Menahem Friedman and Abraham Kandel

Calculus Light

Intelligent Systems Reference Library, Volume 9

Editors-in-Chief

Further volumes of this series can be found on our homepage: springer.com

Vol. 1. Christine L. Mumford and Lakhmi C. Jain (Eds.)
Computational Intelligence: Collaboration, Fusion and Emergence, 2009
ISBN 978-3-642-01798-8

Vol. 2. Yuehui Chen and Ajith Abraham
Tree-Structure Based Hybrid Computational Intelligence, 2009
ISBN 978-3-642-04738-1

Vol. 3. Anthony Finn and Steve Scheding
Developments and Challenges for Autonomous Unmanned Vehicles, 2010
ISBN 978-3-642-10703-0

Vol. 4. Lakhmi C. Jain and Chee Peng Lim (Eds.)
Handbook on Decision Making: Techniques and Applications, 2010
ISBN 978-3-642-13638-2

Vol. 5. George A. Anastassiou
Intelligent Mathematics: Computational Analysis, 2010
ISBN 978-3-642-17097-3

Vol. 6. Ludmila Dymowa
Soft Computing in Economics and Finance, 2011
ISBN 978-3-642-17718-7

Vol. 7. Gerasimos G. Rigatos
Modelling and Control for Intelligent Industrial Systems, 2011
ISBN 978-3-642-17874-0

Vol. 8. Edward H.Y. Lim, James N.K. Liu, and Raymond S.T. Lee
Knowledge Seeker – Ontology Modelling for Information Search and Management, 2011
ISBN 978-3-642-17915-0

Vol. 9. Menahem Friedman and Abraham Kandel
Calculus Light, 2011
ISBN 978-3-642-17847-4

Menahem Friedman and Abraham Kandel

Calculus Light

Prof. Menahem Friedman
Ben Gurion University of the Negev
Beer-Sheva 84105
Israel
E-mail: mlfrid@netvision.net.il

Prof. Abraham Kandel
University of South Florida
4202 E. Fowler Ave. ENB 118
Tampa
Florida 33620
USA
E-mail: kandel@babbage.csee.usf.edu

ISBN 978-3-642-43430-3 ISBN 978-3-642-17848-1 (eBook)

DOI 10.1007/978-3-642-17848-1

Intelligent Systems Reference Library ISSN 1868-4394

Softcover re print of the Hardcover 1st edition 2011

Typeset & Cover Design: Scientific Publishing Services Pvt. Ltd., Chennai, India.

Printed on acid-free paper

9 8 7 6 5 4 3 2 1

springer.com

To our grandchildren

Gal, Jonathan and Tomer

Kfeer, Maya and Riley

with love

Preface

The first question we may face is: "Another Calculus book?" We feel that the answer is quite simple. As long as students find mathematics and particularly calculus a scary subject, as long as the failure rate in mathematics is higher than in all other subjects, except maybe among students who take it as a major in College and as long as a large majority of the people mistakenly believe that only geniuses can learn and understand mathematics and particularly analysis, there will always be room for a new book of Calculus. We call it Calculus Light.

This book is designed for a one semester course in "light" calculus, and meant to be used by undergraduate students without a deep mathematical background who do not major in mathematics or electrical engineering but are interested in areas such as biology or management information systems. The book's level is suitable for students who previously studied some elementary course in general mathematics. Knowledge of basic terminology in linear algebra, geometry and trigonometry is advantageous but not necessary.

In writing this manuscript we faced two dilemmas. The first, what subjects should be included and to what extent. We felt that a modern basic book about calculus dealing with traditional material of single variable is our goal. The introduction of such topics and their application in solving real world problems demonstrate the necessity and applicability of calculus in practically most walks of life.

Our second dilemma was how far we ought to pursue proofs and accuracy. We provided rigorous proofs whenever it was felt that the readers would benefit either by better understanding the specific subject, or by developing their own creativity. At numerous times, when we believed that the readers were ready, we left them to complete part or all of the proof. Certain proofs were beyond the scope of this book and were omitted. However, it was most important for us never to mix intuition and heuristic ideas with rigorous arguments.

We start this book with a historical background. Every scientific achievement involves people and is therefore characterized by victories and disappointments, cooperation and intrigues, hope and heartbreak. All of these elements exist in the story behind calculus and when you add the time dimension – over 2400 years since it all started, you actually get a saga. We hope the reader enjoys reading the first chapter as much as we enjoyed the writing.

In chapters 2-7 we present the topic of single variable calculus and these chapters should be studied in sequential order. The next two chapters provide basic theory and applications of Fourier series and elementary numerical methods. They are expected to motivate the student who is interested in applications and practicality.

The final chapter contains several special topics and combines beauty - the proof that e is irrational, and practicality - the theory of Lagrange multipliers introduced with a short introduction to multi-variable calculus.

Each chapter is divided into sections and at the end of almost every section, variety of problems is given. The problems are usually arranged according to the order of the respective topics in the text. Each topic is followed by examples, simple and complex alike, solved in detail and graphs are presented whenever they are needed. In addition we provide answers to selected problems.

It should be noted that the content of this book was successfully tested on many classes of students for over thirty years. We thank many of them for their constructive suggestions, endless reviews and enormous support.

Tampa, FL 2010

M. Friedman
A. Kandel

Contents

1 Historical Background ... 1
1.1 Prelude to Calculus or the Terror of the 'Infinite' ... 1
1.2 Calculus – Where Do We Start? ... 2
1.3 The Countdown ... 5
1.4 The Birth of Calculus ... 6
1.5 The Priority Dispute ... 8

2 The Number System ... 11
2.1 Basic Concepts about Sets ... 11
2.2 The Natural Numbers ... 14
2.3 Integers and Rational Numbers ... 17
2.3.1 Integers ... 18
2.3.2 Rational Numbers ... 19
2.4 Real Numbers ... 22
2.5 Additional Properties of the Real Numbers ... 30

3 Functions, Sequences and Limits ... 37
3.1 Introduction ... 37
3.2 Functions ... 39
3.3 Algebraic Functions ... 48
3.4 Sequences ... 55
3.5 Basic Limit Theorems ... 62
3.6 Limit Points ... 70
3.7 Special Sequences ... 74
3.7.1 Monotone Sequences ... 75
3.7.2 Convergence to Infinity ... 77
3.7.3 Cauchy Sequences ... 82

4 Continuous Functions ... 87
4.1 Limits of Functions ... 87
4.2 Continuity ... 89
4.3 Properties of Continuous Functions ... 94
4.4 Continuity of Special Functions ... 98
4.5 Uniform Continuity ... 104

5 Differentiable Functions 107
5.1 A Derivative of a Function 107
5.2 Basic Properties of Differentiable Functions 115
5.3 Derivatives of Special Functions 125
5.4 Higher Order Derivatives; Taylor's Theorem 131
5.5 L'Hospital's Rules 141

6 Integration 147
6.1 The Riemann Integral 147
6.2 Integrable Functions 155
6.3 Basic Properties of the Riemann Integral 158
6.4 The Fundamental Theorem of Calculus 166
6.5 The Mean-Value Theorems 171
6.6 Methods of Integration 175
6.7 Improper Integrals 179

7 Infinite Series 183
7.1 Convergence 183
7.2 Tests for Convergence 186
7.3 Conditional and Absolute Convergence 193
7.4 Multiplication of Series and Infinite Products 203
7.5 Power Series and Taylor Series 210

8 Fourier Series 217
8.1 Trigonometric Series 217
8.2 Convergence 225
8.3 Even and Odd Functions 229
8.3.1 Even Functions 229
8.3.2 Odd Functions 230

9 Elementary Numerical Methods 233
9.1 Introduction 233
9.2 Iteration 235
9.3 The Newton - Raphson Method 242
9.4 Interpolation Methods 247
9.4.1 Lagrange Polynomial 247
9.4.2 Cubic Splines 250
9.5 Least – Squares Approximations 252
9.5.1 Linear Least – Squares Method 253
9.5.2 Quadratic Least – Squares Method 254
9.6 Numerical Integration 256
9.6.1 The Trapezoidal Rule 256
9.6.2 Simpson Rule 257
9.6.3 Gaussian Integration 259

10 Special Topics.. 263
10.1 The Irrationality of e 263
10.2 Euler's Summation Formula 264
10.3 Lagrange Multipliers 270
10.3.1 Introduction: Multi-variable Functions 270
10.3.2 Lagrange Multipliers..274

Solutions to Selected Problem...283

Index 297

1 Historical Background

1.1 Prelude to Calculus or the Terror of the 'Infinite'

Zeno, the 5-th century BC Greek philosopher, who is mainly remembered for his paradoxes, never gained the same prestige and admiration as did for example Socrates or Plato. But more than any other philosopher before or after him, Zeno introduced strong elements of uncertainty into mathematical thinking. He stumped mathematicians for over 2000 years, and his paradoxes provided both controversy and stimulation (besides entertainment of course) that inspired new research and ideas, including the development of calculus. This is quite fascinating considering the fact that he was a philosopher and logician, but not a mathematician.

Born around 495 BC in Elea, a Greek colony in southern Italy, Zeno was a scholar of the Eleatic School, established by his teacher Parmenides. He quickly adopted Parmenides' monistic philosophy which stated that the universe consists of a *single* eternal entity (called '*being*'), and that this entity is *motionless*. If our senses suggest otherwise, we must assume that they are illusive and ignore them. Only abstract logical thinking should count as a tool for obtaining conclusions.

Determined to promote his teacher's concept of '*All is One*', Zeno invented a new form of debate – the *dialectic*, where one side supports a premise, while the other attempts to reduce the idea to nonsense. He also presented a series of paradoxes through which he tried to show that any allegation which does not support Parmenides' ideas, leads necessarily to contradiction and is therefore absurd. The keyword in the paradoxes was '*infinite*'. According to Zeno, *any* assertion that legitimizes the concept of "infinity", openly or through the back door, is absurd and therefore false. Consequently, the *opposite* assertion is true.

In the following paradox, Zeno argued that motion is impossible: If a body moves from A to B it must first reach the midpoint B_1 of the distance AB. But prior to this it must reach the midpoint B_2 of the distance AB_1. If the process is repeated indefinitely we obtain that the body must move through an infinite number of distances. This is an absurd and thus the body cannot move.

The most famous of Zeno's paradoxes is that which presents a race between the legendary Greek hero Achilles and the tortoise. The much faster Achilles gracefully allows the tortoise a headstart of 10 meters. But his generosity is going to cost him the race no matter how slow the tortoise is. This astonishing

M. Friedman and A. Kandel: Calculus Light, ISRL 9, pp. 1–9.
springerlink.com

conclusion is easily derived. Achilles must first pass the 10 meters gap. At that time the tortoise travels a shorter distance, say 1 meter. Now, before catching up with the tortoise, Achilles must travel a distance of 1 meter. The tortoise does not wait and advances another 0.1 meter. If the process is repeated we realize that just to *get* to the tortoise, Achilles must travel an infinite number of distances, namely

$$(10+1+\frac{1}{10}+\frac{1}{10^2}+\cdots)\ meters$$

which cannot be done in a finite time.

Most philosophers and scientists were not comfortable with Zeno's paradoxes and did not accept his convictions. Aristotle discarded them as "fallacies" but did not possess any scientific tools to justify his conviction, as the basic concepts of calculus, the 'infinite series' and the 'limit', were yet unknown. Only the rigorously formulated notion of converging series as presented by the French mathematician Cauchy, and the theory of infinite sets developed by the German mathematician Cantor in the 19-th century, finally explained Zeno's paradoxes to some satisfaction.

1.2 Calculus – Where Do We Start?

Some people claim that it was Sir Isaac Newton, the 17-th century English physicist and mathematician who discovered calculus. Others give the credit to the German philosopher and mathematician Baron Gottfried Wilhelm von Leibniz, but most believe that both scientists invented calculus independently. Later we will discuss this particular dispute in length. One fact though is above dispute. All agree that discovery of calculus took place between the years 1665 and 1675 and that the first related publication was Leibniz's article *Nova Methodus pro Maximis et Minimis* ("New Method for the Greatest and the Least") which appeared in *Acta Eruditorum Lipsienium* (Proceedings of the Scholars of Leipzig) 1684.

Yet, to think that this unique mathematical field started from scratch in the second half of the 17-th century is incorrect. Various results closely related to some of the fundamental problems and ideas of calculus, particularly the integral calculus, has been known for thousands of years. The ancient Egyptians for example who were practically involved in almost every important development in culture and science, knew how to calculate the volume of a pyramid and how to approximate the area of a circle.

Then came Eudoxus. Born about 400 BC in Cnidus, southwest Asia Minor, this brilliant Greek studied mathematics and medicine (not an unusual combination for those days), in a school that competed with Hipocrates' famous medical school in the island of Cos. A local rich physician, impressed by the talented youngster paid Eudoxus's way to Plato's Academy in Athens. Eudoxus then spent some time in Egypt where he studied astronomy. Later he was a traveling teacher at northwest of Asia Minor and finally returned to Athens where he became an accomplished legislator until his death (about 350 BC).

The best mathematician of his day, Eudoxus is remembered for two major contributions in mathematics which are extensively discussed in the thirteen books collection *Elements* by Euclid (second half of 4-th century BC). The first is the *Theory of Proportions* which paved the way to the introducing of *Irrational Numbers* and is discussed in Book V. The second is the *Method of Exhaustion*, presented in Book XII, which was developed to calculate areas and volumes bounded by curves or surfaces. The general idea of this method was to approximate an unknown area or a volume using elementary bodies like rectangles or boxes, whose areas or volumes were already known. At each stage, at least half (or other constant fraction) of the remaining area must be replaced by new elementary bodies. This guarantees that the unknown area may be approximated to any desired degree. It was the first time that the concept of *infinitesimally small quantity* was included, over 2000 years before Newton and Leibniz. It is therefore quite legitimate to consider Eudoxus as the discoverer of the conceptual philosophy of integral calculus.

The true successor of Eudoxus and his method of exhaustion was Archimedes of Syracuse (287-212 BC). A first class inventor, the founder of theoretical mechanics and one of the greatest mathematicians of all times, Archimedes studied in Alexandria, capital of Egypt and center of the scientific world at the time. After completing his studies he returned to Syracuse where he remained for the rest of his life. The Greek biographer Plutarch (about 40-120 AD) claims that Archimedes despised everything of practical nature and rather preferred pure mathematical research, whether it was solving a complex problem in geometry or calculating a new improved approximation to the circumference of a circle. Yet, when forced by the circumstances, his beloved city under Roman siege, he put his engineering ingenuity to work and devised war machines for Hieron II, King of Syracuse, which inflicted heavy casualties on the Roman Navy. When the city finally surrendered to the Romans, Archimedes, realizing that everything was lost, went back to his research. Unfortunately there was not much time left for him. While engaged in drawing his figures on the sand he was stabbed to death by a Roman soldier.

As stated above, Archimedes adopted and improved the method of exhaustion originated by Eudoxus and followed by Euclid. One of his applications was to approximate the area of a unit circle (equals to 2π) by using inscribed and circumscribed equilateral polygons to bound the circle from below and above. It is an early example of integration, which led to approximate values of π. A sample of his diagram is given in Fig. 1.1.1.

By increasing the number of the vertices of each polygon, the interval where the exact value of the area is located decreases, and the approximation improved. At some point Archimedes believed that π equals $\sqrt{10}$. However, after using the method of exhaustion, he found the assumption false and by taking 96 vertices showed

$$3\frac{10}{71} < \pi < 3\frac{1}{7}$$

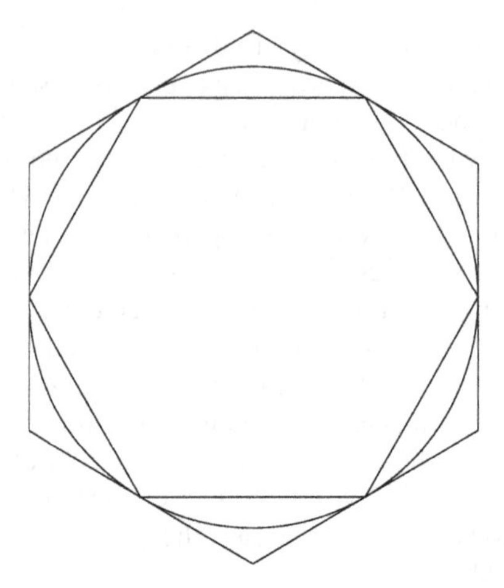

Fig. 1.1.1. Approximating the circle by equilateral polygons.

Around the year 225 BC, already over sixty, Archimedes published his book *Quadrature of the Parabola* where he demonstrated that any portion of a parabola confined by a chord AB (Fig. 1.1.2) is 4/3 of the triangle with base AB and vertex C (obtained by drawing the tangent parallel to AB; the segment connecting the midpoint M of AB with C is parallel to the parabola axis of symmetry). Archimedes constructed a series of increasing areas, each composed of triangles, which approximated the parabola to any desired degree. If A denotes the triangle's area, then at the n-th step, 2^n new triangles of total area $A/4^n$ were added to the $(n-1)$-th approximation.

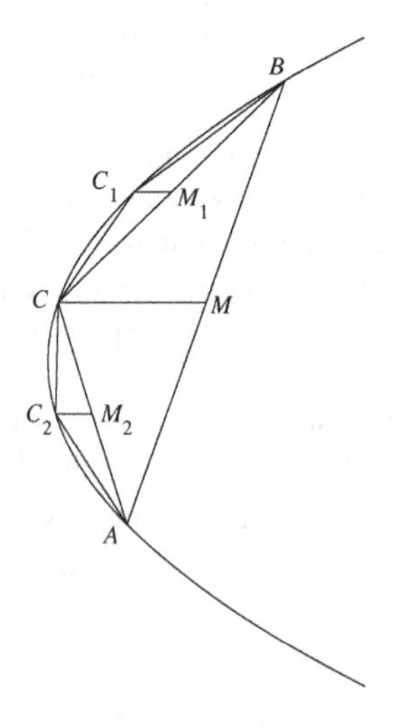

Fig. 1.1.2. $\Delta(ABC)+\Delta(CBC_1)+\Delta(ACC_2)=A+A/8+A/8=A+A/4$

The series of approximations is

$$A, A+\frac{A}{4}, A+\frac{A}{4}+\frac{A}{4^2}, \ldots$$

and it *converges* to $(4/3)A$, the area of the parabola. This was the first known example in history that summation of an infinite series took place and was *documented.*

1.3 The Countdown

No real breakthrough in the development of calculus occurred within the next 1700 years between Archimedes' death and early 16-th century. However, during the late middle ages, mathematicians studied Archimedes' treatises translated from Arabic and the concept of 'the infinite' which confused the Greeks, started to gain some respect and recognition. From late 16-th century on, an increasing number of scientists, often driven by problems in mechanics and astronomy, made significant contributions that led to the creation of calculus several years later.

In his work on planetary motion, J. Kepler (1571-1630), the German astronomer, had to find areas of sectors of an ellipse. He considered an area as a sum of lines – a crude way of integration. Not bothered with rigor he was nevertheless lucky and got the correct answer, after making two canceling errors in his work.

The Italian mathematician B. Cavalieri (1598-1647) stimulated by Euclid's work and a disciple of the famous astronomer Galileo, published his *Geometry by Indivisibles* where he expanded Kepler's work on calculating volumes. He used 'indivisible magnitudes' to calculate the areas under the curves $y = x^n$, $1 \le n \le 9$. His approach though, observing an area like Kepler, as a collection of lines, was not rigorous and drew criticism. He also formulated a theorem (now called Cavalieri's theorem) which determines the volume created by a rotating figure confined between two parallel planes.

The French mathematician G. P. Roberval (1602-1675) considered the same problems as Cavalieri, but conducted a much more rigorous study. He treated the area between a curve and a straight line as an infinite number of 'infinitely narrow' rectangles and then applied his scheme first to approximate the area of x^n from 0 to 1, then to 'obtain' this area as the value $1/(n+1)$, to which the approximates were approaching.

Two other French mathematicians are credited for major contributions during these pre-calculus years. R. Descartes (1596-1650), a philosopher and mathematician, is unanimously considered the inventor of coordinate geometry (today – analytic geometry). His goal was to unify algebra and geometry, mainly to apply algebraic methods for solving problems in geometry. He was the first to classify curves by the equations which *produced* them. An independent discoverer of analytic geometry was P. Fermat (1601-1665) mainly remembered for Fermat's last theorem and other contributions in modern theory of numbers. Fermat calculated the greatest and the least values of algebraic expressions using methods similar to those of modern calculus and finding when the tangents to the associated curves were parallel to the x-axis. This is why some consider Fermat the real inventor of calculus.

An English mathematician, J. Wallis (1616-1703), published *Arithmetica Infinitorum* (Arithmetic of Infinitesimals), a book in which he showed how to use features that hold for finite processes to extrapolate formulas for infinite processes. He also introduced the symbol ∞ for 'infinity'. I. Barrow (1630-1677) who was also Newton's teacher, obtained a method for determining tangents, that was very close to the methods used in calculus. In his book *Lectiones Geometricae* (Geometrical Lectures) published in 1670, Barrow provided a geometrical presentation of the inverse relationship between finding tangents and areas.

By the late 17-th century there were strong indications that calculus was just beyond the horizon.

1.4 The Birth of Calculus

There is a general consensus that Newton and Leibniz contributed more than anyone else to the development of calculus, and are its inventors. It should be noted though, that 150 years after Newton and Leibniz, the foundations of calculus were still shaky. Only in 1821, the French mathematician A.L. Cauchy (1789-1857), finally removed the remaining logical obstacles. His 'theory of limits' did not depend on geometric intuition, but gave a rigorous, logical presentation of basic concepts as continuity, derivative and integral. Fortunately however, the lack of rigorous formalism in its early stages, did not prevent calculus from quickly becoming the most important and effective mathematical tool in the development of modern science.

For both Newton and Leibniz, calculus had not been the sole or even the main subject of interest. Nevertheless, the question who deserved credit for the significant discovery, later became a major source of fury and agony for the two scientists, casting a giant shadow over their remaining days.

Born in 1642 at Woolsthorpe in Lincolnshire, England, Isaac Newton did not enjoy a happy childhood. As if his father's death, three months before he was born, was not enough, his mother soon married an elderly minister and left the 2-year old boy with his grandmother. Only when widowed for the second time did she take him back at the age of eleven. There is no doubt that these traumatic events, could account at least partially, for Newton's general ill temper, his occasional nervous breakdowns and his negative attitude towards women. His scientific genius however, was not affected.

Soon after Newton's reunion with his mother, he was appointed the manager of her estate. Fortunately, the youngster's uncle, Reverend William Ayscough interfered. A Cambridge scholar and a man of authority, Newton's uncle convinced his sister that her son's future should not end at Woolsthorpe. After studying in the grammar school at Grantham Newton arrived in Cambridge and graduated with a BA in 1665. Due to an outburst of the plague, he returned home for almost two years. This period was the most fruitful and creative in Newton's or in any other scientist's life throughout human history.

Within 20 months, Newton laid the foundations to calculus, optics and gravitation. He disclosed his results to close friends but unfortunately did not publish them. Only in 1687, some of Newton's discoveries, including his theory of gravitation, appeared in the first edition of his *Philosophiae Naturalis Principia Mathematica* (Mathematical Principles of the Natural Philosophy). It took him additional 17 years to publish the details of his 'Fluxtional Method', which he used to lay the foundations of calculus. Newton's approach to calculus was dynamic, observing variables as flowing quantities generated by continuous motion of points. He referred to a variable quantity denoted for example by x as 'fluent' and to a variable's rate of change, which he denoted by $\dot{x}$, as 'fluxion'. The two fundamental problems of calculus as regarded by Newton were: Given a relation between two fluents, find the relation between their respective fluxions and vice versa.

Newton's unique contributions to modern science were just one aspect in his life. He was always interested in theology as well and involved also, particularly in later years, with alchemy and mysticism. A professor in Cambridge from 1669, Newton retired in 1696 to become warden and then master of the Royal Mint in London. In 1703 he was elected president of the Royal Society. Knighted by Queen Anne in 1705, it was the first time a scientist was ever awarded this honor.

Unable to confront criticism, accept a colleague's success or share credit for discoveries with anyone, Newton's relations with other scientists were often disrupted permanently. Nothing though compares with the violent crusade he carried against Leibniz, over the glory of inventing calculus. This dispute dominated the last 25 years of his life, until his death in 1727. It is an appalling story which glorifies neither the Newton, nor the Royal Society, for its unconditional surrender to its president's unbalanced behavior. However, before reviewing the details of this dispute, let us present the other contestant: A scientist of Newton's caliber, who also found himself at the center of the most famous, bitter and quite scandalous scientific priority dispute of modern time.

If anyone should be regarded as 'universal genius', it is Gottfried Wilhelm Leibniz, the late 17-th century philosopher and mathematician. He influenced every possible discipline of natural and social sciences. Born in Leipzig, Germany in 1646, Leibniz lost his father, a professor of moral philosophy, at the age of 6 (compare with Newton). At the age of 15 he entered the University of Leipzig as a law student, and received the degree of doctor of law at 20. He turned down an offer of a professor's chair at the University of Altdorf. Instead, Leibniz pursued a political career and became a skillful diplomat at the service of the Archbishop of Mainz.

During four years in Paris (1672-76) Leibniz was intellectually stimulated by constant interaction with leading philosophers and scientists. He particularly benefited from meeting the Dutch scientist C. Huygens, who enriched his mathematical knowledge and thinking. In Paris he also devised a calculating machine. While an adding and subtracting machine was already at hand, Leibniz designed an elegant device which mechanized multiplication – another aspect of his genius. This device was still found in calculating machines in the 1940's!

In 1676 Leibniz returned to Germany and settled in Hanover for the rest of his life as a Counselor of the Duke of Brunswick – Luneberg. His trivial duties, such as composing the family tree of the Brunswicks, left Leibniz plenty of time for more serious research in philosophy and mathematics. His major work in philosophy, published in early 18-th century, describes the universe as a collection of 'monads'. Each monad is a unique, indestructible dynamic substance, distinguished from other monads by its degree of consciousness.

Leibniz's most important contribution to mathematics and to science, in general, was laying the foundations to calculus. The fruitful discussions with Huygens gave him the insight that the operations of summing a sequence and taking its differences sequence, are in a sense inverse to each other. He also realized that quadrature could be observed as summation of equidistant ordinates and that the difference of two consecutive ordinates approximates the local slope of the tangent. Finally, if the distance between successive ordinates becomes 'infinitely small', this approximation becomes exact. Leibniz did not really explain the concept '*infinitely small*', but introduced the notations which were adopted forever by the mathematical world: dx, dy for the 'differentials', dy/dx for the derivative and a long s written as $\int$ for the integral. Leibniz published his discovery first in 1684 followed by a second article about integral calculus in 1686.

Unhappy with his post, the quarrel with Newton over the invention of calculus, only added to Leibniz's bitterness and frustration. Mistreated by the princes whom he served and condemned to remain in the provincial Hanover, he died lonely and forgotten in 1716.

1.5 The Priority Dispute

The scientific world never witnessed such a long, vicious and paranoid battle as that of Newton and Leibniz. This may seem strange, since scientists supposedly portray an image of sincerity, objectivity, moderation, and even modesty. But scientists, like others, struggle for kudos often at the expense of the truth.

Three hundred years after the stormy priority dispute over calculus, it is unanimously agreed that while Newton was the first to lay the foundations of calculus, Leibniz, who discovered it ten years later using a different approach, was the first to publish his results. For years mutual respect existed between the two scientists. But misinterpretation of certain documents, half true hearsay and ignorance about each other's work, led to open accusations and unrestrained attacks. While Leibniz probably never denied Newton's unique contribution, Newton's arrogance and obsession for continued praise, prevented him from crediting Leibniz. Unable to accept the fact that the first scientific publication announcing the birth of calculus was not his, he sent the Royal Society in 1699, a communication accusing Leibniz of plagiarism.

In 1712, the society, totally controlled by Newton, appointed a special committee to look into this issue. Staffed with five Newtonians, the committee not only supported Newton's priority claim, but officially accepted the plagiarism charge. The humiliated Leibniz tried to fight back. For some time he trusted the Berlin Academy to support him, but fighting the Royal Society was unpopular and the academy had neither troops nor desire to do so. Not even Leibniz's death halted Newton's obsession. He once arrogantly told an admirer that "he broke Leibniz's heart with his reply to him". Ironically, it was Leibniz who complemented his opponent: "Taking mathematics from the beginning of the world to the time of Newton, what he has done is much better than half"...

We would like to close the historical background with the following, somewhat related, 'optimistic' episode. Sometime in early 19-th century, in a little pub in Gottingen, Germany, students of Goethe and Schiller, the most prominent German poets, were arguing which of the two was greater. As beer was spilling, the debate was warming up, almost getting violent. Finally, moments before swords were drawn, one peacemaker suggested to bring the issue before Goethe himself, known for his objectivity and trusted by everyone to pass an unbiased judgment. Goethe listened quietly and then, so they say, turned towards the students with a smile, mixed with sadness and irony and said: "Gentlemen, I am amazed. Think how fortunate you all are, to have both Goethe and Schiller in your generation. And here, instead of being thrilled and thankful, all that interests you is who is number one..."

2 The Number System

The basics, which are essential for the study of calculus, are Numbers and Arithmetic Operations. However, we feel that prior to introducing the number system, we should familiarize the reader with some helpful elementary concepts from set theory.

2.1 Basic Concepts about Sets

A *set* is a collection of *distinct* objects, which can be observed as one entity. Each object is called *element* or *member* of the set. Each element of a set is said to *belong* to the set, while the set is *composed* of its elements. Examples of sets are found everywhere: a family, a forest, a class. A family is composed of persons; a forest is composed of trees; a class is composed of students.

Sets are usually denoted by capital letters: $A, B, C, \ldots$; elements are usually denoted by lower-case letters: $a, b, c, \ldots$. The fact that an element x belongs to a set S , is expressed by the notation

$$x \in S$$

If the element x is not a member of S , we write $x \notin S$. The relation between a set and an element is very *crispy*: the element either *belongs* to the set or *does not belong* to it. Another convenient notation is displaying the elements of a set in braces. For example a set whose elements are the numbers $0, -5, 17$ will be written as $\{0, -5, 17\}$.

A set whose elements are all the positive integers from 1 to 1000, will be written as $\{1, 2, 3, \ldots, 1000\}$, and a set consisting of all the positive integers will be written as $\{1, 2, 3, \ldots\}$ or as $\{1, 2, 3, \ldots, n, \ldots\}$. Note that $\{0, -5, 17\}$ may be also written as $\{-5, 17, 0\}$, i.e. the *order* of the elements is irrelevant. What counts is *membership*. This will become clearer as we define the basic relations between sets.

Definition 2.1.1. Two given sets A, B are said to be equal, i.e. $A = B$, if each element of A belongs also to B and if each element of B belongs also to A . This can be also written in short notation as $x \in A \Rightarrow x \in B$ and $x \in B \Rightarrow x \in A$, or even shorter as $x \in A \Leftrightarrow x \in B$. If the two sets are not equal we write $A \neq B$.

M. Friedman and A. Kandel: Calculus Light, ISRL 9, pp. 11–35.
springerlink.com

For example, $\{0,-5,17\}=\{-5,17,0\}$.

Definition 2.1.2. A set A will be called a *subset* of a set B, if every element in A is also in B, i.e. if $x\in A \Rightarrow x\in B$. In this case we write $A\subseteq B$. If there is an $x\in B$ such that $x\notin A$, we write $A\subset B$.

For example, let $A=\{Mom,Dad,Jack,Martha,Tom\}$ and $B=\{Mom,Tom\}$ then B is a subset of A, or $B\subseteq A$. Note that here $B\subset A$ as well.

Theorem 2.1.1. Let A,B be arbitrary sets. Then

$$A=B \quad \textit{if and only if} \quad A\subseteq B \textit{ and } B\subseteq A \tag{2.1.1}$$

The proof is trivial and follows from the previous definitions.

Definition 2.1.3. For arbitrary sets A,B we define the *union* of A and B as the set C, composed of all the elements which belong to either A or B. We write $C=A\cup B$.

Definition 2.1.4. For arbitrary sets A,B we define the *intersection* of A and B as the set C, composed of all the elements which belong to both A and B. We write $C=A\cap B$.

Definition 2.1.5. For arbitrary sets A,B we define the *difference* of A and B as the set C, composed of all the elements which belong to A but not to B. We write $C=A-B$.

Example 2.1.1. A *prime* number is a positive integer which has no factor except itself and 1. Let A denote the set of all the primes between 3 and 18 and let B denote all the odd numbers between 12 and 20, i.e. $A=\{3,5,7,11,13,17\}$ and $B=\{13,15,17,19\}$.

Then

$$A\cup B=\{3,5,7,11,13,15,17,19\},\ A\cap B=\{13,17\}$$
$$A-B=\{3,5,7,11\},\ B-A=\{15,19\}$$

If a set has no elements, it is called the *empty set* and is denoted by $\varnothing$. The sets A and B are called *disjoint* if $A\cap B=\varnothing$. The notations of union, intersection and difference, are illustrated in Fig. 2.1.1.

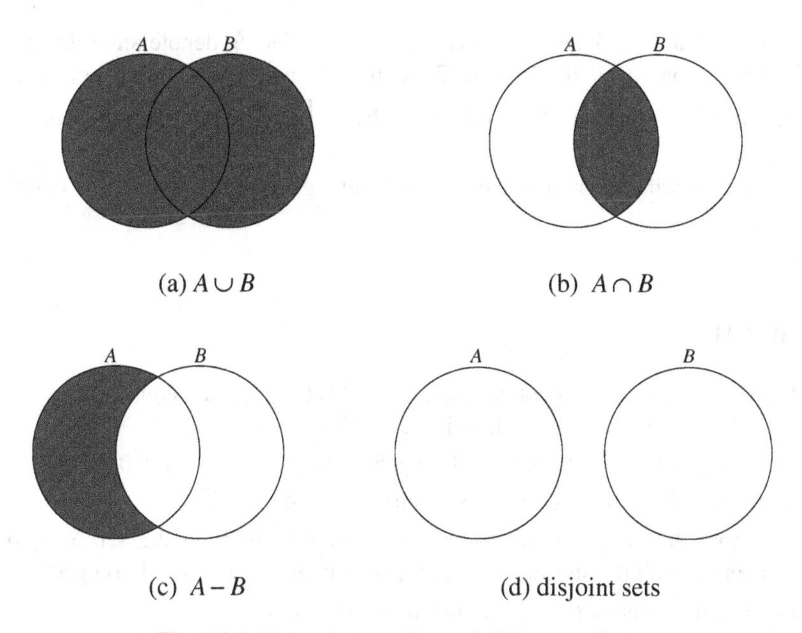

Fig. 2.1.1. Union, intersection and difference of sets.

A popular way for defining a particular set is by writing

$$S = \{x \mid x \text{ satisfies } P\}$$

It means that an element x belongs to S if and only if x satisfies some property P. For example, let $S = \{x \mid x \text{ is a positive integer less than } 4\}$. Then S contains the numbers 1, 2 and 3 and will be written as $S = \{1,2,3\}$. Quite often several properties are needed to define a set. For example, the set of all positive numbers whose square is less than 7 may be presented as $\{x \mid x > 0\,,\, x^2 < 7\}$.

The operations 'union' and 'intersection' are *commutative*, i.e. arbitrary sets A,B satisfy $A \cup B = B \cup A$ and $A \cap B = B \cap A$. They are also *associative*, i.e. for arbitrary sets A,B,C we have $(A \cup B) \cup C = A \cup (B \cup C)$ and $(A \cap B) \cap C = A \cap (B \cap C)$. These results are straightforward. Consider for example the assertion $(A \cap B) \cap C = A \cap (B \cap C)$. We have to show that each side is a subset of the other. Indeed, if $x \in (A \cap B) \cap C$ then $x \in A \cap B$ and $x \in C$. Thus $x \in A$ and $x \in B$ as well. Consequently $x \in B \cap C$ and since $x \in A$ we obtain $x \in A \cap (B \cap C)$. Therefore, the left-hand side of the relation is a subset of its right-hand side. The opposite is shown similarly.

We will now extend the notations of union and intersection for any number of sets.

Definition 2.1.6. Let $\Im$ denote a 'class' of sets and let A denote an arbitrary set in $\Im$. The union of all the sets in $\Im$ is the set which contains every x that belongs to at least one set A. It is denoted by $\bigcup_{A\in\Im} A$. The intersection of the sets in $\Im$ is the set which contains every x that belongs to all A in $\Im$. It is denoted by $\bigcap_{A\in\Im} A$.

PROBLEMS

1. Which of the following collections are sets? Which sets are equal?
   (a) $\{-4,7,3,-3\}$ (b) $\{-3,7,-4,-3\}$ (c) $\{7,3,-4,-3\}$
2. Let $A=\{x\,|\,0<x<4\}$, $B=\{x\,|\,3<x<5\}$. Obtain $A\cup B$, $A\cap B$, $A-B$.
3. Show $(A-B)\cap(B-A)=\varnothing$ for arbitrary sets A,B.
4. (a) Show that $\varnothing$ is a subset of *any* given set. (b) Find the set S whose elements are all the subsets of $\{1,2,3\}$. How many elements did you get?
5. For arbitrary sets A,B,C prove the *distributive* laws:

$$A\cap(B\cup C)=(A\cap B)\cup(A\cap C)$$
$$A\cup(B\cap C)=(A\cup B)\cap(A\cup C)$$

6. For arbitrary sets A,B,C show $A-(B\cap C)=(A-B)\cup(A-C)$.
7. For arbitrary sets A,B,C show $A-(B\cup C)=(A-B)-C$.
8. For arbitrary sets A,B: $A\cup(A\cap B)=A\cap(A\cup B)=A$.
9. Let $\Im$ denote a class of sets and let B denote an arbitrary set. Then

$$B-\bigcup_{A\in\Im} A=\bigcap_{A\in\Im}(B-A)\quad,\quad B-\bigcap_{A\in\Im} A=\bigcup_{A\in\Im}(B-A)$$

2.2 The Natural Numbers

The 19-th century German mathematician Leopold Kronecker once said: "*God created the natural numbers. All the rest is man-made.*" Indeed, it is impossible to imagine a world without the natural numbers. A world where simple questions like "How many children have you got?" or "When do you expect the next full moon" would be so difficult to answer. The rigorous formalism for introducing the set N of natural numbers is based on the following four *axioms* i.e. statements which intuitively are true but we are unable to prove:

1. If the successors of n and m are equal, then $n = m$.
2. Every natural number $n \in N$ has another associated unique natural number $n' \in N$, called the successor of n.
3. There is a number called 1, which is not a successor of any natural number.
4. If a subset M of N is such that the two requests: (a) $1 \in M$; and (b) If $n \in M$ then $n' \in M$ are satisfied, then $M = N$.

These are the Peano axioms and their *acceptance* is the foundation for the natural number system. We denote the successor of 1 by '2', the successor of 2 by '3' and thus obtain the familiar natural numbers.

Axiom 4 is called the *Principle of Mathematical Induction*. It provides a process by which we can determine, whether a statement $P(n)$ associated with the natural number n, is valid for all the natural numbers. The process consists of:

(a) Showing that $P(1)$ is true.
(b) Showing that if $P(n)$ is true, than $P(n')$ is true as well.

If both (a) and (b) are satisfied, then by virtue of Axiom 4, the set of natural numbers for which P holds, is identical to N. If instead of (a) we show that $P(n_0)$ is true for some natural number n_0 and that (b) holds as well, then P must be true at least for all the natural numbers greater or equal to n_0.

Every result associated with natural numbers, follow from the Peano axioms. For example, Axiom 3 guarantees that N is nonempty: at least 1 belongs to N. Another example is the statement that the numbers 1, 2, 3 are all different. This is shown as follows:

(1) $2 \neq 1$: Otherwise $2 = 1 \Rightarrow 1' = 1 \Rightarrow 1$ is a successor of a natural number, contradicting Axiom 3.
(2) $3 \neq 1$: Otherwise $3 = 1 \Rightarrow 2' = 1 \Rightarrow$ again contradicting Axiom 3.
(3) $3 \neq 2$: Otherwise $3 = 2 \Rightarrow 2' = 1' \Rightarrow 2 = 1$ by Axiom 2. However, this is impossible as we have seen in part (1).

By using induction (Axiom 4) we can show that all the natural numbers are different.

In order to 'calculate' with natural numbers, we introduce 'arithmetic operations' between them. We define *addition* (+) and *multiplication* $(\cdot)$ and expect them to possess the following elementary familiar properties:

$$n + 1 = n' \quad , \quad n + m' = (n + m)' \tag{2.2.1}$$

$$n \cdot 1 = n \quad , \quad n \cdot m' = n \cdot m + n \tag{2.2.2}$$

Do such operations exist? This is answered next.

Theorem 2.2.1. There exist unique operations $+, \cdot$ between the natural numbers such that Eqs. (2.2.1-2) are satisfied for all $n, m \in N$.

We usually use the shorter notation nm for multiplication. Addition and multiplication satisfy the basic rules of arithmetic:

$$m+n=n+m \quad , \quad mn=nm \qquad \text{(commutative laws)}$$
$$(m+n)+l=m+(n+l) \quad , \quad (mn)l=m(nl) \qquad \text{(associative laws)}$$
$$(m+n)l=ml+nl \quad , \quad m(n+l)=mn+ml \qquad \text{(distributive laws)}$$

all of which can be verified using induction and Eqs. (2.2.1-2). For example, let us obtain the associative law for addition.

Theorem 2.2.2. For arbitrary $n, m, l \in N$ we have

$$(m+n)+l=m+(n+l) \tag{2.2.3}$$

Proof.

For arbitrary prefixed $n, m \in N$ we will apply induction to show the validity of Eq. (2.2.3) for all $l \in N$.

(a) $l=1$: By virtue of Eq. (2.2.1) $(m+n)+1=(m+n)'=m+n'=m+(n+1)$

(b) The assumption $(m+n)+l=m+(n+l)$ combined with Eq. (2.2.1) yield

$$(m+n)+l'=[(m+n)+l]'=[m+(n+l)]'=m+(n+l)'=m+(n+l')$$

which concludes the proof.

The following result, the *trichotomy* law, defines *order* among the natural numbers.

Theorem 2.2.3. For arbitrary $n, m \in N$ one and only one of the following relations is valid:

(1) $m=n$
(2) $m=n+l$, for some $l \in N$.
(3) $n=m+l$, for some $l \in N$.

If case (1) holds, the numbers are *equal* ; if case (2) holds, m is *greater* than n and we write $m>n$; if case (3) holds, m is *less* than n and we write $m<n$. Theorem 2.2.3 combined with the basic arithmetic rules, leads to many results which we usually take for granted. For example

$$m > n \Rightarrow m + k > n + k \quad \textit{for all} \quad k \in N$$

Indeed, $m > n \Rightarrow m = n + l$ for some $l \in N$. Therefore

$$m + k = (n + l) + k = n + (l + k) = n + (k + l) = (n + k) + l$$

i.e. (by definition) $m + k > n + k$.

PROBLEMS

1. Obtain the commutative law for addition, $m + n = n + m$, by prefixing m and applying induction on n .
2. Obtain the associative law for multiplication, $(mn)l = m(nl)$, by prefixing m, n and applying induction on l .
3. Obtain the commutative law for multiplication, $mn = nm$. Hint: to show $m \cdot 1 = 1 \cdot m$ apply a second induction on m .
4. Show : $m > n\ , k > l \Rightarrow m + k > n + l\ , mk > nl$.
5. Show: $m \geq n\ , k > l \Rightarrow m + k > n + l\ , mk > nl$.
6. Show: $m \geq n\ , k \geq l \Rightarrow m + k \geq n + l\ , mk \geq nl$.
7. Show: (a) $m > n \Rightarrow mk > nk$ (b) $mk > nk \Rightarrow m > n$.
8. Define $n^2 = n \cdot n$ and apply the basic arithmetic rules to show:

$$\text{(a)} \quad (n+1)^2 = n^2 + 2n + 1 \quad \text{(b)} \quad (n+m)^2 = n^2 + 2nm + m^2 .$$

9. Show $n^2 > n + 3$ for all $n > 2$.

2.3 Integers and Rational Numbers

A simple equation such as $3 + x = 7$, where x is some unknown natural number, is solved uniquely by $x = 4$. However, other equations, just as simple, for example $5 + y = 2$, or $6z = 13$, have no solutions, i.e. there are no y or z for which these equations hold. Nevertheless, these particular relations may simulate real 'situations'. Consequently, unless determined to significantly limit our capacity to solve real-world problems, the concept of *number* must be *extended* beyond the set of the natural numbers. In this section we discuss two extensions and introduce the *integers* and the *rational* numbers. However, prior to this, we ought to introduce the concept of 'extension' which is often used throughout this chapter. Let A and B denote sets of 'numbers' and let P_A and P_B be 'arithmetic operations', which are defined over A and B respectively. The set B will be called an *extension* of A , if either

(a) (1) $A \subset B$, (2) $xP_B y = xP_A y$ for all $x, y \in A$; or
(b) There exists a subset $A' \subset B$ such that

(1) There is a one-to-one correspondence between A and A'.
(2) If $x, y \in A$ correspond to $x', y' \in A'$ then $x' P_B y' \in A'$ and $xP_A y$ corresponds to $x' P_B y'$.

In constructing the whole number system we use both types of extensions.

2.3.1 Integers

We add new objects to N : $-n$ (called *minus* n) for each $n \in N$ and 0 (called *zero*).

In order to calculate with them, we extend the arithmetic operations as follows.

Definition 2.3.1. For arbitrary $m, n \in N$

$$\begin{aligned} &m + (-m) = 0 \\ &m + (-n) = -l \quad \text{if } n = m + l \quad , \quad l \in N \\ &m + (-n) = l \quad \text{if } m = n + l \quad , \quad l \in N \end{aligned} \tag{2.3.1}$$

and

$$\begin{aligned} (-n) + m &= m + (-n) \\ (-m) + (-n) &= -(m + n) \end{aligned} \tag{2.3.2}$$

Definition 2.3.2. For arbitrary $m, n \in N$

$$\begin{aligned} m \cdot (-n) &= (-n) \cdot m = -(m \cdot n) \\ (-m) \cdot (-n) &= m \cdot n \end{aligned} \tag{2.3.3}$$

Definition 2.3.3. For arbitrary $n \in N$

$$\begin{aligned} &0 + n = n + 0 = n \\ &0 + (-n) = (-n) + 0 = -n \\ &-(-n) = n \end{aligned} \tag{2.3.4}$$

and

$$0 \cdot n = n \cdot 0 = (-n) \cdot 0 = 0 \cdot (-n) = 0 \tag{2.3.5}$$

Definition 2.3.4. $0 + 0 = 0 \cdot 0 = -0 = 0$.

The elements $-n, n \in N$ are the *negative integers*. We refer to the union of the natural numbers, also called the *positive integers*, the negative integers and 0, as the *integers*. It can be shown that all the integers are different. The set of all the integers is denoted by I. The next result confirms the validity of the basic arithmetic rules for *all* the integers and supports the formalism used for extending N.

Theorem 2.3.1. The commutative, associative and distributive laws for addition and multiplication hold over I.

The equation $m + x = n$ which did not always possess a solution in N, and provided the motivation to legitimate the negative integers and 0, has a unique solution $x = n + (-m)$ for arbitrary $m, n \in I$. We usually write $n + (-m) = n - m$ and call this operation *subtraction*. For example $7 + (-11) = 7 - 11 = -4$. Finally, we leave it for the reader to show, that the procedure for obtaining the integers from the natural numbers, is an extension of type (a).

2.3.2 Rational Numbers

For every ordered pair of integers (m, n) such that $n \neq 0$, we define an object m/n or $\frac{m}{n}$, which we call a *fraction*. It is *positive* if $m > 0, n > 0$ or if $m < 0, n < 0$ and we write $m/n > 0$. It is *negative* and we write $m/n < 0$ if $m > 0, n < 0$ or if $m < 0, n > 0$.

It is *zero* if $m = 0$ and we write $m/n = 0$. Equality and the arithmetic operations between fractions are defined as follows.

Definition 2.3.5. For arbitrary fractions a/b and c/d

$$\frac{a}{b} = \frac{c}{d} \Leftrightarrow ad = bc \tag{2.3.6}$$

$$\frac{a}{b} + \frac{c}{d} = \frac{ad + bc}{bd} \tag{2.3.7}$$

$$\frac{a}{b} \cdot \frac{c}{d} = \frac{ac}{bd} \tag{2.3.8}$$

This definition, which goes back to our early schooldays, is not only intuitively acceptable, but also guarantees the validity of the commutative, associative and distributive laws for fractions. In addition, we can show that for arbitrary fractions r, s one and only one of the following relations holds:

1. $r = s$
2. $r = s + t$, for some fraction $t > 0$.
3. $s = r + t$, for some fraction $t > 0$.

In case (2) we say that r is *greater* than s; in case 3 we say that r is *less* than s.

Note that equality between fractions does not necessarily mean that they are *identical*. For example, $2/3 = 4/6 = 300/450$. Consequently, an equation such as $2/5 + x = 1/2$ has an infinite number of solutions, namely, $1/10, 2/20, (-4)/(-40)$ etc.

Disturbing? Not really! An elegant way to remove this confusion is to create a set, called a *rational number*, whose elements are *all* the different equal fractions. For example the fraction $(-2)/(-6)$ belongs to the rational number $\{m/n \mid m/n = 1/3\}$.

For all practical purposes, such as arithmetic operations, it is irrelevant which fraction of the set is used. We can therefore identify a fraction that solves a given equation, with the rational number to which this fraction belongs. Consequently, there are unique rational numbers which solve

$$\frac{a}{b} + x = \frac{c}{d} \tag{2.3.9}$$

$$\frac{a}{b} \cdot x = \frac{c}{d}, \ a \neq 0 \tag{2.3.10}$$

The equal fractions $(-a)/b$, $a/(-b)$ are denoted by $-(a/b)$. Using this notation we can write the solution to Eq. (2.3.9) as $c/d + [-(a/b)]$ and define the operation 'subtraction' as

$$\frac{c}{d} - \frac{a}{b} = \frac{c}{d} + \left(-\frac{a}{b}\right) \tag{2.3.11}$$

The solution to Eq. (2.3.10) is $(bc)/(ad)$ (check by substituting). This motivates to define the operation 'division' as

$$\frac{\left(\dfrac{c}{d}\right)}{\left(\dfrac{a}{b}\right)} = \frac{bc}{ad} \quad , \quad a \neq 0 \tag{2.3.12}$$

The left-hand side of Eq. (2.3.12) is the *quotient* of the two fractions.

Note that there is one-to-one correspondence, $n \leftrightarrow \frac{n}{1}$, between all the integers n , $n \in I$ and the fractions $\frac{n}{1}$. This correspondence is preserved under addition and multiplication, i.e.

$$m+n \leftrightarrow \frac{m}{1}+\frac{n}{1} \quad , \quad m \cdot n \leftrightarrow \frac{m}{1} \cdot \frac{n}{1}$$

Indeed

$$m+n \leftrightarrow \frac{m+n}{1} = \frac{m \cdot 1 + n \cdot 1}{1 \cdot 1} = \frac{m}{1} + \frac{n}{1}$$

$$m \cdot n \leftrightarrow \frac{m \cdot n}{1} = \frac{m \cdot n}{1 \cdot 1} = \frac{m}{1} \cdot \frac{n}{1}$$

Consequently, each rational number $\frac{n}{1}$ can be replaced by the corresponding integer n in any computational process.

The set of all the rational numbers is denoted by R . It was derived from the set of integers, using an extension of type (b). We will also use R^+ , R^- to denote the sets of all the positive and negative rational numbers, respectively.

PROBLEMS

1. Use induction to show $1+2+3+\ldots+n = [n(n+1)]/2$ for all $n \in N$.
2. Use induction to show $1^3+2^3+3^3+\ldots+n^3 = [n^2(n+1)^2]/4$ for all $n \in N$.
3. Use induction to define 2^n and show $2^n > n$ for all $n \in N$.
4. Define $1! = 1$; $n! = n \cdot (n-1)!$, $n > 1$ and show $n! > n^2$, $n \geq 4$.
5. Show the validity of the commutative laws for fractions.
6. Show the validity of the associative laws for fractions.
7. Let r, s denote fractions such that $r < s$. Find a fraction $t : r < t < s$.
8. Let $r \in R$, $R = \{r \mid r - \mathit{fraction}\}$. Define r^n , $n \in N$ and show

$$1+r+r^2+\ldots+r^n = \frac{r^{n+1}-1}{r-1} \quad , \quad n \in N \quad , \quad r \neq 1$$

9. Show $(1+r)^n \ge 1+nr$, $n \in N$ (Bernoulli's inequality) for arbitrary fraction

$$r > -1.$$

2.4 Real Numbers

By introducing the rational numbers we can now apply the four arithmetic operations over R with the single restriction of dividing by 0. In addition we define *integer powers* of a fraction as

$$r \ne 0 : r^0 = 1 \,;\, r^n = r \cdot r^{n-1} \,,\, n \in N$$

$$r \ne 0 : r^{-n} = \frac{1}{r^n} \,,\, n \in N$$

and $0^n = 0$, $n \in N$. At this stage it may seem that additional extension of the number system is unnecessary, particularly since computers take and provide *only* rational numbers. Yet, a strong indication against this attitude is given simply by showing that some *legitimate numbers* are still missing from R ! For example, the area of the unit circle, denoted by π , is not a rational number, i.e. cannot be expressed as a fraction. The proof of this allegation is very complicated and is beyond the scope of this book. Instead, let us verify an equally important though much simpler statement.

Theorem 2.4.1. There is no fraction x such that $x^2 = 2$.

Proof.

If such a fraction exists, then $m^2/n^2 = 2$ for some integers m,n . We may assume that these integers have no common factor. Since $m^2 = 2n^2$ we conclude that m^2 is even. Therefore m itself is even and we may write $m = 2l$ for some integer l and obtain $2l^2 = n^2$. A similar argument shows that n is even too, which contradicts the assumption that m,n have no common factor. Thus, the integer 2 is not a square of any fraction.

An immediate consequence of Theorem 2.4.1 is that we can either extend the number system to include objects such as $\sqrt{2}$ or π , or maybe deny their existence. The latter choice, though simpler, leads to an illogical situation. Indeed, consider a right angled isosceles triangle whose equal sides are of length 1 (Fig. 2.4.1). By applying Pythagoras theorem we get $AC^2 = AB^2 + BC^2 = 1^2 + 1^2 = 2$, i.e. AC has a length $\sqrt{2}$ or has *no length* at all. If we accept that *every* segment has a length, we must find a way to include $\sqrt{2}$ in our number system. These new objects will be called *irrational* numbers. The set of all the rational and the irrational numbers, is called the *real* number system and is denoted by $\Re$.

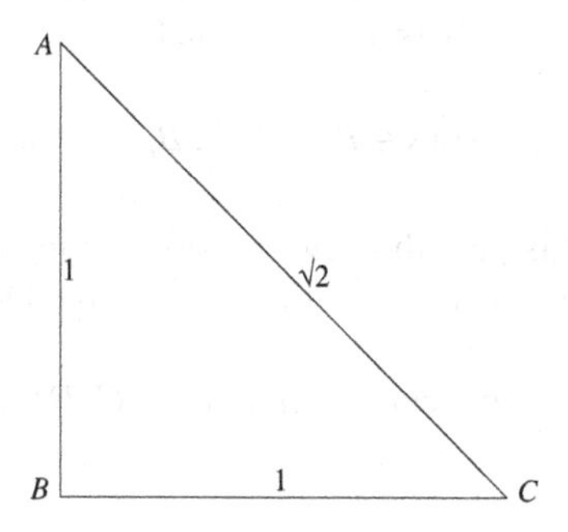

Fig. 2.4.1. Introducing $\sqrt{2}$.

Accepting the 'existence' of irrational numbers is only one step. We still have to formally define the *real* numbers, define arithmetic operations between them and show that the outcome $\Re$ is indeed an *extension* of R. The popular way for doing it, is by using the Dedekind cuts.

Definition 2.4.1. A *Dedekind cut* is an ordered pair (A,B) of sets of rational numbers, which satisfy the following requirements:

1. A and B are nonempty.
2. Every $r \in R$ belongs to either A or B.
3. Every $a \in A$ is less than every $b \in B$.

For example, A may consist of all the rational numbers less than or equal to 3, while B consists of all the rational numbers that are greater than 3. There are four possible types of Dedekind cuts (or simply 'cuts'):

(a) A has a largest number and B does not have a smallest number.
(b) B has a smallest number and A does not have a largest number.
(c) A has a largest number and B has a smallest number.
(d) A does not have a largest number and B does not have a smallest number.

We can immediately dismiss one type. Clearly, case (c) is impossible: If a is the largest number in A and b is the smallest number in B, then $(a+b)/2$ is a rational number greater than a, smaller than b, and thus cannot belong to either A or B. This contradicts part (2) of Definition 2.4.1. Consequently, a cut of type (c) does not exist. We can also disregard cuts of type (b). Indeed, a cut (A,B) of type (b) with a smallest $b \in B$, is merely an alternative representation of the cut

$(A \cup \{b\}, B - \{b\})$ of type (a). Thus, we should consider only two types of cuts: (a) and (d). If a cut is of type (a) it is called a *rational* cut. A type (d) cut is called an *irrational* cut. There is a simple one-to-one correspondence between all the cuts of type (a) and all the rational numbers $r \in R$, namely

$$r \leftrightarrow (A_r, B_r)\,;\; A_r = \{s \mid s \in R,\, s \le r\},\; B_r = \{s \mid s \in R,\, s > r\} \tag{2.4.1}$$

The 'other' cuts, if they exist, are the new 'irrational numbers'. Our first objective is to define equality and arithmetic operations among the Dedekind cuts.

Definition 2.4.2. Two arbitrary cuts (A,B) and (C,D) are called equal if and only if $A = C$ and $B = D$.

Definition 2.4.3. The *sum* of arbitrary cuts (A,B) and (C,D) is the cut

$$(E,F) = (A,B) + (C,D)$$

where $E = \{e \mid e = a + c\,;\, a \in A\,,\, c \in C\}$ and $F = \{f \mid f = b + d\,;\, b \in B\,,\, d \in D\}$.

Showing that (E,F) is a cut, is left as an exercise for the reader.

Definition 2.4.4. The *multiplication* of arbitrary cuts (A,B) and (C,D) is the cut

$$(E,F) = (A,B) \cdot (C,D)$$

where the sets E,F are determined as follows:

(1) If A,C both have nonnegative rationals then

$$F = \{f \mid f = bd\,;\, b \in B,\, d \in D\} \quad , \quad E = R - F$$

(2) If A has nonnegative rationals and C has only negative rationals

$$E = \{e \mid e = bc\,;\, b \in B,\, c \in C\} \quad , \quad F = R - E$$

(3) If A has only negative rationals and C has nonnegative rationals

$$E = \{e \mid e = ad\,;\, a \in A,\, d \in D\} \quad , \quad F = R - E$$

(4) If A,C both have only negative rationals

$$F = \{f \mid f = ac\,;\, a \in A,\, c \in C\} \quad , \quad E = R - F$$

and if F has a smallest rational we transfer it to E and maintain a cut of type (a).

Showing that (E,F) is a cut (of type (a) or (d)), is left as an exercise for the reader.

Given a pair of arbitrary cuts (A,B), (C,D) , we can show that the equation

$$(A,B)+(X,Y)=(C,D)$$

has a unique solution (X,Y) which we denote by $(A,B)-(C,D)$ (*subtraction*). We also have a unique solution to

$$(A,B)\cdot(X,Y)=(C,D)$$

provided that $(A,B)\neq 0$. This solution is denoted by $(C,D)/(A,B)$ (*division*).

We will now define *order* among the real numbers.

Definition 2.4.5. A cut (A,B) is *greater* than a cut (C,D) if and only if there is a rational number r such that $r\in A$, $r\notin C$. If there is a rational number r such that $r\in C$, $r\notin A$ we say that (A,B) is *less than* (C,D).

The cut (A,B) where A contains the rational 0 and all the negative rationals, is the real number 0. A real number is *positive*, if it is greater than 0. It is *negative*, if it less than 0. We denote by $\mathfrak{R}^+, \mathfrak{R}^-$, the sets of the all the positive and the negative real numbers, respectively. Now, consider two positive real numbers a,b such that $0<a<b$. What can be said about the sums $a+a, a+a+a,\ldots,na,\ldots$? It was Archimedes who first realized that no matter how small a is, we can always find some positive integer n (possibly very large) for which $na>b$.

Lemma 2.4.1. (Archimedes' axiom) For arbitrary positive real numbers a,b, there is a positive integer n such that $na>b$ (This lemma is valid also for arbitrary $a,b\in\mathfrak{R}$).

Proof.

Assume that a,b are rational numbers. Let $a=r/s$, $b=p/q$ where r,s,p,q are positive integers. In order to get $na>b$, we should find n such that $n(r/s)>(p/q)$, or $n>(sp/rq)$. Thus, we may choose $n=sp$. The proof for positive real numbers a,b (which are now represented by cuts) is similar and is left as an exercise for the reader.

Corollary 2.4.1. For every arbitrarily small $\varepsilon>0$ there is a positive integer n such that $1/n<\varepsilon$ (substitute $a=\varepsilon$, $b=1$ in Lemma 2.4.1).

It can be seen that under the one-to-one correspondence of Eq. (2.4.1), $\Re$ is an extension of R. We will now show the *existence* of irrational cuts, i.e. of 'irrational numbers'. Consider the cut $\alpha = (A,B)$ where A consists of all the rationals whose square is less than 2, and B - of all the rationals whose square is greater than 2. Since there is no rational number x such that $x^2 = 2$, every rational is in either A or B. Assume that there is a largest rational $a \in A$.. If we define

$$\varepsilon = \min(a, \frac{2-a^2}{8a})$$

then ε is a positive rational which satisfies $a^2 < (a+\varepsilon)^2 < 2$ i.e. $(a+\varepsilon) \in A$ contradicting our assumption. Similarly, there is no smallest rational in B. Hence, (A,B) is an irrational cut. Thus, the set of irrational cuts is nonempty! We will now show that this particular cut solves the equation $x^2 = 2$.

Lemma 2.4.2. Let r, s denote rational numbers such that $0 < r < s$. Then a rational number $t : r < t^2 < s$ can be found.

Proof.

We should consider the following three cases:

(a) Let $r < s \le 1$. We wish to find a positive rational number m/n, $m < n$ such that:

$$(1) \quad \frac{m^2}{n^2} < s \quad , \quad (2) \quad \frac{(m+1)^2}{n^2} \ge s$$

For any prefixed n, we can use Lemma 2.4.1 and find a nonnegative integer m that satisfies (2). Let m_0 be the smallest m for which (2) holds. Then, m_0 and n satisfy (1) as well. Since $s \le 1$ we also must have $m_0 < n$. We will now show how to prefix n so that

$$r < \frac{m_0^2}{n^2}$$

Indeed

$$s - \frac{m_0^2}{n^2} \le \frac{(m_0+1)^2}{n^2} - \frac{m_0^2}{n^2} = \frac{2m_0+1}{n^2} < \frac{3n}{n^2} = \frac{3}{n}$$

and since by corollary 2.4.1 we can find $n = n_0$ that satisfies $(3/n_0) < s - r$, we obtain

$$s - \frac{m_0^2}{n_0^2} < s - r$$

and finally $r < m_0^2/n_0^2$. We therefore choose $t = m_0/n_0$ and the proof is concluded.

(b) Let $r < 1 < s$. Here we simply choose $t = 1$.
(c) Let $1 \le r < s$. This case is left for the reader (transfer to case (a)).

Theorem 2.4.2. The previously defined irrational cut α satisfies $\alpha^2 = 2$.

Proof.

Consider the cut $\beta = (C, D) = \alpha^2$. Every element $s \in D$, can be written as $s = r_1 r_2$, $r_1, r_2 \in B$. Since $r_1^2 > 2$, $r_2^2 > 2$ we also have $s > 2$. The opposite is also true: given a rational number $r > 2$, we can apply Lemma 2.4.2 and find a rational number t such that $2 < t^2 < r$. Consequently $t \in B$ and $t^2 \in D$. Therefore $r \in D$ as well and we conclude that D consists exactly of all the rationals which are greater than 2. Thus, $2 \in C$ and $(C, D) = 2$.

Solving the equation $x^2 = 2$ is a particular case of a more general problem: Given $a \in R^+, n \in N$, find a positive real number $\alpha \in \Re^+$, such that $\alpha^n = a$. It can be shown that this problem has a *unique* solution (see problem 10 below) which we denote by $\sqrt[n]{a}$ or $a^{1/n}$. We can now go further and define;

$$a^{-1/n} = 1/(a^{1/n}),\ a \in R^+, n \in N \tag{2.4.2}$$
$$a^{m/n} = \sqrt[n]{a^m}\ ,\ a \in R^+, n \in N\ , m \in I$$

Finding a solution to $x^n = a$ can be done by using Newton's *Binomial Theorem* or even just Bernoulli's inequality (problem 13 below) combined with a process similar to that applied for solving $x^2 = 2$. We will present the binomial theorem – a powerful mathematical tool, prove it, but leave the rest of solving the equation $x^n = a$ to the interested reader. We start by defining the binomial coefficients.

Definition 2.4.6. For arbitrary nonnegative integers n,k such that $k \le n$, the number $\binom{n}{k} = \frac{n!}{k!(n-k)!}$ is called the *binomial coefficient* n *over* k (0! is defined as 1).

Lemma 2.4.3. The binomial coefficients satisfy

$$\binom{n}{k-1} + \binom{n}{k} = \binom{n+1}{k}$$

The proof, based on Definition 2.4.6, is straightforward and is left for the reader. We now define the power notation and use it combined with mathematical induction to get the binomial theorem.

Definition 2.4.7. For arbitrary real number a and positive integer n:

$$a^1 = a \,,\; a^{n+1} = a^n \cdot a$$

Theorem 2.4.3 (Newton's Binomial Theorem). For arbitrary $a,b \in \Re$, $n \in N$:

$$(a+b)^n = a^n + \binom{n}{1}a^{n-1}b + \binom{n}{2}a^{n-2}b^2 + \ldots + \binom{n}{n-2}a^2b^{n-2} + \binom{n}{n-1}ab^{n-1} + b^n$$
$$= \sum_{k=0}^{n}\binom{n}{k}a^{n-k}b^k \tag{2.4.3}$$

Proof.

(a) $n = 1$. Both sides of Eq. (2.4.3) are $a+b$.

(b) Assume Eq. (2.4.3) to hold for arbitrary $n \ge 1$. Then

$$(a+b)^{n+1} = (a+b)^n(a+b) = \left[\sum_{k=0}^{n}\binom{n}{k}a^{n-k}b^k\right](a+b)$$

$$= \binom{n+1}{0}a^{n+1} + \binom{n}{1}a^n b + \binom{n}{2}a^{n-1}b^2 + \ldots + \binom{n}{n-1}a^2b^{n-1} + \binom{n}{0}ab^n$$

$$+ \binom{n}{0}a^n b + \binom{n}{1}a^{n-1}b^2 + \ldots + \binom{n}{n-2}a^2b^{n-1} + \binom{n}{n-1}ab^n + \binom{n+1}{n+1}b^{n+1}$$

By adding the two last rows and applying Lemma 2.4.3 obtain that Eq. (2.4.3) holds for $n+1$ as well, which concludes the proof.

At this point, it should be emphasized, that during actual calculations, we do not have to present each number as a cut with two infinite sets of rational numbers and perform our calculations on the elements of these sets. This would be correct, but endless, tiresome and *unnecessary*. Instead, we represent each number in the familiar form of a decimal fraction $d_0.d_1d_2\ldots$, finite or infinite, and use all the usual standard techniques to calculate. If the number is rational, the equivalent decimal fraction is either finite, or infinite and periodic. For example, 0.25 represents the rational number $1/4$ while $3/11$ is represented by $0.2727\ldots$. An irrational number is represented by an infinite non-periodic decimal fraction: $\sqrt{3} = 1.7320508\ldots$. Note that a decimal fraction $d_0.d_1d_2\ldots$, which represents a cut (A,B), actually *defines* the set A. Indeed, A is the union of all the sets $D_i = \{r \mid r \in R\,, r \le d_0.d_1d_2\ldots d_i\}$.

The set of the real numbers can be also represented geometrically, as all the points of a straight line called the *real line* or the *real axis* (Figure 2.4.2). The one-to-one correspondence between numbers and points defines the order of the real numbers from left to right.

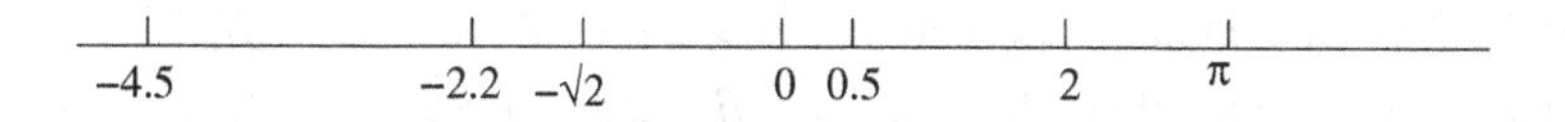

Fig. 2.4.2. The real line.

Those readers, who feel somewhat uncomfortable with the Dedekind's cuts and with the irrational numbers, being located on the real line, may try and visualize a real number as a point where the two disjoint sets A and B meet. This *meeting point*, is the rightmost point of A, if the number is rational. If A has no rightmost point, A and B still meet at *some point*, which now represents an irrational number α (Figure 2.4.3).

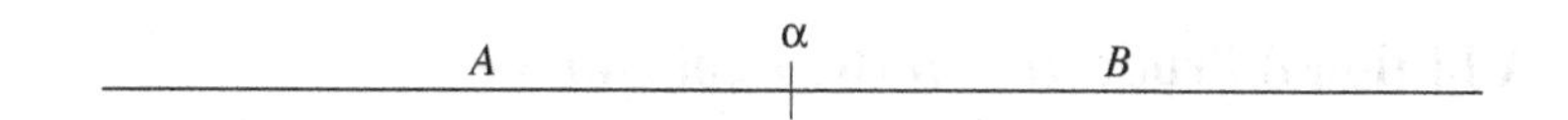

Fig. 2.4.3. An irrational number: A and B meet at α.

PROBLEMS

1. If k is a prime number, then $x^l = k$ has no rational solution for any positive integer $l > 1$.

2. Let α and r be irrational and rational numbers respectively. Then, the real numbers $\alpha+r$, αr are both irrational.
3. Find the solution of $(A,B)+(X,Y)=(C,D)$ for arbitrary cuts (A,B), (C,D).
4. Find the solution of $(A,B)\cdot(X,Y)=(C,D)$, $(A,B)\neq 0$ for arbitrary cuts (A,B), (C,D).
5. Let α,β,γ denote arbitrary cuts. Show: (a) $\alpha\beta=\beta\alpha$ (b) $(\alpha\beta)\gamma=\alpha(\beta\gamma)$.
6. Given rational numbers $r,s\,;r<s$ find an irrational number $\alpha:\ r<\alpha<s$. (Hint: Find a rational number t between r and s ; then use Archimedes' law and obtain an irrational number of the form $t+\sqrt{2}/n$ that does not exceed s).
7. Given a real number $\alpha=(A,B)$, show: (a) $r<\alpha$ for every $r\in A$. (b) $\alpha<r$ for every $r\in B$.
8. Given irrational numbers $\alpha,\beta\,;\alpha<\beta$ find a rational number $t:\alpha<t<\beta$.
9. Given real numbers (rational or irrational) $\alpha,\beta\,;\alpha<\beta$ find a rational number $t:\alpha<t<\beta$ and an irrational number $\gamma:\alpha<\gamma<\beta$.
10. If α,β are positive real numbers such that $\alpha^2=\beta^2=2$, then $\alpha=\beta$.
11. If α,β,γ are real numbers such that $\alpha>\beta$, $\gamma>0$, then $\alpha\gamma>\beta\gamma$.
12. If α,β are real numbers such that $\alpha>\beta$, then $\alpha+\gamma>\beta+\gamma$ for all real γ.
13. (a) Show $\alpha^2\geq 0$ for arbitrary real number α. (b) Prove Bernoulli's inequality: $(1+\alpha)^n\geq 1+n\alpha$ for all positive number n and real number α such that $\alpha>-1$.
14. Prove Lemma 2.4.3.
15. Show the existence of a solution to $x^n=a$ for arbitrary $a\in R^+$, $n\in N$.
16. Show that for arbitrary $n\in N$: $\sum_{k=1}^{n}\binom{n}{k}=2^n$.

2.5 Additional Properties of the Real Numbers

An important notation associated with numbers is the absolute value. We define it, derive the triangle inequality and then use it to get other basic relations, such as the Cauchy – Schwarz inequality.

Definition 2.5.1. For every real number $a\in\Re$, we define the *absolute value* of a as

$$|a|=\begin{cases} a\ , & a\geq 0\\ -a\ , & a<0\end{cases}\tag{2.5.1}$$

For example, $|17| = 17$, $|-13| = 13$, $|0| = 0$.

Some basic properties of the $|\cdot|$ operator, which can be easily derived, are

(1) $-|a| \le a \le |a|$, $a \in \Re$.
(2) $|ab| = |a||b|$, $a,b \in \Re$.
(3) $|a/b| = |a|/|b|$, $a,b \in \Re$, $b \neq 0$.

Less trivial is the next result.

Theorem 2.5.1. The *triangle inequality*

$$|a+b| \le |a| + |b| \tag{2.5.2}$$

is valid for arbitrary real numbers $a,b \in \Re$.

Proof.

Six cases should be considered:

(a) $a \ge 0$, $b \ge 0$: $|a+b| = ||a| + |b|| = |a| + |b|$.
(b) $a \ge 0$, $b < 0$, $a+b \ge 0$: $|a+b| = a+b < a-b = |a| + |b|$.
(c) $a \ge 0$, $b < 0$, $a+b < 0$: $|a+b| = -a-b < a-b = |a| + |b|$.

If change the roles of a and b , we obtain the three remaining cases.
By using induction we derive the more general triangle inequality

$$|a_1 + a_2 + \ldots + a_n| \le |a_1| + |a_2| + \ldots + |a_n|$$

The next result, known as the Cauchy – Schwarz inequality, is most helpful in numerous applications in analysis. It is based on the following lemma.

Lemma 2.5.1. If the inequality $ax^2 + bx + c \ge 0$ holds for some fixed $a,b,c \in \Re$ and all $x \in \Re$, then $b^2 \le 4ac$.

Proof.

If $a = 0$, then $ax^2 + bx + c \ge 0$, $x \in \Re$ implies $bx + c \ge 0$, $x \in \Re$. Consequently $b = 0$ (why?) and $b^2 \le 4ac$ since both sides vanish. If $a > 0$ we rewrite

$$ax^2 + bx + c = a(x + \frac{b}{2a})^2 + c - \frac{b^2}{4a} \geq 0$$

and since the inequality must hold for all $x \in \Re$, it is valid also for $x = -b/(2a)$ which implies $c - b^2/(4a) \geq 0$, i.e. $(4ac - b^2)/(4a) \geq 0$. Since $a > 0$ the numerator is nonnegative and we get $4ac \geq b^2$. The case $a < 0$ cannot occur (why?). This concludes the proof.

We now introduce the *sigma* notation, defined as

$$\sum_{k=1}^{n} a_k = a_1 + a_2 + \ldots + a_k$$

Theorem 2.5.2 (*Cauchy – Schwarz inequality*). Let $n \in N$ and $a_i, b_i \in \Re$ for $1 \leq i \leq n$. Then

$$\left(\sum_{k=1}^{n} a_k b_k\right)^2 \leq \left(\sum_{k=1}^{n} a_k^2\right)\left(\sum_{k=1}^{n} b_k^2\right) \tag{2.5.3}$$

Proof.

Define $f(x) = \sum_{k=1}^{n} (a_k + b_k x)^2$, $x \in \Re$. Then $f(x) \geq 0$, $x \in \Re$, i.e.

$$\left(\sum_{k=1}^{n} b_k^2\right)x^2 + 2\left(\sum_{k=1}^{n} a_k b_k\right)x + \sum_{k=1}^{n} a_k^2 \geq 0 \quad , \quad x \in \Re$$

and by applying Lemma 2.5.1 we conclude Eq. (2.5.3).

In the particular case $b_k = 1$, $1 \leq k \leq n$, the Cauchy-Schwarz inequality implies

$$\left(\sum_{k=1}^{n} a_k\right)^2 \leq n \sum_{k=1}^{n} a_k^2$$

or

$$\left(\frac{\sum_{k=1}^{n} a_k}{n}\right)^2 \leq \frac{\sum_{k=1}^{n} a_k^2}{n} \tag{2.5.4}$$

Definition 2.5.2. For arbitrary positive real numbers a_k , $1 \le k \le n$ we define

$$A(a_1,a_2,\ldots,a_n) = \frac{a_1 + a_1 + \ldots + a_n}{n} \qquad \textit{(arithmetic mean)} \qquad (2.5.5)$$

$$G(a_1,a_2,\ldots,a_n) = \sqrt[n]{a_1 a_2 \ldots a_n} \qquad \textit{(geometric mean)} \qquad (2.5.6)$$

$$H(a_1,a_2,\ldots,a_n) = \frac{1}{\frac{1}{n}\left(\frac{1}{a_1} + \frac{1}{a_2} + \ldots + \frac{1}{a_n}\right)} \qquad \textit{(harmonic mean)} \qquad (2.5.7)$$

$$Q(a_1,a_2,\ldots,a_n) = \sqrt{\frac{a_1^2 + a_2^2 + \ldots + a_n^2}{n}} \qquad \textit{(qudratic mean)} \qquad (2.5.8)$$

By virtue of Eq. (2.5.4) we have $A(a_1,a_2,\ldots,a_n) \le Q(a_1,a_2,\ldots,a_n)$, i.e., the arithmetic mean never exceeds the quadratic mean. For example

$$3 = \frac{1+2+6}{3} < \sqrt{\frac{1^2 + 2^2 + 6^2}{3}} = \sqrt{\frac{41}{3}}$$

We will now show that the geometric mean never exceeds the arithmetic mean.

Lemma 2.5.2. For arbitrary positive real numbers $a,b \in \Re$, $a \le b$ and $m \in N$ we have

$$a^m b \le \left(\frac{ma+b}{m+1}\right)^{m+1}$$

Proof.

By virtue of Bernoulli's inequality we get

$$\left(\frac{ma+b}{m+1}\right)^{m+1} = \left(a + \frac{b-a}{m+1}\right)^{m+1} = a^{m+1}\left(1 + \frac{b-a}{a(m+1)}\right)^{m+1} \ge a^{m+1}\left(1 + \frac{(m+1)(b-a)}{a(m+1)}\right) = a^m b$$

which concludes the proof.

Theorem 2.5.3. For arbitrary positive real numbers a_k , $1 \le k \le n$ we have

$$G(a_1,a_2,\ldots,a_n) \le A(a_1,a_2,\ldots,a_n) \qquad (2.5.9)$$

Proof.

Without any loss of generality, we may assume that a_k , $1 \le k \le n$ are already given in an increasing order, i.e., $a_1 \le a_2 \le \ldots \le a_n$. Replace a_1, a_2 by the two equal numbers $\alpha_2 = (a_1 + a_2)/2$. By applying Lemma 2.5.2 with $m = 1$ we obtain a new set of increasing numbers $\alpha_2, \alpha_2, a_3, \ldots, a_n$ with the *original* arithmetic mean, but with a *non-decreasing* geometric mean. In the next step we replace α_2, α_2, a_3 by the three equal numbers $\alpha_3 = (2\alpha_2 + a_3)$ and leave the remaining numbers unchanged. By using Lemma 2.5.2 with $m = 2$ we get the increasing numbers $\alpha_3, \alpha_3, \alpha_3, a_4, \ldots, a_n$ still maintaining the original arithmetic mean and a non-decreasing geometric mean. After $(n-1)$ steps we obtain n equal numbers $\alpha_n, \alpha_n, \ldots, \alpha_n$ ($\alpha_n = \sum_{k=1}^{n} a_k$). Denote the original arithmetic and geometric means by A, G and the final means by A^*, G^* respectively. Then, $A^* = A = G^* = \alpha_n$, $G \le G^*$ which implies $G \le A$ as stated.

It can be easily derived that $G(a_1, a_2, \ldots, a_n) = A(a_1, a_2, \ldots, a_n)$ if and only if all the numbers are equal. Also, by applying Theorem 2.5.3 on the *inverses* of a_k , $1 \le k \le n$, i.e., to $1/a_1, 1/a_2, \ldots, 1/a_n$ we get $H(a_1, a_2, \ldots, a_n) \le G(a_1, a_2, \ldots, a_n)$. Hence, for any positive real numbers, we have

$$H(a_1, a_2, \ldots, a_n) \le G(a_1, a_2, \ldots, a_n) \le A(a_1, a_2, \ldots, a_n) \le Q(a_1, a_2, \ldots, a_n) \quad (2.5.10)$$

Example 2.5.1. Consider the numbers $1, 3, 4, 7, 10$. The calculated means are $A = 5$, $G = \sqrt[5]{840} \approx 3.845$, $Q = \sqrt{35} \approx 5.916$, $H \approx 2.738$ and they satisfy Eq. (2.5.9) (the symbol $\approx$ means ‘approximately’).

PROBLEMS

1. For arbitrary real numbers $a, b \in \Re$: $|a - b| \ge ||a| - |b||$.
2. For arbitrary real numbers $a, b, c \in \Re$: $|a - b| \le |a - c| + |c - b|$.
3. For arbitrary $a, b \in \Re$: $|ab| = |a||b|$ and if $b \ne 0$: $\left|\frac{a}{b}\right| = \frac{|a|}{|b|}$.
4. If $a < c < b$ for $a, b, c \in \Re$, then $|a - b| = |a - c| + |c - b|$.
5. Find all x for which $|x + 2| < 3$ (Hint: assume first $x + 2 \ge 0$, then $x + 2 < 0$).

6. Find all x which satisfy $|x+5| < |x-7|$.

7. Show $a + \frac{1}{a} \geq 2$ for all $\Re^+$.

8. Show $a > b \Leftrightarrow a^2 > b^2$ for all $a, b \in \Re$.

9. Prove the *Minkowsky Inequality*

$$\left(\sum_{k=1}^{n}(a_k + b_k)^2\right)^{1/2} \leq \left(\sum_{k=1}^{n} a_k^2\right)^{1/2} + \left(\sum_{k=1}^{n} b_k^2\right)^{1/2}$$

for arbitrary real numbers a_k, b_k, $1 \leq k \leq n$ (*Hint*: use the result of problem 6 combined with the Cauchy-Schwarz inequality).

10. Show $n! \leq \left(\frac{n+1}{2}\right)^n$, $n \in N$ without using induction.

11. Show $r^{(n-1)/2} \leq \frac{r^n - 1}{n(r-1)}$, $n \in N$ for all $r \in R^+$, $r \neq 1$ without induction.

12. Let the set $\{a_1, a_2, \ldots, a_n\}$ satisfy $a_i \neq 0$, $1 \leq i \leq n$. Show:

$$n^2 \leq \sum_{i=1}^{n} a_i^2 \sum_{i=1}^{n}\left(\frac{1}{a_i^2}\right).$$

13. Use the result of problem 12 to prove: $n \leq \frac{(n+1)}{2}\left(1 + \frac{1}{2} + \ldots + \frac{1}{n}\right)$, $n \geq 1$.

3 Functions, Sequences and Limits

In this chapter we define the notations of *function* and *sequence* and introduce the most important concept of calculus: the *limit*. This chapter is vital to the understanding of any further reading; in particular, the reader must come to control the subject of limit and *convergence*.

3.1 Introduction

A preliminary notation, necessary for introducing the function concept, is the Cartesian product.

Definition 3.1.1. The *Cartesian product* of given sets A and B is the set $A \times B$ of all the ordered pairs (a,b), where $a \in A$ and $b \in B$:

$$A \times B = \{(a,b) \mid a \in A , b \in B\} \tag{3.1.1}$$

Example 3.1.1. Let $A = \{1,2,4\}$, $B = \{2,3\}$. Then

$$A \times B = \{(1,2),(1,3),(2,2),(2,3),(4,2),(4,3)\}$$

Definition 3.1.2. Any subset of the $A \times B$ is called a *relation* from A to B. The set of all the ordered pairs of a relation present the *graph* or the *curve* of the relation.

Example 3.1.2. In the previous example, the set $\{(1,2),(2,2),(2,3),(4,3)\}$ is a relation from A to B. In order to obtain the graph associated with this relation, we first draw two infinitely long straight lines, perpendicular to each other. These horizontal and vertical lines are called *axes* and are usually denoted by x and y respectively. Secondly, we mark all the points corresponding to the ordered pairs, relative to the intersection point of the axes (Fig. 3.1.1), called *origin* and defined as $(0,0)$.

M. Friedman and A. Kandel: Calculus Light, ISRL 9, pp. 37–86.
springerlink.com

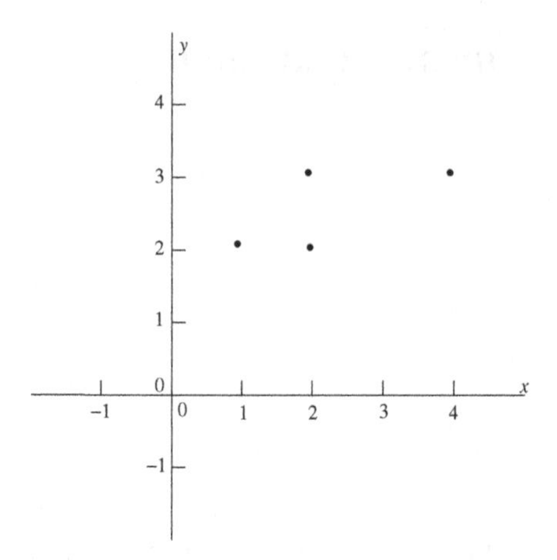

Fig. 3.1.1. The graph of the relation in Example 3.1.2.

Example 3.1.3. The following sets are relations from $\Re$ to $\Re$:

(a) $S_1 = \{(x, y) \mid x \in \Re, y = x^3$ (b) $S_2 = \{(x, y) \mid x, y \in \Re, x^2 + y^2 = 1\}$

In case (a) we calculate x^3 for every $x \in \Re$ and place the ordered pair (x, x^3) in S_1. In case (b), given an arbitrary $x \in \Re$, we find all the y's such that $x^2 + y^2 = 1$ and place each (x, y) in S_2.

Throughout this text we will usually discuss relations where A and B are subsets of $\Re$. The most commonly used subsets are the intervals, which are defined next.

Definition 3.1.3. The following subsets of $\Re$ are called *intervals*:

$[a,b] \equiv \{x \mid a \le x \le b\, ,\, a,b \in \Re\}$ (*closed interval* with endpoints a and b)
$(a,b) \equiv \{x \mid a < x < b\, ,\, a,b \in \Re\}$ (*open interval* with endpoints a and b)
$[a,b) \equiv \{x \mid a \le x < b\, ,\, a,b \in \Re$
$(a,b] \equiv \{x \mid a < x \le b\, ,\, a,b \in \Re$
(*semi open interval* with endpoints a and b)

A semi open interval is also called '*semi closed*' or '*half open* (closed)'. The symbol '$\equiv$' means "defined as".

For example, the set of all the real numbers between 2 and 17.5 (including 2 and 17.5) is the closed interval $[2,17.5]$. We often rewrite the set of all the real numbers, $\Re$, as $\Re = \{x \mid -\infty < x < \infty\}$ or as $\Re = (-\infty, \infty)$. By $(-\infty, a)$ we mean "all the real numbers less than a". By (b, ∞) we mean "all the real numbers greater than b". If the endpoint is included we write $(-\infty, a]$ and $[b, \infty)$. Note that the symbols $-\infty$ (minus infinity) and ∞ (infinity) are not legitimate real numbers.

Example 3.1.4. All the real numbers greater or equal to 5 are denoted by $[5,\infty)$. All the real numbers less than $\sqrt{2}$, are represented by $(-\infty,\sqrt{2})$. The set of all the real numbers can be rewritten as the union $(-\infty,3)\cup[-3,10]\cup(10,\infty)$.

PROBLEMS

1. Let $A=\{x,1\}$, $B=\{2,a,0\}$. Find the Cartesian products $A\times B$ and $B\times B$.
2. Let $A=\{x,y,1\}$, $B=\{2,a,0\}$, $C=\{2,b,y\}$. Find $A\times(B-C)$ and $(A-C)\times B$.
3. Let A,B denote arbitrary sets. Show that $A\times B=B\times A$, if and only if $A=B$.
4. What is the relation $S=\{(x,y)\mid x\in\{2,3\}, -\infty<y<\infty\}$?
5. Consider the relation $S=\{(p,q)\mid p,q,(p+q+1)\ primes\}$. Find the part of the graph of S for which $1<p,q<10$.
6. Define a relation whose graph is the upper positive half of the $x-y$ plane.

3.2 Functions

We are particularly interested in one type of relations, called 'functions'.

Definition 3.2.1. A *function* f, is a relation from A to B, whose graph, denoted by *graph*(f), satisfies: If $(a,b),(a,c)\in graph(f)$ then $b=c$. The set

$$D_f=\{a\mid a\in A \text{ and there exists } b\in B \text{ such that } (a,b)\in graph(f)\} \quad (3.2.1)$$

is called the *domain* of f. The set

$$R_f=\{b\mid b\in B \text{ and there exists } a\in D_f \text{ such that } (a,b)\in graph(f)\} \quad (3.2.2)$$

is called the *range* of f.

The most important feature of a function f is that for every $x\in D_f$, there exists a *unique* $y\in B$, such that $(x,y)\in graph(f)$. We call y the *image* of the *source* x and denote it by $f(x)$. If we restrict ourselves to sets A and B that are subsets of $\Re$, the function f is called a *real-valued function.* Note that f is always *single-valued.* We also say that f is a *mapping of* $D_f\subseteq A$ to (or into) B, or a *mapping from* A to B. The first coordinate x of an ordered pair $(x,y)\in graph(f)$, is called the *independent* coordinate. The second one, y, is the *dependent* coordinate.

Example 3.2.1. The relation $\{(x,y) \mid 0 \le x \le 1,\ y = x^2 + x\}$ is a function from $\mathfrak{R}$ to $\mathfrak{R}$. It is also a mapping of $[0,1]$ to $\mathfrak{R}$. Indeed, for every x between 0 and 1, the image y is uniquely determined. On the other hand, the relation $S = \{(x,y) \mid x, y \in \mathfrak{R},\ x^2 + y^2 = 1\}$ is not a function, since for arbitrary $x, 0 < x < 1$ there is more than one ordered pair (x,y), which belongs to the graph of S. In fact, there are two different y_1 and y_2, such that $(x,y_1),(x,y_2)$ are in *graph*(S), i.e. $x^2 + y_1^2 = x^2 + y_2^2 = 1$. For example $x = 1/2$ has two images $y_1 = \sqrt{3}/2$ and $y_2 = -\sqrt{3}/2$. For an arbitrary x, we can represent S by $y = \pm\sqrt{1-x^2}$, an explicit form of the two images $y_1 = +\sqrt{1-x^2}$ and $y_1 = -\sqrt{1-x^2}$. Usually, by writing $\sqrt{(\cdot)}$ we mean the unique positive square root which is now a function and not just a relation. The graphs of the function $y = x^2 + x$ and of the relation S are given in Fig. 3.2.1.

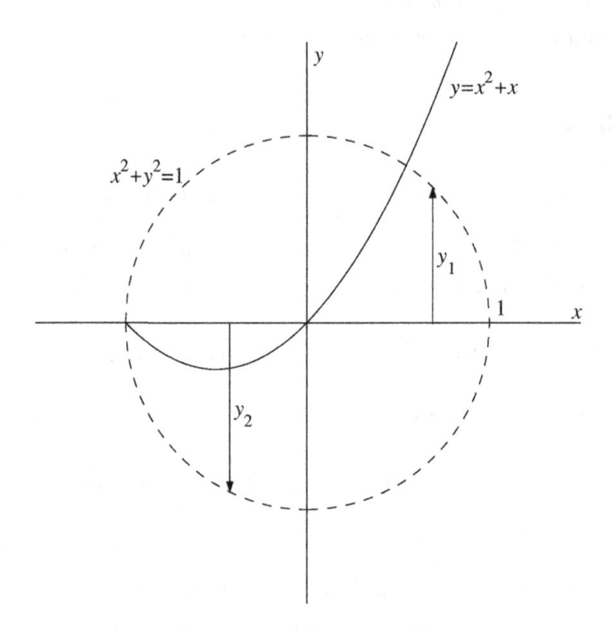

Fig. 3.2.1. A function (solid) vs. a relation (dashed).

Whether a given relation S, is a function, can be determined by the following test: If every vertical line through $(x,0)$, $x \in D_S$ (the relation's domain), intersects the relation's graph at exactly one point, then this relation is a function.

Example 3.2.2. In the previous example, the function $y = x^2 + x$ has the domain $[0,1]$ and we can easily derive that its range is $[0,2]$. However, if we agree to extend the domain to the largest subset of $\Re$, over which the expression $x^2 + x$ is *well defined*, i.e., has a single value, then the domain is clearly $\Re$ itself. In order to decide whether an arbitrary number a is inside the range, we need to solve the equation $x^2 + x = a$ whose formal solutions are given by

$$x_{1,2} = \frac{-1 \pm \sqrt{1+4a}}{2}$$

Since there is no real number, which is the square root of another negative real number, we must have $1+4a \geq 0$. Thus, the *maximum range* of the function, defined as the image of the *maximum domain*, is the interval $[-1/4, \infty)$.

Consider a function f from A to B, denoted by $f: A \to B$, and an arbitrary number $x \in \Re$. If $x \in D_f$, we say that f is *defined* at x and $f(x)$ *exists.* Otherwise, if $x \notin D_f$, then f is *not defined* at x and $f(x)$ *does not exist.* It is usually relevant to consider only the domain of a function rather then whole set A. Therefore, by writing $f: A \to B$ we assume (unless otherwise stated) $A = D_f$ and $R_f \subseteq B \subseteq \Re$ (it may be not be trivial to obtain the exact R_f). For example, if $f(x) = 1 + x^2 + \frac{1}{1+x^2}$ then $f: \Re \to \Re^+$ although a further inquiry reveals $R_f = [2, \infty)$.

A function $f: A \to B$ which satisfies $x_1 \neq x_2 \Rightarrow f(x_1) \neq f(x_2)$ for arbitrary different $x_1, x_2 \in A$, is called *one-to-one* (also: $1-1$). If $R_f = B$, we say that f maps A *onto* B.

Example 3.2.3. Consider the function $y = x^2 + x$ from $\Re$ to $\Re$. It is not one-to-one since for example $x_1 = 3$ and $x_2 = -4$ provide the same image $y = 12$. It is not onto, since the real number -10 is not within the function's range (see Example 3.2.2). On the other hand the function $y = x^3 + 1$ from $\Re$ to $\Re$ is both one-to-one and onto.

Every one-to-one function $f: D_f \to R_f$, can be associated with another one-to-one function $g: R_f \to D_f$, defined as $g(y) = x$, $y \in R_f$, where x is the unique source of y in D_f. Thus, $D_g = R_f$ and $R_g = D_f$. The function g is called the *inverse* of f and is denoted by f^{-1}. For example, if $y = f(x) = 5x + 1$, the inverse function is $x = g(y) = \frac{y-1}{5}$. In order to maintain consistency we will keep using x as the independent coordinate and write $g = f^{-1} = (x-1)/5$.

We now present the concepts of bounded and monotone functions. First, we define a bounded set.

Definition 3.2.2. A set $A \subset \Re$ is called *bounded,* if there is a real number $M > 0$ such that

$$|x| \le M \ , \ x \in A \tag{3.2.3}$$

If there is no such M, the set A is called *unbounded.* If there is a real number M_1 such that $x \le M_1$, $x \in A$, we say that A is *bounded above* and M_1 is an *upper bound* of A. If there is a real number M_2 such that $M_2 \le x$, $x \in A$, we say that A is *bounded below* and M_2 is a *lower bound* of A.

It is easily seen that a set is bounded, if and only if, it is bounded both above and below. Thus, the information that a set A is unbounded, implies that A is *at least* unbounded above, or unbounded below.

Definition 3.2.3. Let A be a bounded above set. If a *smallest* upper bound of A exists, we say that A has a *supremum* or a *least upper bound* and denote it by either $\sup(A)$ or $l.u.b.(A)$. If A is bounded below and a *largest* lower bound of A exists, we say that A has an *infimum* or a *greatest lower bound* and denote it by either $\inf(A)$ or $g.l.b.(A)$.

Example 3.2.4. The set $\Re^+ = (0, \infty)$ is bounded below and has the infimum 0. The set $[2,7)$ is bounded. It has the supremum 7 and the infimum 2. The set of the natural numbers is unbounded below and unbounded above.

The next result confirms the existence of $\sup(A)$ for an arbitrary bounded above set.

Theorem 3.2.1. If A is a nonempty bounded above set, then $\sup(A)$ exists.

Proof.

Let B consist of all the rational numbers, which are upper bounds to A and let $C = R - B$. We will show that the pair (C, B) is a Dedekind cut. Clearly we constructed this pair so that every rational number is in one of its sets. Since A is bounded above, it must have at least one upper limit, say M_1. Find a rational number $r : M_1 < r < M_1 + 1$. Thus $r \in B$, i.e. B is nonempty. The requirement that A itself is nonempty guarantees the existence of some $a \in A$. Let r be a rational number *between* a and $a-1$ such that $a-1<r<a$. Thus, r is not an upper bound of A, i.e. $r \notin B$ and consequently $r \in C$ which confirms that C is also nonempty. It remains to show that every number in C is smaller than every number in B. Indeed, let $r \in C$, $s \in B$. Since r is not an upper bound of A, there is a number $a \in A$ such that $r<a$. Since $a \le s$ we get $r<s$. Hence $\alpha=(C,B)$ is a Dedekind cut.

We now prove that $\alpha=\sup(A)$ and start by showing that α is an upper bound of A. Indeed, if this is false, then for some $a \in A$ we have $\alpha<a$. Find a rational number $r: \alpha<r<a$. Since $\alpha<r$ we must have $r \in B$, i.e. r is a rational upper bound of A which contradicts $r<a$. Next, we show that no upper bound of A, is smaller than α. Otherwise, let $\beta<\alpha$ be an upper bound of A. Find a rational number $r: \beta<r<\alpha$. On one hand $r<\alpha$, i.e. $r \in C$. Consequently, r is not an upper bound of A. On the other hand, $\beta<r$ and since β is an upper bound of A, so is r. This contradicts the previous assumption, i.e. $\alpha=\sup(A)$.

The proof of the next result is left for the reader.

Theorem 3.2.2. If A is a nonempty bounded below set, then $\inf(A)$ exists.

Corollary 3.2.1. Every nonempty bounded set A has both a supremum and an infimum.

If the number $\alpha=\sup(A)$ belongs to A, it is the maximum number in A and we denote $\alpha=\max(A)$. If the number $\alpha=\inf(A)$ belongs to A, it is the minimum number in A and we denote $\alpha=\min(A)$. For example, the set $A=[2,7)$ satisfies $\sup(A)=7$, $\inf(A)=\min(A)=2$. This particular set has no maximum.

We now define bounded and unbounded functions.

Definition 3.2.4. A function $f: A \to B$, $A,B \subseteq \Re$ is called *bounded*, if there is a real number $M>0$ such that

$$|f(x)| \le M \ , \ x \in A \tag{3.2.4}$$

If there is no such M, we say that f is *unbounded*. If there is a real number M_1 such that $f(x) \le M_1$, $x \in A$, we say that f is *bounded above* and M_1 is an *upper bound* of f. If there is a real number M_2 such that $M_2 \le f(x)$, $x \in A$, we say that f is *bounded below* and M_2 is a *lower bound* of f.

For arbitrary function f, we can apply Theorems 3.2.1-2 to the set R_f and conclude that a bounded above f has a smallest upper bound, and a bounded below f has a greatest lower bound.

Definition 3.2.5. Let f be a bounded above function. The *smallest* upper bound of f is called *supremum* or *least upper bound* and is denoted by either $\sup(f)$ or $l.u.b.(f)$. If f is bounded below, the *largest* lower bound of f is called *infimum* or *greatest lower bound* and is denoted by either $\inf(f)$ or $g.l.b.(f)$. If $\sup(f)$ belongs to R_f, it is the maximum of f and we write $\sup(f) = \max(f)$. If $\inf(f)$ belongs to R_f, it is the minimum of f and we write $\inf(f) = \min(f)$.

Example 3.2.5. Consider $f(x) = \dfrac{x}{1+x^2}$ whose domain is $\Re$. Since $|x| \le |2x| \le 1 + x^2$, $x \in \Re$, f is bounded by 1. Further investigation provides $\sup(f) = 0.5$ and $\inf(f) = -0.5$. Since $f(1) = 0.5$ and $f(-1) = -0.5$ we can also write $\max(f) = 0.5$ and $\min(f) = -0.5$.

Example 3.2.6. Let $f(x) = \dfrac{1}{1+x}$, $x \in \Re$ and $x \ne -1$ (Fig. 3.2.2). We will show that f is unbounded above. If this is false, there is $M > 0$ such that $\dfrac{1}{1+x} \le M$, $x \in \Re$. Assume $x > -1$. Then, $0 < \dfrac{1}{1+x} < M$ and consequently $\dfrac{1}{M} < 1 + x$. However, the last inequality fails to hold for $x = -1 + 2/M$. Hence, $f(x)$ is not bounded above. It can be similarly shown that it is not bounded below as well. However, if we restrict the function's domain to $[-0.8, \infty)$, it becomes bounded, with $\sup(f) = 5$, $\inf(f) = 0$.

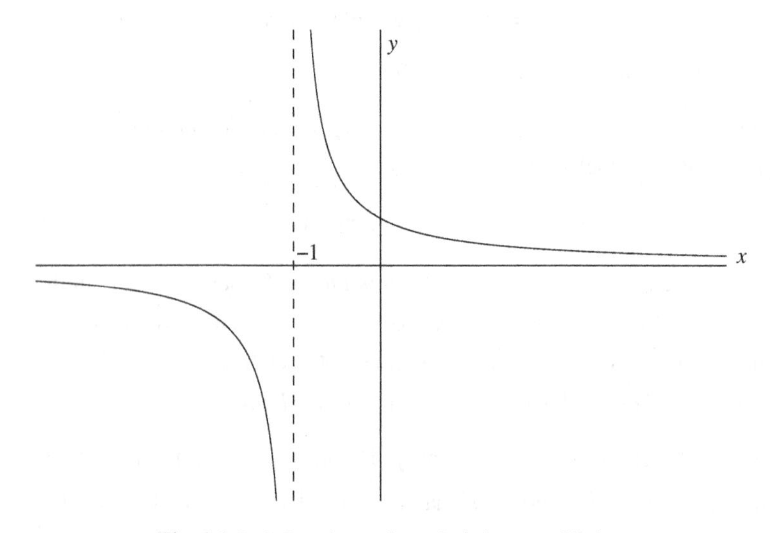

Fig. 3.2.2. A function unbounded above and below.

Example 3.2.7. The function $f(x) = x - |x|$, $x \in \Re$ which satisfies

$$f(x) = \begin{cases} 0, & x \ge 0 \\ 2x, & x < 0 \end{cases}$$

is bounded above by 0 but is not bounded below. Hence, f is unbounded. The function $f(x) = \frac{1}{x^2}$, $x \ne 0$ is bounded below by 0 but unbounded above. It is therefore unbounded.

The next notation is 'monotone functions'.

Definition 3.2.6. A function f is said to be *increasing* if $x_1 < x_2 \Rightarrow f(x_1) \le f(x_2)$ for all $x_1, x_2 \in D_f$. If $x_1 < x_2 \Rightarrow f(x_1) \ge f(x_2)$ for all $x_1, x_2 \in D_f$, the function is called *decreasing*. If $x_1 < x_2 \Rightarrow f(x_1) < f(x_2)$ for all $x_1, x_2 \in D_f$, the function is called *strictly increasing* and if $x_1 < x_2 \Rightarrow f(x_1) > f(x_2)$ for all $x_1, x_2 \in D_f$, the function is *strictly decreasing*. A function f which is either increasing over D_f or decreasing over D_f, is called *monotone*.

Example 3.2.7. The function $f_1(x) = [x]$, $x \in \Re$ defined as the greatest integer less than or equal to x, is increasing over $\Re$, but not strictly increasing over any interval (Figure 3.2.3).The function $f_1(x) = x - |x|$, $x \in \Re$ is also increasing but not strictly increasing. If we restrict it to $x \le 0$ it is strictly increasing. The function $f_2(x) = \frac{1}{x^2}$, $x \ne 0$ is not monotone throughout its whole domain. However, it is clearly strictly increasing over $(-\infty, 0)$ and strictly decreasing over $(0, \infty)$.

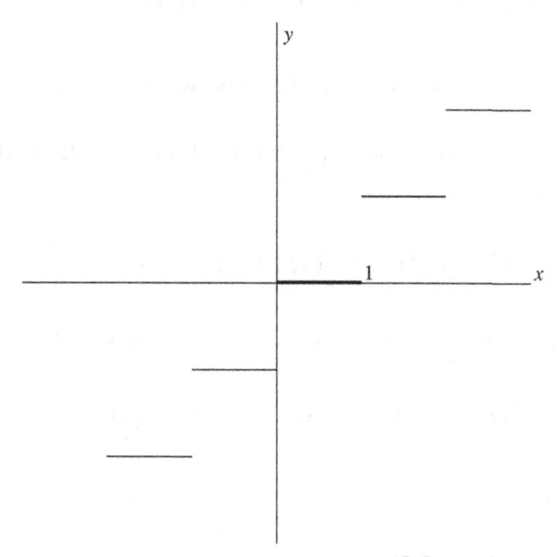

Fig. 3.2.3. Graph of $y = [x]$.

Example 3.2.8. Sometimes, finding the sub-domains where a function is either increasing or decreasing, is less simple than in the previous example. Consider the function $f_3(x) = \dfrac{x^2 - 3x + 1}{x^2 + 1}$, $x \in \Re$ whose graph in Fig. 3.2.4 represents three sub-domains of monotony:

1. $-\infty < x < -1$: strictly increasing.
2. $-1 \le x < 1$: strictly decreasing.
3. $1 \le x < \infty$: strictly increasing.

Finding these domains using elementary algebra, is possible but somewhat tiresome. Later, after introducing differentiation, we will familiarize ourselves with a more elegant procedure for determining the sub-domains where an arbitrary function is increasing or decreasing.

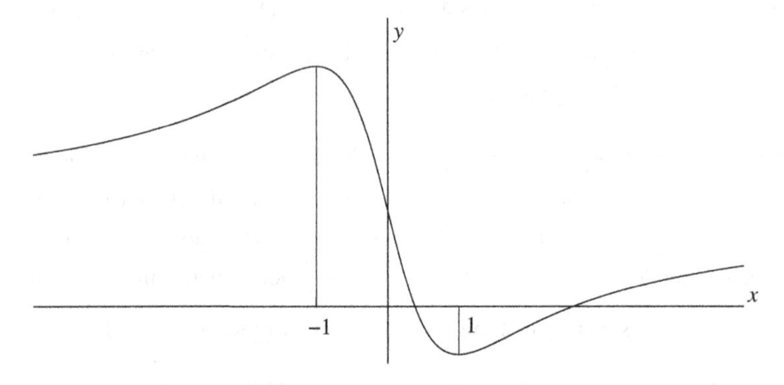

Fig. 3.2.4. Graph of $(x^2 - 3x + 1)/(x^2 + 1)$.

We now define the basic arithmetic operations between functions.

Definition 3.2.7. Let f and g denote arbitrary functions defined over the same domain D . We define

$$(f + g)(x) = f(x) + g(x) \,,\, x \in D \quad \text{(sum)} \tag{3.2.5}$$

$$(f - g)(x) = f(x) - g(x) \,,\, x \in D \quad \text{(difference)} \tag{3.2.6}$$

$$(fg)(x) = f(x)g(x) \,,\, x \in D \quad \text{(multiplication)} \tag{3.2.7}$$

and if $g(x) \ne 0$, $x \in D$

$$\left(\frac{f}{g}\right)(x) = \frac{f(x)}{g(x)}\ ,\ x \in D \qquad \text{(division)} \qquad (3.2.8)$$

Example 3.2.9. Let $f(x) = 3x^2 - \sqrt{x}\ , x \ge 0$ and $g(x) = 1 - x^2, x \in \Re$. Then $(f + g)(x) = 4x^2 - \sqrt{x} + 1\ , x \ge 0$ and $\left(\frac{f}{g}\right)(x) = \frac{3x^2 - \sqrt{x}}{1 - x^2}\ , x \ge 0, x \ne 1$.

Definition 3.2.8. Let f and g denote arbitrary functions such that R_g is a subset of D_f. The function $(f \circ g)(x) = f(g(x))\ , x \in D_g$ is called the *composite* function of f and g.

Example 3.2.10. If $f(x) = x^2$, $g(x) = \sqrt{x+1}$ then $(f \circ g)(x) = (\sqrt{x+1})^2 = x + 1$. Note that every image of g is in $D_f = \Re$ as requested. The domain of $f \circ g$ is $D_g = \{x \mid x \ge -1\}$.

PROBLEMS

1. Find the domains of the following relations from $\Re$ to $\Re$ and determine which of them are functions.

   (a) $\{(x, y) \mid x^2 + y^4 = 1\}$ (b) $\{(x, y) \mid y = \sqrt{x-1}\}$ (c) $\left\{(x, y) \mid \frac{x+y}{2x-y} = 1\right\}$

   (d) $\{(x, y) \mid \sqrt{x} - y^2 = x$

2. Find the domain and the range of $f(x) = \frac{1}{(1+x)^2}$ and draw its graph.

3. Find the domain and the range of $f(x) = \frac{1}{x^2 - 5x + 6}$ and draw its graph.

4. (a) Find the domain and the range of $f(x) = \frac{2x^2 + 1}{x - 1} + \frac{1}{x - 2}$ and draw its graph.
   (b) Compare the graph of part (a) with that of $g(x) = 2x$. Any conclusions?

5. Find which of the following functions are one-to one:

   (a) $17x - 5$ (b) $\sqrt{x+1}$ (c) $x^4 - x^2$ (d) x^3 (e) $x^3 + x$

6. Find the inverse, if it exists, of the following functions:

(a) $y = x^2 + 3$, $x \ge 0$ (b) $y = \dfrac{4x-1}{3x+5}$ (c) $y = x + \dfrac{1}{x}$ (d) $y = x^2 + \dfrac{1}{x}$

(e) $y = \sqrt{x^4 + x^2}$, $x \ge 0$

7. Show that $f(x) = x + \sqrt{x}$, $x \ge 0$ is one-to-one and find its inverse.
8. Find the domains of the following functions; determine which of them are bounded above and obtain upper bounds whenever they exist.

(a) x^2 (b) $2 + \dfrac{1}{x^2+1}$ (c) $x^2 - \sqrt{x^4+1}$ (d) $\sqrt[10]{x}$

9. Find the domains of the following functions; determine which of them are bounded below and obtain lower bounds whenever they exist.
(a) x^4 (b) $x^2 + x$ (c) $\sqrt[3]{x} - \sqrt{x}$ (d) $\dfrac{1}{\sqrt{x+1}}$
10. Determine whether the function is increasing, strictly increasing, decreasing, strictly decreasing or not monotone at all.

(a) $-x^3 - x$ (b) $x + \dfrac{1}{x}$ (c) $\dfrac{1+x^2}{x}$, $x \ge 2$ (d) $x - x^3$, $x \ge \dfrac{1}{\sqrt{3}}$ (e) $\lfloor x \rfloor + \lfloor x \rfloor^2$

11. Let $f(x) = \sqrt{x} + \dfrac{1}{x}$, $x \ge 0$ and $g(x) = x^3 - 7x^2 + 12$. Calculate and find the domains of $f+g$, $f-g$, fg, f/g.
12. Calculate the composite function $f \circ g$ in the following cases:

(a) $f = \sqrt{1-x^2}$, $g = \dfrac{1}{1+x^2}$ (b) $f = x^3 - \dfrac{1}{x}$, $g = \sqrt{1+x^2}$

(b) $f = \sqrt{2-x^2}$, $g = \sqrt{x^2-2}$ (d) $f = \dfrac{1}{x^2}$, $g = \dfrac{1}{x}$

13. (a) Let $f = x^2$, $g = \sqrt{x}$. Calculate $f \circ g$. (b) Use (a) to obtain a general rule for calculating $f \circ f^{-1}$.

3.3 Algebraic Functions

In most applications, a function is defined over an interval or several intervals by a 'formula', for example $y = x^3 - x$, $y = \dfrac{1}{\sqrt{x-1}}$ or $y = \sin(x)$. A particular class of functions, which are most commonly used in all scientific disciplines are the polynomials.

Definition 3.3.1. The function

$$p(x) = a_n x^n + a_{n-1} x^{n-1} + \ldots + a_1 x + a_0 \;,\; x \in \Re \tag{3.3.1}$$

where $a_i \in \Re \,, 0 \le i \le n \,; a_n \ne 0$, is called a real *polynomial* of *order* or *degree* n. The numbers a_i , $0 \le i \le n$ are the polynomial's *coefficients.*

Example 3.3.1. The function $y = -3x^2 + 5x + 2$ is polynomial of order 2, while $y = x^7 - 1$ is a polynomial of order 7. The *constant* function $y = 5$ is a polynomial of order 0.

The polynomials are a small fraction of a larger class of functions, called *algebraic functions*, which we define next.

Definition 3.3.2. A function $y = f(x)$, $x \in D_f$ which satisfies

$$p_n(x) y^n + p_{n-1}(x) y^{n-1} + \ldots + p_1(x) y + p_0(x) = 0 \,,\, x \in D_f \tag{3.3.2}$$

where $p_i(x)$, $0 \le i \le n$ are polynomials, is called an *algebraic function.*

Example 3.3.2. The functions x^2 , $\frac{2+x}{1-x^2}$, $\sqrt{1-x}$ are algebraic. Indeed, the first two satisfy Eq. (3.3.2) with $n = 1$. The function $y = \sqrt{1-x}$ satisfies $y^2 + x - 1 = 0$ which has the form of Eq. (3.3.2) with $n = 2$.

Every function that is not algebraic, is called *transcendental.* A function that satisfies Eq. (3.3.2) with $n = 1$ is called *rational.* Thus, a rational function is given by $y = \frac{p(x)}{q(x)}$, where $p(x)$ and $q(x)$ are polynomials.

In many applications, it is important to find a *root* r of a given function $f(x)$, i.e., a real number r such that $f(r) = 0$. A function may not possess any real roots. For example, the polynomial $x^2 + 1$ is always positive and therefore has no roots. On the other hand the polynomial $x^2 - 3x + 2$ has two real roots, 1 and 2.

If $r \in \Re$ is a root of an n-th order polynomial $p(x)$, then $p(x) = (x - r)q(x)$, where $q(x)$ is an $(n-1)$-th polynomial. This result is an immediate consequence of the following Lemma.

Lemma 3.3.1. For arbitrary $a, b \in \Re$ and $n \in N$

$$a^n - b^n = (a-b)(a^{n-1}b + a^{n-2}b^2 + \ldots + a^2b^{n-2} + ab^{n-1})$$

The proof using induction, is left for the reader.

Theorem 3.3.1. An arbitrary polynomial $p(x)$ which has a root $r \in \Re$, can be rewritten as $p(x) = (x-r)q(x)$ where $q(x)$ is another polynomial.

Proof.

Let $p(x) = \sum_{k=0}^{n} a_k x^k$. Since $p(r) = 0$ we have $\sum_{k=0}^{n} a_k r^k = 0$. Therefore

$$p(x) = \sum_{k=0}^{n} a_k x^k - \sum_{k=0}^{n} a_k r^k = \sum_{k=0}^{n} a_k (x^k - r^k)$$

By virtue of Lemma 3.3.1 we get $x^k - r^k = (x-r)q_k(x)$ for some polynomial $q_k(x)$, which implies

$$p(x) = \sum_{k=0}^{n} a_k (x^k - r^k) = (x-r)\sum_{k=0}^{n} a_k q_k(x)$$

where $q(x) = \sum_{k=0}^{n} a_k q_k(x)$ is a polynomial. This concludes the proof.

If a polynomial $p(x)$ has a root r and can be factorized as $p(x) = (x-r)^k q(x)$ where $k > 1$ and $q(r) \neq 0$, we say that r is a *multiple root* of multiplicity k. If r is not a multiple root, we say that r is a *simple root*. For example, $p(x) = x^3 - 3x^2 + x + 1$ has the simple root 1 and can be factorized as $p(x) = (x-1)(x^2 - 2x - 1)$. The polynomial $p(x) = x^3 - 3x^2 + 4$ can be rewritten as $p(x) = (x-2)^2(x+1)$. Thus, it has one simple root $r_1 = -1$ and one multiple root $r_2 = 2$ with multiplicity 2.

In order to calculate the roots of a function, *a priori* knowledge of their approximate locations is very helpful. Unfortunately, this is not always the case. If however, the function is a polynomial, one could use the following result of S. Gerschgorin to determine a bound for all the real roots.

Theorem 3.3.2. Let $p(x)=\sum_{k=0}^{n} a_k x^k$, $a_n \neq 0$. If $p(r)=0$, $r \in \Re$ then

$$|r| \le \frac{|a_0|+|a_1|+\ldots+|a_n|}{|a_n|} \tag{3.3.3}$$

Proof.

If $|r| \le 1$, Eq. (3.3.3) holds since its right-hand side is always no less than 1. We may therefore assume $|r|>1$. Since $\sum_{k=0}^{n} a_k r^k = 0$, one can rewrite

$$r = -\frac{a_{n-1}r^{n-1}+\ldots+a_1 r + a_0}{a_n r^{n-1}}$$

which leads to

$$|r| \le \frac{|a_{n-1}|}{|a_n|}+\frac{|a_{n-2}|}{|a_n||r|}+\ldots+\frac{|a_1|}{|a_n||r|^{n-2}}+\frac{|a_0|}{|a_n||r|^{n-1}} \le \frac{|a_{n-1}|}{|a_n|}+\ldots+\frac{|a_1|}{|a_n|}+\frac{|a_0|}{|a_n|}$$

Therefore

$$|r| \le \frac{|a_{n-1}|+\ldots+|a_1|+|a_0|}{|a_n|} \le \frac{|a_n|+|a_{n-1}|\ldots+|a_1|+|a_0|}{|a_n|}$$

which completes the proof.

Example 3.3.3. Let $p(x)=x^3-3x^2-16x-12$. The absolute values of polynomial's roots are bounded by 1+3+16+12=32. Clearly this is not a great bound since the actual roots of this polynomial are $6,-1$ and -2. Yet, if no indication about the roots' locations is available, Gerschgorin's theorem may provide useful information.

Definition 3.3.3. A real number r which is a root of a polynomial

$$p(x)=a_n x^n + a_{n-1}x^{n-1}+\ldots+a_1 x + a_0 \text{ , } a_n \neq 0$$

where the coefficients are all integers, is called an *algebraic number*. Any real number that is not algebraic is called a *transcendental number*.

Thus, the real numbers which can be divided to rational and irrational numbers, can be also divided to algebraic and transcendental numbers. Clearly, every rational number is also algebraic. The opposite is not true. For example, the number $\sqrt{2}$ is algebraic ($x^2 - 2 = 0$) but not rational. The number π is irrational and also transcendental. The proof of these facts is beyond the scope of this book.

The main reason for the popularity of the polynomials is that they are easily manipulated and at the same time may be used to *approximate* transcendental or complex algebraic functions. For example the transcendental function $\cos(x)$ can be approximated by the quadratic polynomial $p(x) = 1 - x^2/2$ for 'small' x . How small? Once the reader controls the concepts of differentiation and infinite series, he should be able to show that by approximating $\cos(x)$ by $p(x)$ we obtain a 'relative error' of less 0.3%, provided that $|x| \le 0.5$, namely,

$$\left|\frac{p(x) - \cos(x)}{\cos(x)}\right| \le 3 \cdot 10^{-3} \ , \ |x| \le 0.5$$

The quadratic $p(x)$ is plotted vs. $\cos(x)$ in Fig. 3.3.1. over the interval $[-1,1]$. The maximum deviation is at the endpoints, but even there it does not exceed over 7%.

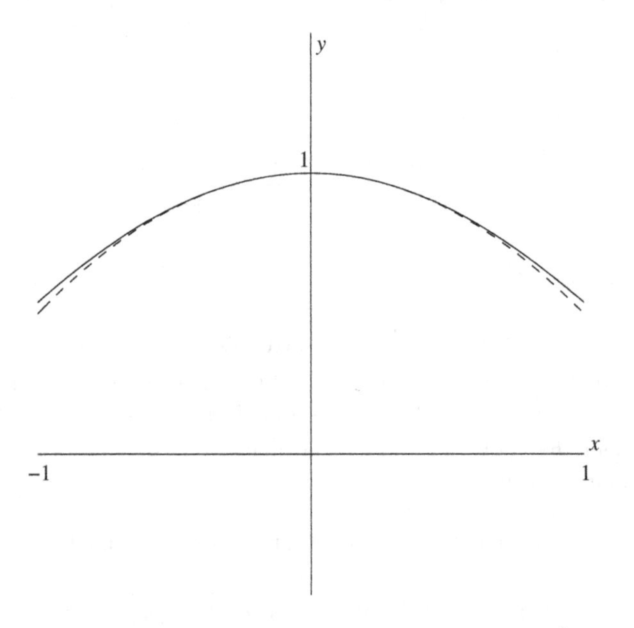

Fig. 3.3.1. Approximating $\cos(x)$ (solid) by $p(x) = 1 - 0.5x^2$ (dashed).

The most commonly used transcendental functions are the trigonometric functions such as $\sin(x), \cos(x), \tan(x), \cot(x)$, the exponential functions such as $10^x, e^x$ (the number e will be defined later in this chapter) and their inverses. Other functions that are not algebraic, are functions such as $[x]$, $|x|$ and

$$f(x) = \begin{cases} 0, & x \text{ rational} \\ 1, & x \text{ irrational} \end{cases} \tag{3.3.4}$$

Definition 3.3.4. A function $f(x)$, defined over the real axis, is called *periodic*, if there exists a real number T such that

$$f(x+T) = f(x), \; x \in \Re \tag{3.3.5}$$

The number T is a *period* of $f(x)$.

It is easily seen that if T is a period of $f(x)$, so is $-T$.

Example 3.3.4. $f(x) = \sin(x)$ is periodic (Fig. 3.3.2) with a period 2π, i.e. $\sin(x + 2\pi) = \sin(x)$ for all x. Clearly, 4π is another period of $\sin(x)$ and so is any integer multiple of 2π. However, the number 2π is the *smallest* of all the periods of $\sin(x)$.

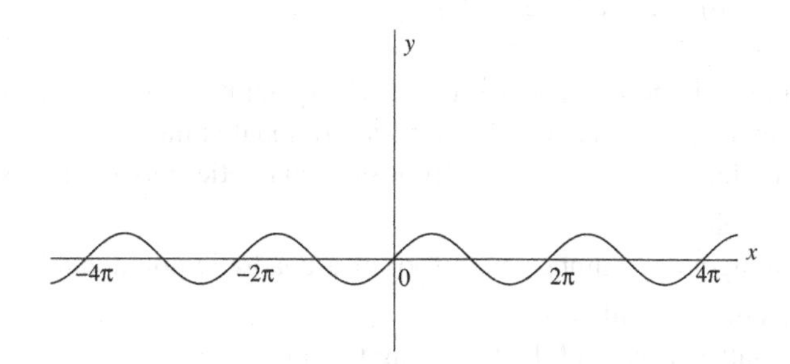

Fig. 3.3.2. The periodic function $\sin(x)$.

Polynomials, unlike trigonometric functions, are never periodic.

Definition 3.3.5. Consider a function $f(x), -a \le x \le a$. If $f(x) = f(-x)$ for all $x \in D_f$, the function is called an *even function* and if $f(x) = -f(-x)$ for all $x \in D_f$, the function is called an *odd function*.

Example 3.3.5. The functions $x^2, \cos(x), |x|$ are even functions, while $x^3, \tan(x)$ are odd functions. The functions 2^x and $x^5 - x^2$ are neither even nor odd functions. The function $\sqrt{x}$ is not defined for $x < 0$ and is therefore neither an even nor an odd function. The function defined by Eq. (3.3.4) is even since a real number x is rational if and only if $-x$ is rational.

PROBLEMS

1. Let x_1, x_2 be the two roots of the quadratic polynomial $ax^2 + bx + c$. Show

$$x_1 + x_2 = -\frac{b}{a}\ ,\ x_1 x_2 = \frac{c}{a}$$

2. Show that $f(x) = \sqrt{x} + x$ is an algebraic function.
3. Show that $f(x) = \sqrt{x} + \sqrt[3]{x}$ is an algebraic function.
4. Use induction to prove Lemma 3.3.1.
5. Find a bound to the roots (absolute value) of the polynomial $100x^3 + x - 1$.
6. Find a bound to the roots of the polynomial $x^3 - x - 100$.
7. Show that the following are algebraic numbers:

   (a) $\sqrt{2} + \sqrt{3}$ (b) $\sqrt{2} + \sqrt[3]{2}$ (c) $\sqrt{2}\sqrt[3]{3}$

8. Let α, an algebraic number, be a root of a quadratic polynomial. Show that the number $\alpha + r$ is algebraic for arbitrary rational number r.
9. Let α, an algebraic number, be a root of a quadratic polynomial. Show that α^2 is algebraic as well.
10. Let α, an algebraic number, be a root of a quadratic polynomial. Show that α^{2n} is algebraic for all $n > 1$.
11. Find the smallest period of the following functions:

    (a) $\cos(3x+1)$ (b) $\tan(x/2)$ (c) $\sin(x) + \tan(x)$

12. Find a domain where the function $y = x^2 + x^6 + \sqrt{x}$ is even.
13. Show that the sum and difference of two even (odd) functions is also even (odd) functions.
14. Show that the product of two even (odd) functions is also an even function.
15. Let $f(x)$, $g(x)$ be even and odd functions respectively. What can you say about the functions: (a) $f(x) + g(x)$ (b) $f(x) - g(x)$ (c) $f(x) \cdot g(x)$.

3.4 Sequences

In this section we introduce the basic notations that distinguish calculus from other subjects in mathematics: sequence, limit and convergence.

Definition 3.4.1. The range of a function $f : D_f \to \Re$, whose domain is an infinite subset of N - the set of natural numbers, is called a *sequence*.

If the domain of f is all of N, the sequence is composed of $f(1), f(2),\ldots$ which are the *terms* or *elements* of the sequence. The *general term* is denoted by $f(n)$ and the sequence is written as $\{f(1), f(2),\ldots, f(n),\ldots\}$ or, using subscripts, as $\{f_1, f_2,\ldots, f_n,\ldots\}$. Representing a sequence using subscripts emphasizes the existence of *order* among the sequence's terms: $f_1 -$ first term, $f_2 -$ second term, ..., $f_n -$ n-th term and so on. If we change the order of the terms we get a different sequence. We usually assume one of two cases: (a) $D_f = N$ or (b) $D_f = N - \{1,2,\ldots,m-1\}$ for some integer $m > 1$. This implies that the subscripts are consecutive positive integers. A sequence $\{a_1, a_2,\ldots, a_n,\ldots\}$ is denoted by $\{a_n\}_{n=1}^{\infty}$ or simply by $\{a_n\}$. A sequence $\{a_m, a_{m+1},\ldots, a_n,\ldots\}$ is denoted by $\{a_n\}_{n=m}^{\infty}$.

Example 3.4.1. The function $a(n) = n^2 + 1$, $n \ge 1$ defines the sequence $\{2,5,10,17,\ldots, n^2+1,\ldots\}$ also written as $\{n^2+1\}_{n=1}^{\infty}$ or as $\{n^2+1\}$. The function $a(n) = \frac{n+3}{n-2}$, $n \ge 3$ defines the sequence $\left\{6, \frac{7}{2}, \frac{8}{3},\ldots, \frac{n+3}{n-2},\ldots\right\}$ which we can write as $\left\{\frac{n+3}{n-2}\right\}_{n=3}^{\infty}$. The function $a(n) = 1 + (-1)^n$, $n \ge 1$ defines the sequence $\{0,2,0,2,\ldots,1+(-1)^n,\ldots\}$.

Consider now the sequence $\{a_n\}_{n=3}^{\infty} = \left\{\frac{n+3}{n-2}\right\}_{n=3}^{\infty}$. If we calculate 'enough' elements, we see that a_n 'approaches' 1 as the n increases. In fact, the difference between a_n and 1 is $\frac{n+3}{n-2} - 1 = \frac{5}{n-2}$. In the previous chapter, we showed, that for *any* given number $\varepsilon > 0$, we can find a natural number $n_0(\varepsilon)$ (i.e. n_0 depends on ε) such that $\frac{1}{n} < \varepsilon$ for all $n > n_0(\varepsilon)$. Similarly we can show (but leave it for the reader), that for *any* given number $\varepsilon > 0$, we can find a natural number $n_0(\varepsilon)$ such that $\frac{5}{n-2} < \varepsilon$ for all $n > n_0(\varepsilon)$. In other words, our sequence's elements $a_n = \frac{n+3}{n-2}$ get *as close as we want* to the number 1. Thus, by letting n increase *indefinitely*, the elements of the sequence 'converge' to 1.

We now present a rigorous definition of the 'convergence' notation for sequences.

Definition 3.4.2. A sequence $\{a_n\}$ *converges* to A as $n \to \infty$, if for *every* $\varepsilon > 0$, there exists a positive integer $n_0(\varepsilon)$ such that

$$n > n_0(\varepsilon) \Rightarrow |a_n - A| < \varepsilon \tag{3.4.1}$$

The number A is called the *limit* of $\{a_n\}$. A sequence that converges is called a *convergent sequence*. Otherwise, the sequence *diverges* and is called a *divergent sequence*.

Example 3.4.2. Let $a_n = \frac{n^2 - 1}{2n^2 + n + 1}$. By dividing both the numerator and the denominator by n^2, we obtain

$$a_n = \frac{1 - \frac{1}{n^2}}{2 + \frac{1}{n} + \frac{1}{n^2}}$$

The numerator and the denominator 'seem' to approach 1 and 2 respectively as n increases. Consequently, we are motivated to believe that $A = \frac{1}{2}$ is limit of the sequence. In fact, we easily get

$$\left|a_n - \frac{1}{2}\right| = \left|\frac{-\frac{1}{n} - \frac{3}{n^2}}{2\left(2 + \frac{1}{n} + \frac{1}{n^2}\right)}\right| = \frac{\frac{1}{n} + \frac{3}{n^2}}{2\left(2 + \frac{1}{n} + \frac{1}{n^2}\right)} < \frac{\frac{1}{n} + \frac{4}{n}}{4} = \frac{5}{4n}$$

Now, given an arbitrary $\varepsilon > 0$ we can guarantee $\left|a_n - \frac{1}{2}\right| < \varepsilon$, provided that $\frac{5}{4n} < \varepsilon$ or $n > \frac{5}{4\varepsilon}$. This suggests the choice $n_0(\varepsilon) = \left[\frac{5}{4\varepsilon}\right] + 1$ and the sequence indeed converges to $\frac{1}{2}$.

If a sequence $\{a_n\}$ converges to A, we denote it by $\lim_{n\to\infty} a_n = A$. For example, in the previous example we proved

$$\lim_{n\to\infty} \frac{n^2-1}{2n^2+n+1} = \frac{1}{2}$$

The convergence notation can be illustrated geometrically as follows. Consider the $x-y$ plane where the natural numbers are marked along the x- axis and the sequence's elements – along the y- axis (Fig. 3.4.1). The sequence $\{a_n\}$ converges to the number A, if and only if, for *every* $\varepsilon > 0$, except for a *finite* number of elements, all a_n are inside the strip $A-\varepsilon < y < A+\varepsilon$. The number of the 'outsiders' may increase when ε decreases, but it is *always* finite. The sequence that in fact produces Fig. 3.4.1, is $a_n = [n+\sin(2n)]/[2n+\cos(n)]$ which converges to $1/2$. The illustrated case is $\varepsilon = 0.05$ for which only 7 elements are out of the strip.

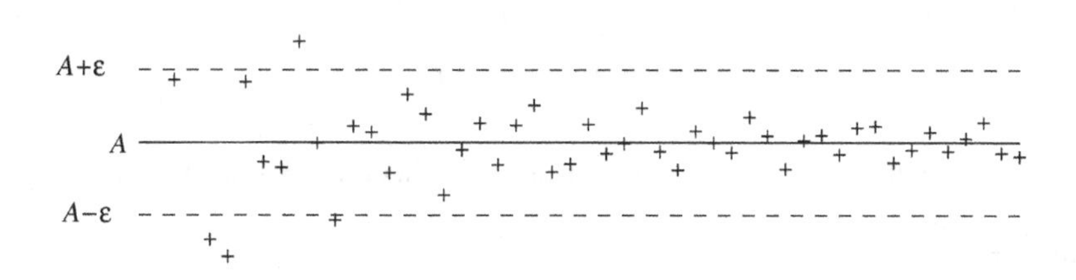

Fig. 3.4.1. Illustrating 'convergence'.

The next result follows directly from Definition 3.4.2 and its proof is left as an exercise for the reader.

Theorem 3.4.1. A sequence $\{a_n\}$ converges to A, if and only if, the sequence $\{b_n\} = \{a_n - A\}$ converges to 0, i.e. $\lim_{n\to\infty} a_n = A \Leftrightarrow \lim_{n\to\infty}(a_n - A) = 0$.

The question whether a limit of a sequence is unique, is answered clearly by the next theorem.

Theorem 3.4.2. Consider a sequence $\{a_n\}$. If $\lim_{n\to\infty} a_n = A$ and $\lim_{n\to\infty} a_n = B$ then $A = B$.

Proof.

Assume $A \neq B$. For any given $\varepsilon > 0$ there are $n_1(\varepsilon), n_2(\varepsilon)$ such that $|a_n - A| < \varepsilon$, $n > n_1(\varepsilon)$ and $|a_n - B| < \varepsilon$, $n > n_2(\varepsilon)$. Therefore, $|a_n - A| < \varepsilon$, $|a_n - B| < \varepsilon$ for $n > n_0 = \max[n_1(\varepsilon), n_2(\varepsilon)]$ and consequently, by using the triangle's inequality

$$|A - B| = |A - a_n + a_n - B| \le |A - a_n| + |a_n - B| < 2\varepsilon \ , \ n > n_0$$

The particular choice $\varepsilon = \frac{|A - B|}{3}$ leads to contradiction, i.e. $A = B$.

Example 3.4.3. Consider the sequence $a_n = \sqrt{1 + 1/n} - 1$, $n \ge 1$. If we multiply and divide the general term by $\sqrt{1 + 1/n} + 1$, we obtain

$$a_n = \left(\frac{1}{n}\right) \Big/ \left(\sqrt{1 + \frac{1}{n}} + 1\right)$$

and consequently $\lim_{n\to\infty} a_n = 0$. By virtue of Theorem 3.4.1, we also have $\lim_{n\to\infty} \sqrt{1 + 1/n} = 1$.

Definition 3.4.3. A sequence $\{a_n\}$ is called *bounded*, if $|a_n| \le M$, $n \in N$ for some $M > 0$.

The sequence $\{a_n\} = \{n\}$ is not bounded, while $\{a_n\} = \left\{\frac{n+1}{n}\right\}$ is bounded by 2. A bounded sequence may not have a limit. For example, the sequence $\{1,0,1,0,\ldots\}$ does not converge. On the other hand, the opposite is true.

Theorem 3.4.3. A sequence which has a limit, must be bounded.

Proof.

Let $\lim_{n\to\infty} a_n = A$. Then, for $\varepsilon = 1$, an $n_0(1)$ can be found such that $|a_n - A| < 1$ for all $n > n_0(1)$. Therefore, $|a_n| < 1 + |A|$, $n > n_0(1)$. The choice

$$M = \max[|a_1|, |a_2|, \ldots, |a_{n_0(1)}|, 1 + |A|]$$

guarantees $|a_n| \le M$, $n \in N$.

Example 3.4.4. Consider the sequence

$$a_n = \begin{cases} n^2, 1 \le n \le 10^6 \\ \dfrac{n}{2n+1}, n > 10^6 \end{cases}$$

The sequence converges to 0.5 and is therefore bounded. It does consist of some large elements such as $a_{10000} = 10^8$ but is still bounded, for example, by $M = 10^{12}$.

Example 3.4.5. The sequence $\{\sqrt{n}\}$ is not bounded. Indeed, if this were false, one could find a number $M > 0$ such that $\sqrt{n} \le M$ for every natural number $n \in N$. However, this last inequality does not hold for $n = ([M] + 1)^2$, i.e. the sequence is not bounded.

Definition 3.4.4. A sequence $\{a_n\}$ is called *bounded away from zero*, if there exist a number $\rho > 0$ and a positive integer m, such that $|a_n| \ge \rho$ for all $n > m$.

Example 3.4.6. The sequence $\left\{1, 0, 0, \frac{3}{4}, \frac{5}{8}, \frac{9}{16}, \ldots\right\}$ is bounded away from 0, since $|a_n| \ge \frac{1}{2}$, $n > 3$. The sequence $\left\{\frac{1}{n}\right\}$ is not bounded away from 0 and so is the sequence $\left\{1, \frac{1}{2}, 1, \frac{1}{3}, \ldots, 1, \frac{1}{n}, \ldots\right\}$ (why?).

A *sufficient* condition for a sequence to be bounded away from 0, is given next.

Theorem 3.4.4. A sequence $\{a_n\}$ that converges to $A \neq 0$, is bounded away from 0.

Proof.

Let $\varepsilon = \frac{|A|}{2}$. Since $\{a_n\}$ converges, we can find $n_0(\varepsilon) > 0$ such that $|a_n - A| < \frac{|A|}{2}$ for all $n > n_0(\varepsilon)$. The triangle's inequality implies $|A| - a_n \leq |a_n - A| < \frac{|A|}{2}$ for $n > n_0(\varepsilon)$. Consequently $a_n > \frac{|A|}{2} > 0$, $n > n_0(\varepsilon)$, i.e. $\{a_n\}$ is bounded away from 0.

The next results are very useful in determining whether a given sequence converges.

Theorem 3.4.5. Consider the sequences $\{a_n\}$ and $\{b_n\}$ where $0 \leq a_n \leq b_n$, $n \geq 1$. If $\lim_{n\to\infty} b_n = 0$, then $\lim_{n\to\infty} a_n = 0$.

Theorem 3.4.6. If a sequence $\{a_n\}$ converges to A, then $\{ca_n\}$ converges to cA for all $c \in \Re$.

Theorem 3.4.7. If the $\{|a_n|\}$ converges to 0, then $\{a_n\}$ converges to 0 as well.

The proofs of these theorems are left as an exercise for the reader.

Example 3.4.7. Let $a_n = q^n$, $n \geq 1$ where $0 < q < 1$. Then: $\lim_{n\to\infty} a_n = 0$.

Indeed, rewrite $q = \frac{1}{1+p}$ where $p = \frac{1}{q} - 1 > 0$. Using Bernoulli's inequality we get

$$(1+p)^n \geq 1 + np > 0$$

Since $(1+p)^n > 0$ and $1 + np > 0$ we also have

$$0 \leq \frac{1}{(1+p)^n} \leq \frac{1}{1+np} < \frac{1}{np}$$

The sequence $\left\{\frac{1}{n}\right\}$ converges to 0. Thus, by virtue of Theorems 3.4.5-6, $\left\{\frac{1}{np}\right\}$, $\left\{\frac{1}{1+np}\right\}$ and $\{a_n\}=\left\{\frac{1}{(1+p)^n}\right\}$ all converge to 0 as well. By implementing Theorem 3.4.7, we get $\lim\limits_{n\to\infty} q^n = 0$ for $|q|<1$.

One of the most important features of the convergence notation is the following result.

Theorem 3.4.8. If the sequences $\{a_n\}$ and $\{b_n\}$ satisfy $a_n = b_n$, $n > n_0$ for some positive integer n_0 and $\lim\limits_{n\to\infty} a_n = A$, then $\lim\limits_{n\to\infty} b_n = A$ as well.

We thus conclude, that the *head* of a sequence, defined as the first n_0 elements where n_0 is any fixed positive integer, plays no role in deciding whether or not the sequence converges. The convergence or divergence of the sequence is determined *solely* by *remaining* elements.

The proof of Theorem 3.4.8 is quite trivial and is left for the reader.

PROBLEMS

1. Find which of the following sequences converge and obtain their limits.

   (a) $\{1,-2,3,-4,5,-6,\ldots\}$ (b) $\left\{\frac{n+10}{n}\right\}$ (c) $\left\{\frac{2\cdot 3}{1\cdot 2},\frac{3\cdot 4}{2\cdot 3},\frac{4\cdot 5}{3\cdot 4},\ldots\right\}$

   (d) $\left\{1,-1,\frac{3}{2},-1,\frac{7}{4},-1,\frac{15}{8},\ldots\right\}$ (e) $\left\{\frac{n-\sqrt{n}}{n}\right\}$ (f) $\left\{\frac{n+(-1)^n(n/2)}{n}\right\}$

   (g) $\left\{\frac{n^2+n-1}{2n^2+2n-5}\right\}$

2. Let $a_n = \frac{n-3}{7n+2}$. Find n_0 such that $\left|a_n - \frac{1}{7}\right| < 10^{-3}$, $n > n_0$. Does finding n_0 guarantee that $\{a_n\}$ converges to $\frac{1}{7}$? Converges to any limit?

3. Let $a_n = \frac{\sqrt{n}-10\sqrt[3]{n}}{1-3\sqrt{n}}$. Find n_0 such that $\left|a_n + \frac{1}{3}\right| < 10^{-3}$, $n > n_0$.

4. Let $a_n = 1 + q + \ldots + q^n$, $|q| < 1$. Find $\lim_{n\to\infty} a_n$ if it exists.
5. Show that the following sequences converge to 0:

$$\text{(a)} \left\{\frac{n^2}{n!}\right\} \quad \text{(b)} \left\{\frac{n^3}{n!}\right\} \quad \text{(c)} \left\{\frac{n!}{n^n}\right\}$$

6. Let $\lim_{n\to\infty} a_n = A$ and define $b_n = \frac{a_n + a_{n+1}}{2}$, $n \in N$. Prove: $\lim_{n\to\infty} b_n = A$.
7. Let $a_n = 1 + \frac{1}{2} + \frac{1}{3} + \ldots + \frac{1}{n}$. Show that $\{a_n\}$ is unbounded.
8. Let $a_n = 1 + \frac{1}{2!} + \frac{1}{3!} + \ldots + \frac{1}{n!}$. Show that $\{a_n\}$ is bounded.
9. Use the binomial theorem and problem 8 to show that the sequence $\left\{\left(1 + \frac{1}{n}\right)^n\right\}$ is bounded.
10. Let $a_n = 1 + \frac{1}{1\cdot 2} + \frac{1}{2\cdot 3} + \ldots + \frac{1}{(n-1)\cdot n}$, $n > 1$. Show that $\lim_{n\to\infty} a_n = 2$.
11. Let $a_n = \frac{1}{2\cdot 3} + \frac{4}{5\cdot 6} + \ldots + \frac{3n-2}{(3n-1)\cdot(3n)}$. Use the result of problem 7 to show that $\{a_n\}$ is unbounded.
12. Let $\{a_n\}, \{b_n\}$ denote convergent and divergent sequences respectively. Show an example where: (a) $\{c_n\}, c_n = a_n b_n$ is convergent (b) $\{c_n\}, c_n = a_n b_n$ is divergent.
13. Let $\{a_n\}, \{b_n\}$ denote convergent and divergent sequences respectively. What can you expect from the sequence $\{c_n\}, c_n = a_n + b_n$?

3.5 Basic Limit Theorems

The following theorems significantly extend our ability to determine whether a given sequence has a limit. We start with a limit theorem related to basic arithmetic operations between sequences.

Theorem 3.5.1. Let $\lim_{n\to\infty} a_n = A$ and $\lim_{n\to\infty} b_n = B$. Then

(a) $\lim_{n\to\infty} (a_n + b_n) = A + B$

(b) $\lim_{n\to\infty} (a_n - b_n) = A - B$

(c) $\lim_{n\to\infty} (a_n b_n) = AB$

(d) $\lim\limits_{n\to\infty}\left(\dfrac{a_n}{b_n}\right)=\dfrac{A}{B}$, provided that $B\neq 0$ and $b_n\neq 0\,,\,n\in N$.

Theorem 3.5.1 yields that the sum, difference, product and quotient sequences, of two arbitrary *convergent* sequences, *converge* as well. The only restriction is in part (d), where we must assume that neither the elements of the sequence $\{b_n\}$ nor its limit may vanish. However, if $B\neq 0$ there exists some $m\in N$ such that $b_n\neq 0\,,\,n\geq m$ (why?). Consequently, we may replace part (d) with

$$\lim_{n\to\infty}\left(\frac{a_n}{b_n}\right)_{n=m}^{\infty}=\frac{A}{B}\ \text{, where } B\neq 0 \text{ and } b_n\neq 0\,,\,n\geq m\,.$$

Proof.

(a) Given an arbitrary $\varepsilon>0$, let us define $\varepsilon_1=\dfrac{\varepsilon}{2}$. Since $\{a_n\}$ converges to A , we can find $n_1(\varepsilon_1)$ (denoted by n_1) such that $|a_n-A|<\varepsilon_1\,,\,n>n_1$. Similarly, we can find $n_2(\varepsilon_1)$ (denoted by n_2) such that $|b_n-B|<\varepsilon_1\,,\,n>n_2$. Consequently

$$|(a_n+b_n)-(A+B)|=|(a_n-A)+(b_n-B)|\leq|a_n-A|+|b_n-B|<2\varepsilon_1=\varepsilon$$

provided that $n>n_0=\max(n_1,n_2)$. By virtue of Definition 3.4.2 part (a) is valid.

(b) The proof is almost identical to that of part (a) and is left for the reader.

(c) In order to show $\lim\limits_{n\to\infty}a_nb_n=AB$ we rewrite

$$a_nb_n-AB=(a_nb_n-Ab_n)+(Ab_n-AB)$$

and get

$$|a_nb_n-AB|\leq|a_nb_n-Ab_n|+|Ab_n-AB|=|b_n||a_n-A|+|A||b_n-B|$$

Since a convergent sequence is also bounded, we can find $M>0$ such that $|b_n|\leq M\,,\,n\in N$. Consequently

$$|a_nb_n-AB|\leq M|a_n-A|+|A||b_n-B|$$

Consider an arbitrary $\varepsilon > 0$ and let $\varepsilon_1 = \dfrac{\varepsilon}{2M}$ and $\varepsilon_2 = \dfrac{\varepsilon}{2|A|}$. Since $\lim_{n\to\infty} a_n = A$ we can find $n_1(\varepsilon_1)$ such that $|a_n - A| < \varepsilon_1$ is valid for $n > n_1$. Similarly, since $\lim_{n\to\infty} b_n = B$ we can find $n_2(\varepsilon_2)$ such that $|b_n - B| < \varepsilon_2$ is valid for $n > n_2$. Therefore, by taking $n_0 = \max(n_1, n_2)$, we obtain

$$|a_n b_n - AB| < M\varepsilon_1 + |A|\varepsilon_2 = M\frac{\varepsilon}{2M} + |A|\frac{\varepsilon}{2|A|} = \varepsilon \ , \ n > n_0$$

and the proof is completed.

(d) Rewrite

$$\frac{a_n}{b_n} - \frac{A}{B} = \frac{a_n B - b_n A}{b_n B} = \frac{a_n B - AB + AB - b_n A}{b_n B} = \frac{a_n - A}{b_n} + \frac{A}{b_n B}(B - b_n)$$

which implies

$$\left|\frac{a_n}{b_n} - \frac{A}{B}\right| \le \frac{|a_n - A|}{|b_n|} + \frac{|A|}{|b_n B|}|B - b_n|$$

Since $\{b_n\}$ converges to $B \ne 0$ and since none of the elements vanishes, the sequence $\{b_n\}$ is bounded away from zero, i.e. we can find $M > 0$ such that $|b_n| \ge M$ for all n.

Consequently

$$\left|\frac{a_n}{b_n} - \frac{A}{B}\right| \le \frac{|a_n - A|}{M} + \frac{|A|}{M|B|}|B - b_n|$$

Now, consider an arbitrary $\varepsilon > 0$. Define $\varepsilon_1 = \dfrac{M\varepsilon}{2}$ and $\varepsilon_2 = \dfrac{M|B|\varepsilon}{2|A|}$. Choose $n_1(\varepsilon_1)$ and $n_2(\varepsilon_2)$ such that $|a_n - A| < \varepsilon_1 , n > n_1$ and $|b_n - B| < \varepsilon_2 , n > n_2$. This is feasible since $\{a_n\}$ converges to A and $\{b_n\}$ converges to B. Thus,

$$\left|\frac{a_n}{b_n} - \frac{A}{B}\right| \le \frac{|a_n - A|}{M} + \frac{|A|}{M|B|}|B - b_n| < \frac{\varepsilon_1}{M} + \frac{|A|\varepsilon_2}{M|B|} = \frac{\varepsilon}{2} + \frac{\varepsilon}{2} = \varepsilon$$

for $n > n_0 = \max(n_1, n_2)$ and this completes the proof of part (d).

It should be noted that the particular choices of ε_1, ε_2 in the proofs of parts (c) and (d), were neither lucky guesses nor a demonstration of special expertise. It is merely a simple, standard procedure, which adds a touch of elegance to the proof. For example, consider the proof of part (c). After obtaining the inequality

$$|a_n b_n - AB| \le M|a_n - A| + |A||b_n - B|\ ,\ n \in N$$

we continue the proof differently as follows. For a given $\varepsilon > 0$ we have to show the existence of $n_0(\varepsilon)$ such that $|a_n b_n - AB| < \varepsilon\ ,\ n > n_0$. However, for an arbitrary $\varepsilon^* > 0$ (it is irrelevant whether we use the symbol ε or ε^*), we can find $n_1(\varepsilon^*)$ such that $|a_n - A| < \varepsilon^*$ is valid for $n > n_1$. Similarly, we can find $n_2(\varepsilon^*)$ such that $|b_n - B| < \varepsilon^*$ is valid for $n > n_2$. Therefore, by taking $n_0 = \max(n_1, n_2)$, we obtain

$$|a_n b_n - AB| < M\varepsilon + |A|\varepsilon^* = (M\varepsilon^* + |A|\varepsilon^*) = (M + |A|)\varepsilon^*\ \ ,\ \ n > n_0$$

Since this is true for *arbitrary* $\varepsilon^* > 0$, it is true also for the particular choice $\varepsilon^* = \dfrac{\varepsilon}{M + |A|}$. In other words, there exists n_0 such that

$$|a_n b_n - AB| < (M + |A|)\varepsilon^* = (M + |A|)\frac{\varepsilon}{(M + |A|)} = \varepsilon,\ \ n > n_0$$

and the proof is concluded. Thus, instead of having to pre-calculate 'clever' choices for ε_1 and ε_2, we obtain an inequality such as $|u_n - U| < c\varepsilon^*\ ,\ n > n_0$ and apply it for $\varepsilon^* = \dfrac{\varepsilon}{c}$.

Theorem 3.5.2. Let $a_n \ge 0\ ,\ n \in N$ and $\lim\limits_{n\to\infty} a_n = A$. Then $\lim\limits_{n\to\infty} \sqrt[m]{a_n} = \sqrt[m]{A}$ for arbitrary $m \in N$.

Proof.

Clearly $A \ge 0$ (why?). Define $b_n = \sqrt[m]{a_n}\ ,\ n \in N$ and $B = \sqrt[m]{A}$. We distinguish between two cases.

(a) $A > 0$ which implies $B > 0$. Rewrite

$$a_n - A = b_n^m - B^m = (b_n - B)(b_n^{m-1} + b_n^{m-2}B + \ldots + b_n B^{m-2} + B^{m-1})\,.$$

Since $B > 0$ and $b_n \ge 0$, $n \in N$ we obtain

$$|a_n - A| = |b_n - B|\left|b_n^{m-1} + b_n^{m-2}B + \ldots + b_n B^{m-2} + B^{m-1}\right| \ge B^{m-1}|b_n - B|$$

Therefore $|b_n - B| \le \frac{|a_n - A|}{B^{m-1}}$ which implies $\lim_{n\to\infty} b_n = B$ (why?).

(b) $A = 0$ which implies $B = 0$. We have to show $\lim_{n\to\infty} b_n = 0$, i.e., that for arbitrary $\varepsilon > 0$ there exists $n_0 \in N$ such that $|b_n| < \varepsilon$ for $n > n_0$. Define $\varepsilon_1 = \varepsilon^m$. Since $\lim_{n\to\infty} a_n = 0$, one can find $n_0 \in N$ such that $|a_n| < \varepsilon_1$, $n > n_0$. Thus $\sqrt[m]{a_n} < \sqrt[m]{\varepsilon_1} = \varepsilon$ for $n > n_0$. This concludes the proof.

Example 3.5.1. Let $a_n = \sqrt{n^2 + 2n} - \sqrt{n^2 + 1}$, $n \ge 1$. In order to find whether the sequence converges, rewrite

$$a_n = \frac{(\sqrt{n^2 + 2n} - \sqrt{n^2 + 1})(\sqrt{n^2 + 2n} + \sqrt{n^2 + 1})}{\sqrt{n^2 + 2n} + \sqrt{n^2 + 1}} = \frac{2n - 1}{\sqrt{n^2 + 2n} + \sqrt{n^2 + 1}}$$

which, after dividing the numerator and the denominator by n, leads to

$$a_n = \frac{2 - \frac{1}{n}}{\sqrt{1 + \frac{2}{n}} + \sqrt{1 + \frac{1}{n^2}}}$$

By Theorem 3.5.1 part (b), the numerator converges to 2. Indeed, it is the difference sequence of $\{2,2,2,\ldots\}$ and $\left\{\frac{1}{n}\right\}$ that converges to $2 - 0 = 2$. The denominator converges to 2 as well. First, $\lim_{n\to\infty} \frac{1}{n^2} = \left(\lim_{n\to\infty} \frac{1}{n}\right)^2 = 0$. Therefore $\lim_{n\to\infty}\left(1 + \frac{1}{n^2}\right) = 1$ leading, by Theorem 3.5.2 to $\lim_{n\to\infty} \sqrt{1 + \frac{1}{n^2}} = 1$. Similarly, $\lim_{n\to\infty} \sqrt{1 + \frac{2}{n}} = 1$ and the whole denominator converges to 2. Therefore, $\lim_{n\to\infty} a_n = 1$.

The following results are common knowledge to the reader who is familiar with elementary trigonometry:

$$\text{(R1)}\ \lim_{n\to\infty}\sin\left(\frac{1}{n}\right)=0\quad \text{(R2)}\ \lim_{n\to\infty}n\sin\left(\frac{1}{n}\right)=1\quad \text{(R3)}\ \lim_{n\to\infty}\cos\left(\frac{1}{n}\right)=1$$

Result (R2) follows from the inequality $\sin(x)<x\,, 0<x<\pi/2$ (proved in Section 5.3) while (R1) and (R3) follow straightforward from (R2).

Example 3.5.2. Let $a_n=\dfrac{(n+3)(\sqrt{n+\sqrt{n}}-\sqrt{n})}{(n-1)(n-2)\sin(1/n)}$. To find if $\lim_{n\to\infty}a_n$ exists we first rewrite $a_n=\dfrac{(n+3)}{(n-1)}\dfrac{(\sqrt{n+\sqrt{n}}-\sqrt{n})}{(n-2)\sin(1/n)}$. The fraction $\dfrac{n+3}{n-1}$, rewritten as $\dfrac{1+3/n}{1-1/n}$, converges to 1 (Theorem 3.5.1). The next expression is

$$\sqrt{n+\sqrt{n}}-\sqrt{n}=\frac{(\sqrt{n+\sqrt{n}}-\sqrt{n})(\sqrt{n+\sqrt{n}}+\sqrt{n})}{\sqrt{n+\sqrt{n}}+\sqrt{n}}=\frac{\sqrt{n}}{\sqrt{n+\sqrt{n}}+\sqrt{n}}=\frac{1}{\sqrt{1+1/\sqrt{n}}+1}$$

which by virtue of Theorems 3.5.1 and 3.5.2, converges to 0.5. The last expression $(n-2)\sin(1/n)=n\sin(1/n)-2\sin(1/n)$, converges (results R1, R2 and Theorem 3.5.1) to 1. Consequently, $\lim_{n\to\infty}a_n=0.5$.

Example 3.5.3. Let $a_n=\sqrt[n]{a}\,, n\in N$ for a fixed $a>1$. Clearly $a_n>1$ for all $n\in N$, i.e. $a_n=1+\alpha_n\,, \alpha_n>0$. Therefore, by Bernoulli's inequality: $a=(1+\alpha_n)^n\ge 1+n\alpha_n$ and consequently $0<\alpha_n\le\dfrac{a-1}{n}$. Using previous result we obtain $\lim_{n\to\infty}\alpha_n=0$ and finally $\lim_{n\to\infty}a_n=1$. If $0<a<1$ we have $\lim_{n\to\infty}\sqrt[n]{\dfrac{1}{a}}=\dfrac{1}{\sqrt[n]{a}}=1$ (why?) and therefore $\lim_{n\to\infty}\sqrt[n]{a}=1$ as well. Thus, $\lim_{n\to\infty}\sqrt[n]{a}=1$ for arbitrary $a>0$.

Example 3.5.4. Consider the sequence

$$a_n=\sqrt{\frac{n}{\cos(1/n)}}\,\frac{\sqrt{n+\sqrt[3]{n}}-\sqrt{n}}{\sqrt{n+\sqrt{n}}-\sqrt{n}}\sin\left(\frac{1}{\sqrt[3]{n}}\right)$$

Multiply and divide the midterm's numerator by $\sqrt{n+\sqrt[3]{n}}+\sqrt{n}$ and its denominator by $\sqrt{n+\sqrt{n}}+\sqrt{n}$. The midterm can be therefore rewritten as

$$\frac{\sqrt{n+\sqrt[3]{n}}-\sqrt{n}}{\sqrt{n+\sqrt{n}}-\sqrt{n}}=\frac{\sqrt[3]{n}\,(\sqrt{n+\sqrt{n}}+\sqrt{n})}{\sqrt{n}\,(\sqrt{n+\sqrt[3]{n}}+\sqrt{n})}$$

which implies

$$a_n=\frac{\sqrt[3]{n}\sin(1/\sqrt[3]{n})}{\sqrt{\cos(1/n)}}\frac{\sqrt{n+\sqrt{n}}+\sqrt{n}}{\sqrt{n+\sqrt[3]{n}}+\sqrt{n}}=\frac{\sqrt[3]{n}\sin(1/\sqrt[3]{n})}{\sqrt{\cos(1/n)}}\frac{\sqrt{1+1/\sqrt{n}}+1}{\sqrt{1+1/\sqrt[3]{n^2}}+1}$$

A multiple use of Theorems 3.5.1 and 3.5.2 shows $\lim_{n\to\infty} a_n = 1$. The details are left as an exercise for the reader.

Theorem 3.5.3. If the sequences $\{a_n\}$, $\{b_n\}$, $\{c_n\}$ satisfy $\lim_{n\to\infty} a_n = \lim_{n\to\infty} b_n = A$ and $a_n \le c_n \le b_n$, $n \in N$ then $\lim_{n\to\infty} c_n = A$ as well.

Proof.

For arbitrary $n \in N$ we have

$$|c_n - A| = |c_n - a_n + a_n - b_n + b_n - A| \le |c_n - a_n| + |a_n - b_n| + |b_n - A| \le 2|a_n - b_n| + |b_n - A|$$

The right-hand side converges to 0 which implies $\lim_{n\to\infty}|c_n - A| = 0$ and concludes the proof.

Example 3.5.5. Let $a_n = 1$, $b_n = 1+\frac{1}{n}$, $c_n = 1+\frac{1}{n^2}$. Here $\lim_{n\to\infty} a_n = \lim_{n\to\infty} b_n = 1$ and $1 < 1+\frac{1}{n^2} < 1+\frac{1}{n}$. Therefore, $\lim_{n\to\infty}\left(1+\frac{1}{n^2}\right) = 1$.

PROBLEMS

1. Find the limits of the following sequences:

a. $a_n = \frac{1}{n} + \frac{\sqrt{n}}{n+\sqrt{n}}$ (b) $a_n = \frac{\sin(n)}{\sqrt{n}}$ (c) $a_n = (n^2+n)\sin^2\left(\frac{2}{n}\right)$

(d) $a_n = \frac{n\sqrt{n}+\sqrt[3]{n}}{2\sqrt{n}+5}\tan\left(\frac{2}{n+\sqrt{n}}\right)$

2. Let $\lim_{n\to\infty} a_n = A$, $\lim_{n\to\infty} b_n = B$, $\lim_{n\to\infty} c_n = C$. Prove $\lim_{n\to\infty}(a_n b_n c_n) = ABC$.
3. Extend the result of problem 2 to m convergent sequences for arbitrary $m \in N$.
4. If a sequence $\{a_n\}$ does not converge to A, there must some $\varepsilon = \varepsilon_0$ such that for *any* positive integer m, there is another positive integer $m_0 > m$ for which $\left|a_{m_0} - A\right| \ge \varepsilon_0$. Use this argument and the binomial theorem to show $\lim_{n\to\infty} \sqrt[n]{n} = 1$.
5. Use Example 3.5.3 to calculate $\lim_{n\to\infty} \sqrt[n]{3^{n+3}}$.
6. Let $0 \le \alpha < \beta$. Define $a_n = \sqrt[n]{\alpha^n + \beta^n}$, $n \in N$ and prove $\lim_{n\to\infty} a_n = \beta$. Does this also hold when $\alpha = \beta$?
7. Prove $\lim_{n\to\infty} \frac{a^n}{n!} = 0$ for arbitrary $a \in \Re$.
8. Let $\lim_{n\to\infty} a_n = A$ where $a_n \ge 0$, $n \in N$. Show that for an arbitrary fraction $\frac{m}{k}$ such that $m, k > 0$, the sequence $\{\sqrt[k]{a_n^m}\}$ converges to $A^{\frac{m}{k}}$.
9. Let $\lim_{n\to\infty} a_n = A$. Define $b_n = \frac{a_n + a_{n+1} + a_{n+2}}{3}$ and find $\lim_{n\to\infty} b_n$ if it exists.
10. Let $\lim_{n\to\infty} a_n = A$, $\lim_{n\to\infty} b_n = B$. When does $\lim_{n\to\infty} \sqrt{\frac{a_n b_n}{a_{n+10} + b_{n+10}}}$ exists? If it exists, calculate it.
11. Let $a_1 = a_2 = 1$ and for $n > 2$ define $a_n = \frac{1 + a_{n-1}}{a_{n-2}}$ and assume $\lim_{n\to\infty} a_n = A$. Find A.
12. Let $\{a_n\}, \{b_n\}$ denote two divergent sequences. Is the sequence $\{a_n + b_n\}$ divergent?
13. Show that for arbitrary rational number r there is a sequence of irrational numbers that converges to r.
14. Let $\{a_n\}, \{b_n\}$ denote two divergent sequences. Is the sequence $\{a_n b_n\}$ divergent?
15. The sequence $a_1 = 1$; $a_{n+1} = \cfrac{1}{1 + \cfrac{1}{1 + a_n}}$, $n \ge 1$ is a typical continued fraction (the name refers to the structure of a_{n+1}, once we express a_n in terms of a_{n-1}

etc.). Find $\lim_{n \to \infty} a_n$, if it exists, then calculate the first 10 terms of the sequence and compare.

16. Show that the sequence in problem 15 is decreasing and apply Theorem 3.7.1 (Section 3.7 below) to show that the sequence converges.

3.6 Limit Points

Definition 3.6.1. Let S denote an arbitrary set of real numbers. A number $s \in \Re$ will be called a *limit point* (also *accumulation point* or *cluster point*), if for every $\varepsilon > 0$ there exists a number $r \in \Re$ such that $|r - s| < \varepsilon$, $r \neq s$.

The requirement $|r - s| < \varepsilon$, $r \neq s$ can be replaced by the requirement $0 < |r - s| < \varepsilon$ since $0 < |r - s| \Leftrightarrow r \neq s$. An immediate consequence of Definition 3.6.1 is the following theorem.

Theorem 3.6.1. Let $s \in \Re$ be a limit point of S. Then S is infinite, and there exists a sequence $\{r_n\}$ such that:

(a) $r_n \in S$, $n \in N$

(b) $k \neq l \Rightarrow r_k \neq r_l$

(c) $\lim_{n \to \infty} r_n = s$

Proof.

Since s is a limit point of S we can find $r_1 \in S$ such that $|r_1 - s| < 1$, $r_1 \neq s$. Define $\varepsilon_1 = \frac{|r_1 - s|}{2}$. Clearly $\varepsilon_1 > 0$ and therefore there exists $r_2 \in S$ such that $|r_2 - s| < \varepsilon_1$, $r_2 \neq s$. Since $|r_2 - s| < \frac{|r_1 - s|}{2}$, we must have $r_1 \neq r_2$. In general we choose $\varepsilon_n = \frac{|r_n - s|}{2} > 0$ and find $r_{n+1} \in S$ such that $|r_{n+1} - s| < \varepsilon_n$, $r_{n+1} \neq s$. We thus constructed an infinite sequence which satisfies (a) and (b) (therefore S is infinite). By using induction we get $|r_{n+1} - s| < \frac{|r_1 - s|}{2^n}$, $n \in N$ and (c) is concluded as well.

Example 3.6.1. The set $S_1 = \{1, 2, \ldots, 17\}$ is finite and therefore has no limit points. The set $S_2 = \left\{1, \frac{1}{2}, 2, \frac{1}{3}, 3, \frac{1}{4}, 4, \ldots\right\}$ has a single limit point $s = 0$. A sequence, which is a subset of S_2, whose elements are mutually distinct and which converges to 0, is for example $\left\{\frac{1}{2},\frac{1}{3},\frac{1}{4},\ldots\right\}$. The set $S_3=\left\{\frac{1}{2},\frac{3}{4},\frac{1}{3},\frac{7}{8},\frac{1}{4},\frac{15}{16},\ldots\right\}$ has two limit points, namely 0 and 1.

By definition, the elements of a set are mutually distinct. This is necessarily true for a sequence. Therefore the concept of 'limit point' for a sequence is defined differently. We first introduce the 'subsequence' notation.

Definition 3.6.2. A sequence $\{b_n\}$ is called a *subsequence* of a sequence $\{a_n\}$ if

$$b_k = a_{n_k}\ ,\ k\in N\quad ;\quad n_1<n_2<\ldots<n_k<n_{k+1}<\ldots \tag{3.6.1}$$

Definition 3.6.3. A number $s\in\Re$ is called a *limit point* (also *accumulation point* or *cluster point*) of a sequence $\{a_n\}$ if there exists a subsequence $\{a_{n_k}\}$ of $\{a_n\}$ such that $\lim\limits_{k\to\infty} a_{n_k}=s$.

Example 3.6.2. The sequence $\{a_n\}=\{0,1.9,0,1.99,0,1.999,\ldots\}$ has two limit points. The first is 0 which is the limit of the convergent subsequence $\{0,0,\ldots\}$. The second is 2 which is the limit of the subsequence $\{1.9,1.99,1.999,\ldots\}$.

Theorem 3.6.2. If s is a limit point of a set S, bounded above by M, then $s\le M$.

Proof.

If $s>M$ then $s=M+\alpha$, $\alpha>0$. Since s is a limit point of S, we should be able to find $r\in S$ which satisfies $0<|r-s|<\frac{\alpha}{2}$. Therefore, $s<r+\frac{\alpha}{2}<r+\alpha\le M+\alpha$ which contradicts $s=M+\alpha$.

Corollary 3.6.1. The set K of the limit points of a bounded above set S, is also bounded (by the same bound) and therefore possesses a *supremum*.

Definition 3.6.4. If the set K of the limit points of a bounded above set S is nonempty, its supremum M is called the *limit superior* of S and is denoted by

$$M=\limsup S=\overline{\lim}S \tag{3.6.2}$$

Similarly, the set K of the limit points of a bounded below set, is also bounded below (by the same bound) and therefore possesses an *infimum*.

Definition 3.6.5. If the set K of the limit points of a bounded above set S is nonempty, its infimum L is called the *limit inferior* of S and is denoted by

$$L = \liminf S = \underline{\lim} S \tag{3.6.3}$$

Theorem 3.6.2 and Definitions 3.6.4 and 3.6.5 can be extended for sequences. This is left as an exercise for the reader. The next result however, does not hold for sequences.

Theorem 3.6.3. Let s be a limit point of a set S. Then, for every $\varepsilon > 0$ the interval $(s-\varepsilon, s+\varepsilon)$ contains an infinite number of elements from S.

The proof is straightforward and is left for the reader.

Example 3.6.3. Consider the set $S_1 = \{-5, (-4, -1), 0, 2, [3, 5], 7, 8, 1000\}$ which consists of six discrete integers, the open interval from -4 to -1 and the closed interval from 3 to 5. The set is bounded and its infimum and supremum are -5 and 1000 respectively. Each number within the two intervals is a limit point and

$$\underline{\lim} S_1 = -4 \ , \ \overline{\lim} S_1 = 5$$

Now, consider the sequence $S_2 = \{-2, 3, 1, -1, 1, -1, 1, -1, \ldots\}$. Here the infimum and the supremum are -2 and 3 respectively. There are only two limit points and

$$\underline{\lim} S_2 = -1 \ , \ \overline{\lim} S_2 = 1$$

The following results can be easily derived from Definitions 3.6.4-3.6.5 and their proofs are left for the reader.

Theorem 3.6.4. Let S be an infinite set and M a real number. Then, $M = \overline{\lim} S$ if and only if, for every $\varepsilon > 0$ there is only a finite number of elements in S that are greater than $M + \varepsilon$ while there is an infinite number of elements in S that are greater than $M - \varepsilon$.

Theorem 3.6.5. Let S be an infinite set and M a real number. Then, $M = \underline{\lim} S$ if and only if, for every $\varepsilon > 0$ only there is a finite number of elements in S that are less than $M - \varepsilon$ while there is an infinite number of elements in S that are less than $M + \varepsilon$.

The next result presents an important relation between a sequence and the class of its subsequences.

Theorem 3.6.6. A sequence $\{a_n\}$ converges to A, if and only if *all* its subsequences converge to A.

Proof.

Let $A = \lim_{n\to\infty} a_n$ and let $\{b_k\} = \{a_{n_k}\}$ denote an arbitrary subsequence of $\{a_n\}$. For a given $\varepsilon > 0$ we can find n_0 such that $|A - a_n| < \varepsilon$, $n > n_0$. Using Eq. (3.6.1) we easily obtain $n_k \ge k$ for all k. In particular, $n_{n_0} \ge n_0$ and consequently, $|A - b_n| < \varepsilon$ for $n > n_0$. Thus, every subsequence of $\{a_n\}$ is convergent and converges to A. Now, assume that *every* subsequence of $\{a_n\}$ converges to A. Since $\{a_n\}$ is also a subsequence of itself we have $A = \lim_{n\to\infty} a_n$ and the proof is completed.

PROBLEMS

1. Find limit points of the set

$$S = \left\{-1, \frac{1}{2}, 1 - 1, -\frac{1}{3}, \frac{1}{4}, 2, -2, -\frac{1}{5}, \frac{1}{6}, 3, -3, \ldots\right\}.$$

   Observe S as a sequence and determine whether this sequence is convergent.

2. Find all the limit points of the sequence $\{1,2,5,1,2,5,1,2,5,\ldots\}$. Does the set which consists of all the mutually distinct elements of the sequence have a limit point?
3. Find the supremum, infimum, limit superior and limit inferior of the *sets*:

   (a) $S_1 = \{0,1,0,2,0,3,\ldots\}$
   (b) $S_2 = \{4,3,2,\ldots\}$
   (c) $S_3 = \left\{\frac{2}{3}, -1, \frac{4}{3}, -1, \frac{4}{5}, -1, \frac{6}{5}, \ldots\right\}$
   (d) $S_4 = \{-5, [-4, 0.1], 6, 7, 10, (-8, -6)\}$

4. Find the supremum, infimum, limit superior and limit inferior of the *sequences*:

(a) $S_1 = \{0,1,0,2,0,3,\ldots\}$

(b) $S_2 = \left\{\frac{2}{3},-1,\frac{4}{3},-1,\frac{4}{5},-1,\frac{6}{5},\ldots\right\}$

(c) $S_3 = \{1,5,1,5,1,5,\ldots\}$

(d) $S_4 = \left\{\frac{1}{2},-\frac{1}{2},\frac{2}{3},-\frac{1}{4},\frac{4}{9},-\frac{1}{8},\frac{8}{27},\ldots\right\}$

5. Let $\{a_n\},\{b_n\}$ be bounded above sequences. Show

$$\overline{\lim}(a_n + b_n) \le \overline{\lim}(a_n) + \overline{\lim}(b_n)$$

if both sides exist.

6. Let $\{a_n\},\{b_n\}$ be bounded below sequences. Show

$$\underline{\lim}(a_n + b_n) \ge \underline{\lim}(a_n) + \underline{\lim}(b_n)$$

if both sides exist.

7. Let $\{a_n\},\{b_n\}$ be bounded above sequences such that $a_n \le b_n$, $n > n_0$ for some fixed n_0 . Show $\overline{\lim} a_n \le \overline{\lim} b_n$ if both sides exist.
8. Make a necessary change in Theorems 3.6.4-3.6.5 and extend them for sequences.
9. Consider the sequence $\left\{1,0,\frac{1}{2},0,\frac{1}{3},0,\ldots\right\}$. Find two subsequences that converge to the same limit. Is this sufficient to guarantee the convergence of the sequence?
10. Let $S = \{1.1,10^{-6},1.01,1.001,10^{-6},1.0001,1.00001,1.000001,10^{-6},\ldots\}$. Does the sequence converge? If not, find $\sup S$, $\overline{\lim} S$, $\inf S$, $\underline{\lim} S$.
11. Let $S = \{0.9,\sqrt{2},0.99,\sqrt[3]{3},0.999,\sqrt[4]{4},\ldots\}$. Find $\sup S$, $\overline{\lim} S$, $\inf S$, $\underline{\lim} S$. You may need to use the result of problem 3.5 and Example 3.7.2 (Section 3.7 below).

3.7 Special Sequences

In this section we introduce and investigate special types of sequences, such as monotone sequences, sequences that converge to infinity and Cauchy sequences.

3.7.1 Monotone Sequences

Definition 3.7.1. A sequence $\{a_n\}$ is called *monotone increasing*, if $a_n \le a_{n+1}$ for all n. If $a_n < a_{n+1}$ for all n, it is called *strictly monotone increasing*.

Definition 3.7.2. A sequence $\{a_n\}$ is called *monotone decreasing*, if $a_n \ge a_{n+1}$ for all n. If $a_n > a_{n+1}$ for all n, it is called *strictly monotone decreasing*.

Clearly, every strictly monotone increasing (decreasing) sequence, is also monotone increasing (decreasing).

Example 3.7.1. The sequences $\{1,1,2,2,3,3,\ldots\}$, $\{1,1,1,2,3,4,5,\ldots\}$ are monotone increasing. The sequence $\{2^{-n}\}$ is strictly monotone decreasing. The sequence $\{(-1)^n\}$ is not monotone. The sequence $\{a_n\} = \{\underbrace{1,-1,1,-1,\ldots,1,-1}_{1000},2,4,6,8,\ldots\}$ is strictly monotone increasing for *sufficiently large* n ($n > 1000$).

The next result is useful in numerous applications.

Theorem 3.7.1. A monotone increasing (decreasing) sequence which is bounded above (below), is convergent.

Proof.

Consider a monotone increasing bounded above sequence $\{a_n\}$ and let M denote its least upper bound. Given an arbitrary $\varepsilon > 0$ we can find an element a_{n_0} of the sequence which satisfies $M - \varepsilon < a_{n_0}$. Due to the sequence's monotonity we also have $M - \varepsilon < a_n$, $n > n_0$ and since $a_n \le M$ for all n, we obtain $|a_n - M| < \varepsilon$, $n > n_0$. Since this is valid for arbitrary $\varepsilon > 0$, we obtain $\lim_{n\to\infty} a_n = M$. The proof for decreasing sequences is similar and is left for the reader.

Example 3.7.2. Consider the sequence $a_n = \left(1+\frac{1}{n}\right)^n$, $n \ge 1$. By using the binomial theorem we get

$$a_n = \sum_{k=0}^{n} \binom{n}{k} 1^{n-k} \left(\frac{1}{n}\right)^k = \sum_{k=0}^{n} \frac{n(n-1)\cdots(n-k+1)}{k!} \frac{1}{n^k}$$

and consequently

$$a_n = 1 + \sum_{k=1}^{n} \frac{1}{k!}\left(1-\frac{1}{n}\right)\left(1-\frac{2}{n}\right)\cdots\left(1-\frac{k-1}{n}\right)$$

Similarly

$$a_{n+1} = 1 + \sum_{k=1}^{n} \frac{1}{k!}\left(1-\frac{1}{n+1}\right)\left(1-\frac{2}{n+1}\right)\cdots\left(1-\frac{k-1}{n+1}\right) + \frac{1}{(n+1)^{n+1}}$$

and since $1-\frac{j}{n} < 1-\frac{j}{n+1}$, $1 \le j \le k-1$ we obtain $a_n < a_{n+1}$, i.e. the sequence is monotone increasing. Also, $a_n < 1+1+\frac{1}{2!}+\frac{1}{3!}+\ldots+\frac{1}{n!}$ (why?) and by using the inequality $\frac{1}{n!} \le \frac{1}{2^{n-1}}$, $n \ge 2$ we get

$$a_n < 2 + \frac{1}{2} + \frac{1}{2^2} + \ldots + \frac{1}{2^{n-1}} = 2 + \frac{1}{2}\frac{1-(1/2)^{n-1}}{1-1/2} < 3 .$$

Thus, the sequence is bounded above and by virtue of Theorem 3.7.1 it is convergent. The limit of the sequence is denoted by the letter e (after the famous Swiss mathematician Leonard Euler (1707 – 1783), who was the first to use it as the base of the natural logarithm).

It is left for the reader to show $e = \lim_{n\to\infty}\left(1+\frac{1}{n}\right)^n = \lim_{n\to\infty} b_n$ where

$$b_n = 2 + \frac{1}{2!} + \frac{1}{3!} + \ldots + \frac{1}{n!}, \; n \ge 2$$

The number e is transcendental, i.e., e is not a root of any polynomial with integer coefficients. The proof of this fact is complicated and is beyond the scope of this book. A proof that e is irrational, is much simpler to obtain but is postponed until we introduce the basics of Infinite Series.

Example 3.7.3. Consider the sequence $\{a_n\} = \{q^n\}$ where $0 < q < 1$. It is monotone decreasing and bounded below by 0. Therefore, it converges to some $A > 0$. Now, define $\{b_n\} = \{q^{n+1}\}$. This sequence clearly converges to A as well. On the other hand it may be observed as the product of two sequences: $\{q^n\}$ with limit A and $\{q,q,\ldots,q,\ldots\}$ with limit q. Thus, $qA = A$ which implies $(q-1)A = 0$. Since $q \neq 1$ we must have $A = 0$, i.e. $\lim_{n\to\infty} q^n = 0, 0 < q < 1$.

Example 3.7.4. Does the sequence $a_1 = 1$; $a_{n+1} = \sqrt{2+a_n}$, $n \geq 1$ converge? Let us examine monotonity, using induction. Since $a_2 = \sqrt{3}$ we have $a_1 < a_2$. Assume $a_n \leq a_{n+1}$. Then, $a_{n+2} = \sqrt{2+a_{n+1}} \geq \sqrt{2+a_n} = a_{n+1}$ which implies that $\{a_n\}$ is monotone increasing. Is the sequence bounded? We will show $a_n < 3$. This clearly holds for a_1. Assume $a_n < 3$. Then $a_{n+1} = \sqrt{2+a_n} < \sqrt{2+3} = \sqrt{5} < 3$. Thus $\{a_n\}$ is bounded by 3. Consequently, $\lim_{n\to\infty} a_n$ exists. Denote it by A. Then, by using basic limit theorems we get $\lim_{n\to\infty} a_{n+1} = \sqrt{2 + \lim_{n\to\infty} a_n}$, which implies $A = \sqrt{2+A}$ or $A^2 - A - 2 = 0$. The two solutions are $A_1 = 2$ and $A_2 = -1$ but A_2 is obviously exclude (why?), i.e. $\lim_{n\to\infty} a_n = 2$.

As previously stated, *any* sequence that does not converge is called *divergent.* There is however, a particular class of sequences, that do not converge in the usual sense, i.e. they do not satisfy the requirements of Definition 3.4.2, yet, they are still referred to as 'convergent sequences'. Instead of converging to finite numbers they converge to 'infinity'. In the next section we introduce the notations of 'convergence to infinity' and 'infinite limits'.

3.7.2 Convergence to Infinity

Definition 3.7.3. A sequence $\{a_n\}$ *converges to infinity* if for every $M \in \Re$ there is $n_0 \in N$, such that $a_n > M$, $n > n_0$. We then write $\lim_{n\to\infty} a_n = \infty$ (or $\lim_{n\to\infty} a_n = +\infty$). The sequence *converges to minus infinity*, if for every $M \in \Re$ there is $n_0 \in N$, such that $a_n < M$, $n > n_0$. In this case we write $\lim_{n\to\infty} a_n = -\infty$. If a sequence does not converge to a finite limit and also does not converge to $\pm\infty$, the sequence *oscillates*.

Theorem 3.7.2. An arbitrary monotone increasing sequence, which is not bounded above, converges to ∞.

The proof of this result is left for the reader.

Example 3.7.5. The sequence $\{\sqrt{n}\}$ converges to ∞. The sequence $\left\{\frac{n^2+3}{n\sqrt{n}+n+1}\right\}$ converges to ∞ as well (why?). Less trivial is the behavior of $\{a_n\}$ where $a_n = \sum_{k=1}^{n} \frac{1}{k}$. Clearly, the sequence is strictly monotone increasing, and each term a_m, where $m = 2^n$, can be represented as

$$a_m = 1 + \frac{1}{2} + \left(\frac{1}{3} + \frac{1}{4}\right) + \left(\frac{1}{5} + \frac{1}{6} + \frac{1}{7} + \frac{1}{8}\right) + \ldots + \left(\frac{1}{2^{m-1}+1} + \frac{1}{2^{m-1}+2} + \ldots + \frac{1}{2^m}\right)$$

$$> 1 + \frac{1}{2} + \frac{2}{4} + \frac{4}{8} + \ldots + \frac{2^{m-1}}{2^m} = 1 + \frac{m}{2}$$

Thus, the sequence is not bounded above and by Theorem 3.7.2 must converge to ∞.

Theorem 3.7.3. Let $\{a_n\}, \{b_n\}$ be two sequences such that $\lim_{n\to\infty} a_n = \infty$ and $a_n \le b_n$ for sufficiently large n. Then, $\lim_{n\to\infty} b_n = \infty$.

The proof is straightforward and is left for the reader.

Theorem 3.7.4. Consider a sequence $\{a_n\}$ where $a_n > 0$, $n \in N$. Then, $\lim_{n\to\infty} a_n = \infty$ if and only if $\lim_{n\to\infty} \frac{1}{a_n} = 0$.

Proof.

Assume $\lim_{n\to\infty} a_n = \infty$. Given an arbitrary $\varepsilon > 0$ define $M = \frac{1}{\varepsilon}$. Then, find n_0 such that $a_n > M$ for $n > n_0$. Thus, $a_n > \frac{1}{\varepsilon}$ or $\left|\frac{1}{a_n} - 0\right| = \frac{1}{a_n} < \varepsilon$ for $n > n_0$, which implies $\lim_{n\to\infty} \frac{1}{a_n} = 0$. The opposite is just as simple and is left as an exercise for the reader.

Sometimes it is important to determine the *rate* by which a sequence converges to infinity. For example, the sequence $\{\sqrt{n}\}$ 'seems' to converge to ∞ *slower* than $\{n^2\}$. The sequence $\{n^3\}$ 'seems' to converge to ∞ faster than $\{n\}$. What can be said with regard to $\{n^2\}$ vs. $\{2n^2\}$? The following definition provides the exact definition of *convergence rate*.

Definition 3.7.4. Let $\{a_n\}, \{b_n\}$ be two sequences such that $\lim_{n\to\infty} a_n = \lim_{n\to\infty} b_n = \infty$.

If $\lim_{n\to\infty} \frac{a_n}{b_n} = \alpha > 0$, then the sequences are said to increase at the *same rate*. If

$\lim_{n\to\infty} \frac{a_n}{b_n} = \infty$, then $\{a_n\}$ increases at a *faster rate* than $\{b_n\}$. If $\lim_{n\to\infty} \frac{a_n}{b_n} = 0$, then $\{a_n\}$ increases at a *slower rate* than $\{b_n\}$.

Note that if several elements of $\{b_n\}$ are zero, Definition 3.7.4 is still meaningful. Indeed, since $b_n > 0$, $n > n_0$ for some fixed n_0, we simply disregard a_n, b_n for $n \le n_0$ as they do not have any impact on the limits or the rate of convergence.

Example 3.7.6. The sequences $\{n^2\}$ and $\{2n^2\}$ increase at the same rate since $\lim_{n\to\infty} \frac{n^2}{2n^2} = \frac{1}{2} > 0$. The sequence $\{a_n\} = \left\{ \frac{\sqrt{n^3+n+1}}{n+10\sqrt{n}} \right\}$ increases to infinity at a faster rate than $\{b_n\} = \{\sqrt[3]{n}\}$. Indeed, by using the basic limit theorems we get $\lim_{n\to\infty} \frac{b_n}{a_n} = 0$. Therefore, by Theorem 3.7.4 $\lim_{n\to\infty} \frac{a_n}{b_n} = \infty$, i.e. $\{a_n\}$ increases faster than $\{b_n\}$.

The next result is another useful tool in determining whether an arbitrary sequence converges.

Theorem 3.7.5. Consider a sequence $\{a_n\}$ with nonzero elements such that

$$\lim_{n\to\infty} \left| \frac{a_{n+1}}{a_n} \right| = A \tag{3.7.1}$$

Then

$$\begin{aligned} &\lim_{n\to\infty} a_n = 0 \quad \textit{if } A < 1 \\ &\lim_{n\to\infty} |a_n| = \infty \quad \textit{if } A > 1 \end{aligned} \tag{3.7.2}$$

Proof.

(a) Let $A<1$. For $\varepsilon=\frac{1-A}{2}<1$ we can find n_0 such that $\left|A-\left|a_{n+1}/a_n\right|\right|<\varepsilon\,, n>n_0$. Thus, $\left|a_{n+1}/a_n\right|<A+\varepsilon=\frac{1+A}{2}=q<1$ for $n>n_0$. In particular, $\left|a_{n_0+2}\right|<q\left|a_{n_0+1}\right|$ and by using induction we obtain $\left|a_{n_0+k}\right|<q^{k-1}\left|a_{n_0+1}\right|, k\geq 2$. Since $0<q<1$ we have $\lim\limits_{n\to\infty} a_n=0$.

(b) Let $A>1$. In this case we get $\left|a_{n+1}/a_n\right|>\frac{1+A}{2}=q>1$ for $n>n_0$ and since $\lim\limits_{n\to\infty} q^n=\infty$ we can apply Theorem 3.7.3 and obtain $\lim\limits_{n\to\infty}\left|a_n\right|=\infty$.

Example 3.7.7. Let $a_n=\frac{n^r}{q^n}$, $n\in N$ where $r>0$, $q>1$. In this case

$$\lim_{n\to\infty}\left|\frac{a_{n+1}}{a_n}\right|=\lim_{n\to\infty}\frac{(n+1)^r q^n}{n^r q^{n+1}}=\lim_{n\to\infty}\frac{1}{q}\left(1+\frac{1}{n}\right)^r=\frac{1}{q}<1$$

which implies $\lim\limits_{n\to\infty}\frac{n^r}{q^n}=0$. Next, consider the sequence $a_n=\frac{n^n}{n!}$, $n\in N$. Here

$$\lim_{n\to\infty}\left|\frac{a_{n+1}}{a_n}\right|=\lim_{n\to\infty}\frac{(n+1)^{n+1}n!}{n^n(n+1)!}=\lim_{n\to\infty}\left(1+\frac{1}{n}\right)^n=e>1$$

and consequently, $\lim\limits_{n\to\infty}\frac{n^n}{n!}=\infty$.

We previously introduced the limit point notation and examined basic properties of such points. We now present several results related to the existence of such points. We will distinguish between a set whose elements are mutually distinct and a sequence for which this property is not necessarily valid.

Theorem 3.7.6 (Bolzano-Weierstrass theorem for sets). Every infinite bounded set S has at least one limit point.

Proof.

Since S is bounded, we can find a closed interval $[a_1,b_1]$ which contains all the points of S. Define $c_1=\frac{a_1+b_1}{2}$. Then, at least one of the intervals

$[a_1,c_1]$, $[c_1,b_1]$ must contain infinitely many points of S. Denote it by $[a_2,b_2]$ and define $c_2 = \frac{a_2+b_2}{2}$. Again, at least one of the intervals $[a_2,c_2]$, $[c_2,b_2]$ contains infinitely many points of S. Using induction, we thus create a set of intervals $\{[a_1,b_1],[a_2,b_2],\ldots,[a_n,b_n],\ldots\}$ such that $\{a_n\}$ and $\{b_n\}$ are monotone increasing and decreasing sequences respectively, $\{a_n\}$ is bounded above, $\{b_n\}$ is bounded below and *each* interval contains infinitely many points of S. Therefore $\lim_{n\to\infty} a_n = A$, $\lim_{n\to\infty} b_n = B$ for some real numbers A and B, but since $0 \le b_n - a_n = \frac{(b_1-a_1)}{2^{n-1}}$ we get $A = B$. Finally, A is a limit point of S. To prove it we need to show that every interval $(A-\varepsilon, A+\varepsilon)$, $\varepsilon > 0$ contains a point α such that $\alpha \in S$ and $\alpha \ne A$. Indeed, given an arbitrary $\varepsilon > 0$ one can find n_1, n_2 such that $A-\varepsilon < a_n \le A$, $n > n_1$ and $A \le b_n < A+\varepsilon$, $n > n_2$. Choose $n_0 = \max(n_1,n_2)$. Then $A-\varepsilon < a_n \le A \le b_n < a+\varepsilon$ for $n > n_0$. Consequently, the particular interval $[a_m,b_m]$, $m = n_0+1$ is inside $(A-\varepsilon, A+\varepsilon)$ and contains *infinitely many points* of S. Let $\alpha \ne A$ be *one* of them. Then $|\alpha - A| < \varepsilon$ and the proof is completed.

A similar result can be derived for sequences.

Theorem 3.7.7 (Bolzano-Weierstrass theorem for sequences). Every bounded sequence $\{a_n\}$ has at least one limit point.

Proof.

What we have to show is the existence of a subsequence that *converges*. If some point α appears infinitely many times in the sequence, i.e., if there is a subsequence $\{a_{n_k}\}$ such that $a_{n_k} = \alpha$, $k \ge 1$, then α is a limit of this subsequence, and hence a limit point of the whole sequence. If there is no such α, construct the following subsequence of $\{a_n\}$. Let $n_1 = 1$ and find $n_2 > n_1$ such that $a_{n_2} \ne a_{n_1}$. This is possible or else a_{n_1} appears infinitely many times. At each step we can find $n_{k+1} > n_k$ such that $a_{n_{k+1}} \ne a_{n_l}$, $1 \le l \le k$, or else one of the numbers a_{n_l}, $1 \le l \le k$ appears infinitely many times. We thus obtain a subsequence $\{a_{n_k}\}$ with mutually distinct elements. This subsequence is therefore a *set*, and by virtue of the previous theorem it has, *as a set*, at least one limit point α. Namely, there is an infinite number of mutually distinct elements $a_{m_1}, a_{m_2}, \ldots, a_{m_j}, \ldots$ such that $a_{m_j} \in \{a_{n_k}\}$, $j \ge 1$ and $\lim_{j\to\infty} a_{m_j} = \alpha$. Since there is no guarantee that $m_1 < m_2 < \ldots < m_j < \ldots$, the number α is not yet a limit of subsequence of $\{a_n\}$.

The final step is therefore, showing how to construct a subsequence of $a_{m_1}, a_{m_2}, \ldots, a_{m_j}, \ldots$ (which must also converge to α - why?) which is also a subsequence of $\{a_n\}$. We leave this step as an exercise for the reader.

In the next subsection we discuss the derivation of properties of a sequence $\{a_n\}$, using information about $|a_n - a_m|$, where n, m are *sufficiently large*.

3.7.3 Cauchy Sequences

Definition 3.7.5. A sequence $\{a_n\}$ is called a *Cauchy sequence*, if for an arbitrary $\varepsilon > 0$ we can find $n_0 \in N$ (which depends on ε), such that

$$|a_n - a_m| < \varepsilon \quad , \quad n, m > n_0 \tag{3.7.3}$$

Namely, a Cauchy sequence is characterized by the fact that for an arbitrary $\varepsilon > 0$, as small as desired, there is a class whose members are *almost all* the sequence's terms, and the distance between *each two* members is less than ε.

Theorem 3.7.8. A Cauchy sequence $\{a_n\}$ is always bounded.

Proof.

The choice $\varepsilon = 1$ guarantees the existence of some n_0 such that $|a_n - a_m| < 1$ as long as $m, n > n_0$. Thus, if we take $m = n_0 + 1$ we get $|a_n - a_{n_0+1}| < 1$, $n > n_0$. By virtue of the triangle inequality, this yields $|a_n| < 1 + |a_{n_0+1}|$, $n > n_0$. Therefore

$$|a_n| < \max\{|a_1|, |a_2|, \ldots, |a_{n_0}|, 1 + |a_{n_0+1}|\}, \ n \ge 1$$

and the sequence is bounded.

The following result presents the equivalence between the notations of a Cauchy sequence and a convergent sequence.

Theorem 3.7.9. A sequence $\{a_n\}$ converges if and only if it is a Cauchy sequence.

Proof.

(a) Assume $\lim_{n \to \infty} a_n = A$ and consider an arbitrary fixed $\varepsilon > 0$. Define $\varepsilon_1 = \varepsilon/2$.

Then, we can find n_0 such that $|a_n - A| < \varepsilon_1$, $n > n_0$. Let $m, n > n_0$. Then

$$|a_n - a_m| = |a_n - A + A - a_m| \le |a_n - A| + |A - a_m| < 2\varepsilon_1 = \varepsilon \quad , \quad m,n > n_0$$

which implies that a convergent sequence is also a Cauchy sequence.

(b) Let $\{a_n\}$ be a Cauchy sequence. It is therefore bounded and thus, by Bolzano-Weierstrass theorem for sequences, $\{a_n\}$ has a limit point A. We will show $\lim_{n\to\infty} a_n = A$. Indeed, there is a subsequence $\{a_{n_k}\}$ such that $\lim_{k\to\infty} a_{n_k} = A$. Let $\varepsilon > 0$. Find n_1 such that $|a_n - a_m| < \varepsilon$, $n,m > n_1$ and find k_0 such that $|a_{n_k} - A| < \varepsilon$, $k > k_0$. Let n_0 the first n_k , $k > k_0$, greater than n_1. Then

$$|a_n - A| = |a_n - a_{n_0} + a_{n_0} - A| \le |a_n - a_{n_0}| + |a_{n_0} - A|$$

for all $n \in N$. Since $n_0 > n_1$ then $n > n_0$ implies $n, n_0 > n_1$ and consequently $|a_n - a_{n_0}| < \varepsilon$. Also, since $n_0 = n_l$ for some $l > k_0$, we have $|a_{n_0} - A| < \varepsilon$. Therefore, $|a_n - A| < 2\varepsilon$, $n > n_0$, i.e., $\lim_{n\to\infty} a_n = A$ and the proof is completed.

Example 3.7.8. If we replace the requirement given by Eq. (3.7.3) by a weaker requirement such as $|a_{n+1} - a_n| < \varepsilon$, $n > n_0$, we may get a sequence that does not converge. For example, the sequence $a_n = \sum_{k=1}^{n} \frac{1}{k}$ is divergent although it satisfies $|a_{n+1} - a_n| = \frac{1}{n}$ and $\lim_{n\to\infty} \frac{1}{n} = 0$.

Definition 3.7.6. A sequence $\{a_n\}$ is called a *contractive sequence* if

$$|a_{n+1} - a_n| < q|a_{n+1} - a_n| , n \in N$$

for some $q : 0 < q < 1$.

Example 3.7.9. Consider a sequence $\{a_n\}$ such that $a_{n+1} = 2 + \frac{a_n}{2}$. In this case we have $a_{n+2} - a_{n+1} = 2 + \frac{a_{n+1}}{2} - 2 - \frac{a_n}{2} = \frac{1}{2}(a_{n+1} - a_n)$, i.e. a contractive sequence with $q = 1$. The conclusion is the same for $a_{n+1} = A + qa_n$; $A \in \Re$, $q \in (0,1)$.

Theorem 3.7.10. Every contractive sequence is a Cauchy sequence.

Proof.

Let $\{a_n\}$ be a contractive sequence. Then, for every $n \in N$ we have

$$|a_{n+2} - a_{n+1}| \le q|a_{n+1} - a_n| \le \ldots \le q^n |a_2 - a_1|$$

In order to evaluate $|a_n - a_m|$ we will assume $m > n$ and rewrite

$$|a_m - a_n| = |a_m - a_{m-1} + a_{m-1} - \ldots + a_{n+2} - a_{n+1} + a_{n+1} - a_n|$$

$$\le |a_m - a_{m-1}| + |a_{m-1} - a_{m-2}| + \ldots + |a_{n+2} - a_{n+1}| + |a_{n+1} - a_n|$$

$$\le q^{m-2}|a_2 - a_1| + q^{m-3}|a_2 - a_1| + \ldots + q^n|a_2 - a_1| + q^{n-1}|a_2 - a_1|$$

$$= q^{n-1}(1 + q + \ldots + q^{m-n-1})|a_2 - a_1| = q^{n-1}\frac{1-q^{m-n}}{1-q}|a_2 - a_1|$$

$$\le \frac{q^{n-1}}{1-q}|a_2 - a_1|$$

Since $\lim_{n\to\infty} q^{n-1} = 0$, given $\varepsilon > 0$ there is n_0 such that $q^{n-1}|a_2 - a_1| < \varepsilon$ for $n > n_0$. Consequently, $|a_n - a_m| < \varepsilon$ for $m > n > n_0$ which implies that $\{a_n\}$ is a Cauchy sequence.

Corollary 3.7.1. Every contractive sequence is convergent.

The proof follows from Theorems 3.7.9-10.

Theorem 3.7.11. Let $\{a_n\}$ be a contractive sequence which converges to A and has a contractive factor q. Then

$$|a_n - A| \le \frac{q^{n-1}}{1-q}|a_2 - a_1|,\ n \in N$$

Proof.

We previously obtained (proof of Theorem 3.7.10) $|a_m - a_n| < \frac{q^{n-1}}{1-q}|a_2 - a_1|$ for fixed n and all $m > n$. Therefore, $\lim_{m\to\infty}|a_m - a_n| = |A - a_n| \le \frac{q^{n-1}}{1-q}|a_2 - a_1|$ as stated.

This theorem provides a useful tool for bounding (for a contractive sequence) the distance between a given a_n and the limit A. The bound depends on the contraction factor and the distance between the first two terms of the sequence.

Example 3.7.10. Let $a_{n+1} = \sqrt{1+a_n}$, $a_1 = 1$. The sequence is monotone increasing and converges to $A = \frac{1+\sqrt{5}}{2}$ (why?). Also

$$a_{n+2} - a_{n+1} = \sqrt{1+a_{n+1}} - \sqrt{1+a_n} = \frac{a_{n+1} - a_n}{\sqrt{1+a_{n+1}} + \sqrt{1+a_n}} \le \frac{a_{n+1} - a_n}{2\sqrt{2}}$$

which, by Theorem 3.7.11, implies $\left|a_n - \frac{1+\sqrt{5}}{2}\right| \le \frac{\sqrt{2}-1}{(2\sqrt{2})^{n-1}(1-1/2\sqrt{2})}$. For example, if $n = 10$ we get $\left|a_{10} - \frac{1+\sqrt{5}}{2}\right| \le 5.53 \cdot 10^{-5}$. The actual distance between a_{10} and $\frac{1+\sqrt{5}}{2}$ is about $1.74 \cdot 10^{-5}$, i.e. about a third of the calculated bound.

Theorem 3.7.12. Let $\{a_n\}$ be a contractive sequence which converges to A and has a contractive factor q. Then

$$|a_n - A| \le \frac{q}{1-q}|a_n - a_{n-1}|,\ n > 1$$

Proof.

The proof is based on the previous technique (Theorems 3.7.10-11) and is left for the reader.

Example 3.7.11. Consider the sequence from the previous example. For $n = 10$ we get the bound $\frac{q}{1-q}(a_{10} - a_9) \approx 2.13 \cdot 10^{-5}$ ($\approx$ means about or approximately) which is a better estimate for $A - a_{10}$ than before.

PROBLEMS

1. Let $a_{n+1} = 1 + \sqrt{a_n}$, $a_1 = 1$. Show that the sequence $\{a_n\}$ is monotone increasing and bounded above. Find its limit.

2. Let $a_{n+1} = \frac{a_n}{1+a_n}$, $a_1 = 2$. Show that the sequence $\{a_n\}$ is monotone decreasing and bounded below. Find its limit.
3. Let $a_{n+1} = a_n + \sqrt{a_n}$, $a_1 = 1$. Show that the sequence $\{a_n\}$ is monotone increasing but does not have a limit.
4. A sequence $\{a_n\}$ which is monotone for $n \geq n_0$ is called *eventually monotone*. Show that the following sequences are eventually monotone and determine their type of monotonity: (a) $\frac{n^2}{2^n}$ (b) $\frac{3^n}{n^2+n-1}$ (c) $\frac{3^n}{3n+n!}$
5. Show that the sequence $a_n = \sum_{k=1}^{n} \frac{1}{k^2}$ has a limit A which satisfies $1 < A < 2$.
6. Let $\{a_n\}$ and $\{b_n\}$ denote two monotone increasing sequences. Show that $\{c_n\} = \{a_n + b_n\}$ is monotone increasing. Is $\{d_n\} = \{a_n - b_n\}$ monotone increasing?
7. Let $a_n = \frac{1 \cdot 3 \cdot 5 \cdots (2n-1)}{2 \cdot 4 \cdot 6 \cdots (2n)}$, $b_n = \frac{2 \cdot 4 \cdot 6 \cdots (2n)}{1 \cdot 3 \cdot 5 \cdots (2n+1)}$, $n \in N$. Show:

   a. $\lim_{n\to\infty} a_n b_n = 0$ (b) $\lim_{n\to\infty} a_n = 0$ (c) $\lim_{n\to\infty} b_n = 0$

8. Show that $a_n = \frac{\sqrt{n^3 - n + n}}{n + 50\sqrt{n}}$ converges to infinity.
9. Show that $a_n = \frac{n^n}{n!}$ converges to infinity.
10. Let $\{a_n\}, \{b_n\}$ be sequences that satisfy $\lim_{n\to\infty} a_n = 0$ and $\lim_{n\to\infty} b_n = \infty$. What can be said about the sequence $\{c_n\} = \{a_n b_n\}$?
11. Change Eq. (3.7.3) and request $|a_m - a_n| < \varepsilon$ for $n, m > n_0$, $|m-n| \leq k$ where k is a *prefixed* positive integer. Is the sequence necessarily convergent? Explain!
12. Let $\lim_{n\to\infty} a_n = A$. Show that the sequence of the averages converges to A, i.e.

$$\lim_{n\to} \frac{a_1 + a_2 + \ldots + a_n}{n} = A$$

13. Prove $\lim_{n\to\infty} \frac{1}{n}\left(1 + \frac{1}{2} + \frac{1}{3} + \ldots + \frac{1}{n}\right) = 0$ (Hint: First show monotonity).
14. Show $\lim_{n\to\infty} (1+n+n^2)^{1/n} = 1$.

4 Continuous Functions

In this chapter we focus on a special class of functions called *continuous functions*. Basically, a continuous function $y = f(x)$ is characterized by the fact that it consists of no jumps, i.e., a small change in x causes a small change in y. An exact definition of *continuity* is given only in Section 4.2. Until then we suggest the phrase given by some unknown mathematician: "When you draw a continuous function, the chalk never separates from the blackboard".

4.1 Limits of Functions

Definition 4.1.1. A function $f(x)$, $x \in D \subseteq \Re$ is said to have the limit A at a (or, "as x approaches a") if:

1. a is a limit point of D.
2. For every $\varepsilon > 0$ we can find a $\delta > 0$ such that

$$0 < |x - a| < \delta\,,\ x \in D \Rightarrow |f(x) - A| < \varepsilon \tag{4.1.1}$$

In this case we write $\lim_{x \to a} f(x) = A$.

Note, that a *does not have* to belong to the function's domain. Also, the requirement that a is a limit point of D, implies the existence of a sequence $\{x_n\}$, $x_n \in D$ with mutually distinct elements such that $\lim_{n \to \infty} x_n = a$ and by virtue of the second requirement we have $\lim_{n \to \infty} f(x_n) = A$. This assertion is left as an exercise for the reader.

Example 4.1.1. Let $f(x) = x^3$, $x \in \Re$. Clearly, every real number is a limit point within $\Re$. We will show that the limit of $f(x)$ at arbitrary $x_0 \in \Re$ is x_0^3. Thus, given an $\varepsilon > 0$ we will find a $\delta > 0$, such that $|x - x_0| < \delta$ implies $|x^3 - x_0^3| < \varepsilon$.

M. Friedman and A. Kandel: Calculus Light, ISRL 9, pp. 87–106.
springerlink.com

Simple algebra yields

$$\left|x^3 - x_0^3\right| = \left|(x - x_0)(x^2 + xx_0 + x_0^2)\right| \le \left|x - x_0\right|\left(\left|x\right|^2 + \left|x\right|\left|x_0\right| + \left|x_0\right|^2\right) \le \left|x - x_0\right|\left(\left|x\right| + \left|x_0\right|\right)^2$$

Since we are interested only at a close neighborhood of x_0, we confine ourselves to x: $\left|x - x_0\right| < 1$. Therefore $\left|x\right| < 1 + \left|x_0\right|$ which leads to

$$\left|x^3 - x_0^3\right| \le \left|x - x_0\right|\left(1 + 2\left|x_0\right|\right)^2$$

If we also require that

$$\left|x - x_0\right| < \frac{\varepsilon}{\left(1 + 2\left|x_0\right|\right)^2}$$

we obtain $\left|x^3 - x_0^3\right| < \varepsilon$. Thus, the choice $\delta = \min\left(1, \dfrac{\varepsilon}{\left(1 + 2\left|x_0\right|\right)^2}\right)$ *guarantees*

$$\left|x - x_0\right| < \delta \Rightarrow \left|x^3 - x_0^3\right| < \varepsilon$$

Example 4.1.2. The function $f(x) = \sin(x)$ is defined for all $x \in \Re$. We will show that the limit of $f(x)$ at arbitrary $x_0 \in \Re$ is $\sin(x_0)$. Indeed,

$$\sin(x) - \sin(x_0) = 2\sin[(x - x_0)/2]\cos[(x + x_0)/2]$$

Simple trigonometry implies that $\sin(\alpha) < \alpha < \tan(\alpha)$ for all $0 < \alpha < \dfrac{\pi}{2}$ (see Section 5.3). Thus, by confining ourselves to $\left|x - x_0\right| < \dfrac{\pi}{2}$ and using the inequality $\left|\cos(\alpha)\right| \le 1$ which holds for all α, we get

$$\left|\sin(x) - \sin(x_0)\right| = 2\left|\sin[(x - x_0)/2]\right|\left|\cos[(x + x_0)/2]\right| \le \left|x - x_0\right|$$

The rest is straightforward.

In Examples 4.1.1- 4.1.2 we dealt with a particular case where the limit of $f(x)$ and the value $f(x_0)$ exist. We applied a simple principle: Consider a function $f(x)$, $x \in D \subseteq \Re$ and let $x_0 \in D$ be a limit point of D. If there is a constant $\rho > 0$ such that the requirement

$$x \in D, \left|x - x_0\right| \le \rho \Rightarrow \left|f(x) - f(x_0)\right| \le C\left|x - x_0\right| \tag{4.1.2}$$

holds, then $\lim\limits_{x \to x_0} f(x) = f(x_0)$. The proof is simple: Given an arbitrarily small $\varepsilon > 0$ define $\delta = \min\{\rho, \varepsilon/C\}$. If $\left|x - x_0\right| < \delta$ then

$$\left|f(x)-f(x_0)\right| \le C\left|x-x_0\right| < C\cdot\frac{\varepsilon}{C}=\varepsilon$$

which concludes the proof. Thus, if Eq. (4.1.2) holds even for a very small but prefixed ρ, the limit of $f(x)$ as $x\to x_0$ exists and equals $f(x_0)$.

Example 4.1.3. Let $f(x)=\sin(1/x)$, $x\ne 0$. The function does not have a limit at $x=0$. Indeed, it is equal to 1 and -1 at $x_n=\dfrac{1}{(2n+1/2)\pi}$ and $y_n=\dfrac{1}{(2n+3/2)\pi}$ respectively, while both sequences of points approach 0 as n increases.

PROBLEMS

1. Show that the function $f(x)=\cos(x)$ has a limit at all $x:-\infty<x<\infty$.
2. Show that the function $f(x)=\sqrt{x}$ has a limit at all $x:x\ge 0$. Does it have a limit at $x=-1$?
3. Consider the function

$$f(x)=\begin{cases} x^2\ , & x\ne\pm1 \\ 0\ , & x=1 \end{cases}$$

   Does the function has a limit at $x=-1,0,1$? What are the differences between the three points?
4. Let $f(x)=x\sin(1/x)$, $x\ne 0$. Does the function has a limit at $x=0$?
5. Let $f(x)=\dfrac{1}{(x-a)(x-b)(x-c)}$. For which x the function has no limit?
6. Let

$$f(x)=\begin{cases} 1+\sqrt{x}\ , & x\le 0 \\ \sin\left(\dfrac{1}{x}\right), & x>0 \end{cases}$$

   Does $f(x)$ have a limit at $x=0$? If not, can you redefine it so that the limit will exist?

4.2 Continuity

The class C of continuous functions is a very small segment of the set of all functions but is nonetheless very useful. The reason is that most processes in nature can be simulated using continuous functions or functions that are continuous *almost everywhere*. Actually the most applicable functions in the various sciences are even a smaller subset – functions that possess *derivatives*.

These will be treated in detail in Chapter 5. In this chapter we concentrate on the concept of continuity and the properties of the continuous functions.

Definition 4.2.1. A function $f(x)$, $x \in D \subseteq \Re$ is called *continuous* at $x = a \in D$ if a is a limit point of D and if

$$\lim_{x \to a} f(x) = f(a) \quad (4.2.1)$$

If a is an isolated point in D, the function is not continuous at a. A detailed version of this definition would be: for arbitrarily small $\varepsilon > 0$ we can find $\delta > 0$ such that

$$|x - a| < \delta \, , \, x \in D \Rightarrow |f(x) - f(a)| < \varepsilon \quad (4.2.2)$$

If $f(x)$ is not continuous at $x = a$ it is called *discontinuous* at this point.

The domain D of $f(x)$ is *usually* an interval, closed, open, semi-infinite, the whole real line or a union of such intervals. However, Definition 4.2.1 considers the most general domain D.

Example 4.2.1. Let $f(x) = 2x$ over the domain $D = [-1,0] \cup \{1, 1/2, 1/3, \ldots, 1/n, \ldots\}$, composed of a single closed interval and a sequence of points which converges to 0 from the right. It is easily seen that the function is continuous over the interval, including at 0, and is not continuous at each $x_n = 1/n$ which is an isolated point.

Definition 4.2.2. A function $f(x)$, $x \in D \subseteq \Re$ is called *continuous* over D if it is continuous at all $x : x \in D$.

Example 4.2.2. The functions $x^2, \sin(x), \cos(x)$ are continuous for all $x : -\infty < x < \infty$. This is derived using the same algebra as in the previous section and is left for the reader.

Example 4.2.3. Let

$$f(x) = \begin{cases} x^2 \, , \, x \neq \pm 1 \\ 0 \, , \, x = \pm 1 \end{cases}$$

The function is not continuous at $x = \pm 1$. However, by altering its definition, the continuity may be extended to hold there as well. This can be done by simply redefining $f(1) = f(-1) = 1$.

Some discontinuities can never be removed. For example the function $f(x) = \sin(1/x)$, $x \neq 0$ oscillates between 1 and -1 as we approach $x = 0$ and no matter how we define the function at this point, it is impossible to obtain continuity since the basic requirement of Eq. (4.2.1) is not satisfied, simply because the left part of that equation does not exist, i.e., the function *does not converge* as $x \to 0$.

Another type of irremovable discontinuity occurs when a function is not bounded. For example, the function

$$f(x) = \begin{cases} 1/x\,, & x \neq 0 \\ A\,, & x = 0 \end{cases}$$

is not continuous at $x = 0$ for all finite A. The proof is left for the reader.

A *removable* type of discontinuity occurs only in the case when the left-hand side of Eq. (4.2.1) exists. Then, if $f(a)$ is not defined we may take

$$f(a) = \lim_{x \to a} f(x)$$

and if the function is defined but satisfies $f(a) \neq \lim_{x \to a} f(x)$ we simply change and redefine $f(a) = \lim_{x \to a} f(x)$. This last case is called *jump discontinuity*. It disappears via redefining the function.

Example 4.2.4. Consider the function

$$f(x) = \begin{cases} 1\,, & x \text{ rational} \\ 0\,, & x \text{ irrational} \end{cases}$$

This function is discontinuous for all x, rational and irrational alike, since at any arbitrarily small interval there is always at least one point of each type. The exact details are left as an exercise for the reader.

Example 4.2.5. Let each rational number be represented as m/n where m, n have no common divisor. Define

$$f(x) = \begin{cases} 1/n\,, & x = m/n \\ 0\,, & \text{irrational} \end{cases}$$

The function is clearly discontinuous at all rational points since at arbitrarily small interval about any rational number r there is at least one irrational number where the function vanishes. This contradicts the requirement of Eq. (4.2.1) since the function's value at $r = m/n$ is $1/n \neq 0$. On the other hand, at an irrational point ξ, the function is zero and for a given $\varepsilon > 0$ we can find a sufficiently small interval $[\xi - \delta, \xi + \delta]$ at which the function values at the rational points within this interval, are less than ε. All we need to do is to choose an interval that contains only rational numbers m/n for which $n > 1/\varepsilon$ (why is that possible?). Consequently, $f(x)$ is continuous at ξ.

Since the domain of definition of a given function is quite often an interval, we introduce the notation $C[a,b]$ - all the continuous functions defined over a closed interval $[a,b]$. Similarly we define $C(a,b)$ etc.

Definition 4.2.3. A function $f(x)$ is said to be right-continuous at a if

$$\lim_{x \to a^+} f(x) = f(a) \tag{4.2.3}$$

where by $x \to a^+$ we mean $x = a + h$, $h > 0$, $h \to 0$. Left continuity is defined similarly.

Example 4.2.6. Let

$$f(x) = \begin{cases} x^2 , x \leq 1 \\ 2 , x > 1 \end{cases}$$

The function, illustrated in Fig. 4.2.1, is defined everywhere, continuous at $x \neq 1$, has a jump continuity at $x = 1$ but is left-continuous there. The discontinuity is not removable since the left and right limits are different.

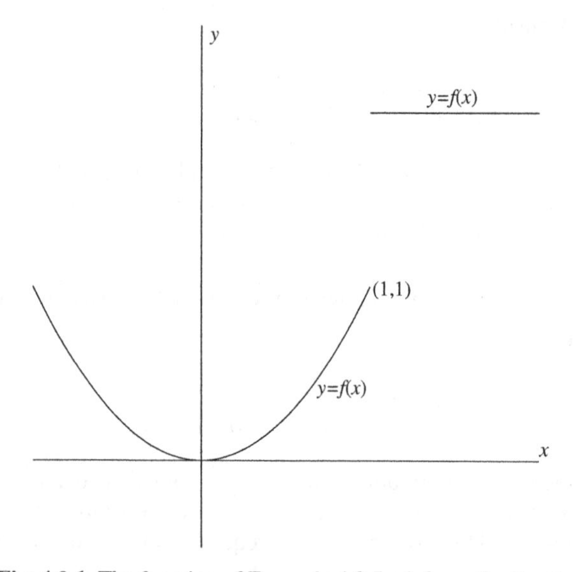

Fig. 4.2.1. The function of Example 4.2.5 – left continuity at $x = 1$.

Theorem 4.2.1. If $f(x)$ is right and left continuous at an arbitrary point $x = a$, then $f(x)$ is continuous at a.

The proof is left as an exercise for the reader.

Theorem 4.2.2. A necessary condition for the continuity of $f(x)$ at an arbitrary $x=a$, is that $f(x)$ is bounded, i.e., there exist $\delta>0$ and $M: 0<M<\infty$ such that

$$|x-a|<\delta \Rightarrow |f(x)|<M$$

Proof.

If $f(x)$ is continuous at a but not bounded there, we can find x_1 within the interval $I_1=[a-1,a+1]$ such that $|f(x_1)|>1$, otherwise $f(x)$ is bounded in I_1 which contradicts our assumption. Next, we can find x_2 within $I_2=[a-1/2,a+1/2]$ such that $|f(x_1)|>2$, otherwise $f(x)$ is bounded at a. We are thus able to construct a sequence $\{x_n\}$ for which $a-1/n \le x_2 \le a+1/n$ and $|f(x_n)|>n$. However, while $\lim_{n\to\infty} x_n = a$, the values $\{f(x_n)\}$ approach infinity rather than $f(a)$, which contradicts the continuity assumption.

PROBLEMS

1. Show that the following functions are continuous for all $x: -\infty<x<\infty$:
   (a) $f(x)=x^2+x$ (b) $f(x)=|x|$ (c) $f(x)=1/[2+\sin(x)]$.
2. Consider the function

$$f(x)=\begin{cases} 0\,, x=0 \\ 3\,, x=-1 \\ \sin(1/x)+\cos[1/(x+1)]\,, x\neq 0,-1 \end{cases}$$

   Find the points of discontinuity. Is it possible to remove these discontinuities by redefining the function?
3. Let $f(x)$, $a\le x\le b$ satisfy the Lipschitz condition $|f(x)-f(y)|<L|x-y|$ for arbitrary $x,y: a\le x,y\le b$. Show that $f(x)$ is continuous over $[a,b]$.
4. Let $f(x)$ be defined everywhere, satisfy $f(x+y)=f(x)+f(y)$ for all x,y and continuous at $x=0$. Show that $f(x)$ is continuous everywhere.
5. Show that $f(x)=\dfrac{1}{\sqrt{x^2+1}+\sqrt{x}}$ is continuous for all $x>0$.
6. Find the points of continuity and discontinuity of

$$f(x)=\begin{cases} \sqrt{-x}\,, -\infty<x\le 0 \\ 1/1000\,, 0<x\le 1 \\ 1/1000+\sqrt{x-1}\,, 1<x<\infty \end{cases}$$

7. Find the points of continuity and discontinuity of

$$f(x)=\begin{cases} x^3 \,, -\infty < x \le 1 \\ 1/(x-1)\,, 1 < x \le 10 \\ 1/8\,, 10 < x < \infty \end{cases}$$

4.3 Properties of Continuous Functions

Throughout the rest of this chapter we mainly deal with the set of all continuous functions. Our next result shows that continuity is preserved under the four basic arithmetic operations.

Theorem 4.3.1. Let $f(x), g(x)$ denote arbitrary continuous functions defined over the same domain. Then, $f(x) \pm g(x)$, $f(x)g(x)$ are continuous as well, and $f(x)/g(x)$ is continuous at all x for which $g(x) \ne 0$.

Proof.

The proof is derived in a similar manner to that of Theorem 3.5.1. The proof of the first three claims are left for the reader while that of the fourth will be given here in detail.

Define $H(x) = f(x)/g(x)$. Assume $g(a) \ne 0$ and let $\varepsilon > 0$. We will show the existence of $\delta > 0$ such that

$$|x-a| < \delta \Rightarrow |H(x) - H(a)| < \varepsilon$$

Clearly,

$$H(x) - H(a) = \frac{f(x)}{g(x)} - \frac{f(a)}{g(a)} = \frac{f(x)g(a) - f(a)g(x)}{g(x)g(a)}$$
$$= \frac{f(x)g(a) - f(a)g(a) + f(a)g(a) - f(a)g(x)}{g(x)g(a)}$$

Hence

$$|H(x) - H(a)| \le \frac{|f(x) - f(a)|}{|g(x)|} + \frac{|f(a)||g(x) - g(a)|}{|g(x)g(a)|}$$

Since $g(a) \ne 0$ and $g(x)$ is continuous, we can find (how?) $\delta_1 > 0$ such that

$$|x-a| < \delta_1 \Rightarrow |g(x)| > \frac{|g(a)|}{2}$$

The continuity of $f(x)$ at a guarantees that given the particular choice $\varepsilon_1 = \frac{\varepsilon|g(a)|}{4}$ we can find $\delta_2 > 0$ such that $|x-a| < \delta_2 \Rightarrow |f(x) - f(a)| < \varepsilon_1$. Also, due to the continuity of $g(x)$ at a, if given the choice $\varepsilon_2 = \frac{\varepsilon|g(a)|^2}{4[1+|f(a)|]}$, we can find $\delta_3 > 0$ such that

$$|x-a| < \delta_3 \Rightarrow |g(x)-g(a)| < \varepsilon_2$$

Therefore, by taking $\delta = \min(\delta_1, \delta_2, \delta_3)$ we obtain

$$|H(x)-H(a)| < \frac{\varepsilon|g(a)|}{4} \Big/ (|g(a)|/2) + \frac{2|f(a)|}{|g(a)|^2} \frac{\varepsilon|g(a)|^2}{4[1+|f(a)|]} = \varepsilon$$

provided that $|x-a| < \delta$. This completes the proof.

Example 4.3.1. Since a constant function and the function $f(x) = x$ are continuous everywhere, so is an arbitrary polynomial $f(x) = a_0 + a_1 x + \ldots + a_n x^n$ which can be obtained from these two using addition and multiplication.

Example 4.3.2. The function $\tan(x) = \sin(x)/\cos(x)$ is continuous except at $x = (n+1/2)\pi$ where n is an arbitrary integer. Indeed, $\sin(x)$ and $\cos(x)$ are continuous everywhere and the denominator vanishes only at $x = (n+1/2)\pi$.

Let $g(x)$ denote a continuous, defined over an interval I and let the set $S = \{y \mid y = g(x), x \in I\}$, i.e., the image of $g(x)$, be a subset of an interval J over which another continuous function $f(x)$. Clearly, the function $f(g(x))$, called the *composite function* and denoted by $(f \circ g)(x)$, is defined over I.

Theorem 4.3.2. The composite function $(f \circ g)(x)$ is continuous over I.

Proof.

Given a within the interior of I, we have to show that for an arbitrary $\varepsilon > 0$ there exists $\delta > 0$ such that

$$|x-a| < \delta, x \in I \Rightarrow |f(g(x)) - f(g(a))| < \varepsilon$$

The continuity of $f(x)$ guarantees that for some $\delta_1 > 0$ we have

$$|y-g(a)| < \delta_1, y \in J \Rightarrow |f(y) - f(g(a))| < \varepsilon$$

Also, since $g(x)$ is continuous over I, we can find $\delta > 0$ such that

$$|x-a| < \delta, x \in I \Rightarrow |g(x)-g(a)| < \delta_1$$

The rest is straightforward:

$$|x-a|<\delta, x\in I \Rightarrow |g(x)-g(a)|<\delta_1 \Rightarrow |f(g(x))-f(g(a))|<\varepsilon$$

and the composite function is indeed continuous at a.

The next result is one of several that relate to an arbitrary continuous function over a closed interval.

Theorem 4.3.3. A continuous function $f(x)$ over a closed interval $I=[a,b]$ is bounded, i.e., there exists $M>0$ such that $|f(x)|<M$ for all $x\in I$.

Proof.

If $f(x)$ is not bounded, then a sequence $\{x_n\}$, $x_n\in I$ such that $|f(x_n)|>n$, can be found. Since the sequence is bounded, then by virtue of Bolzano-Weierstrass theorem for sequences (Section 3.7), it has a limit point ξ. Thus, there is a subsequence $\{x_{n_k}\}$ of $\{x_n\}$ that converges to ξ. The continuity of $f(x)$ implies $\lim_{k\to\infty} f(x_{n_k})=f(\xi)$ which leads to contradiction since $|f(x_{n_k})|>n_k\to\infty$. Thus, $f(x)$ must be bounded which concludes the proof.

By virtue of Theorems 3.2.1 and 4.3.3, a continuous function $f(x)$ over a closed interval $I=[a,b]$ has a supremum, i.e., there exists a number M such that

$$f(x)\le M \ , x\in I \tag{4.3.1}$$

and for an arbitrary $\varepsilon>0$ we can find $x_\varepsilon\in I$ that satisfies $f(x_\varepsilon)>M-\varepsilon$. Similarly, $f(x)$ has an infimum m, i.e., a number for which

$$f(x)\ge m \, , x\in I \tag{4.3.2}$$

and such that for an arbitrary $\varepsilon>0$ we can find $x_\varepsilon\in I$ that satisfies $f(x_\varepsilon)<m+\varepsilon$.

The next result shows that these supremum and infimum are attained by the function, i.e., they are the maximum and minimum of $f(x)$, respectively.

Theorem 4.3.4. Let $f(x)$, a continuous function over the interval $I=[a,b]$, have a supremum M. Then, there exists a point $\xi\in I$ such that $f(\xi)=M$.

Proof.

Since M is a supremum of $f(x)$, we should be able to construct a sequence $\{x_n\}$, $x_n\in I$ such that $f(x_n)>M-1/n$ which implies that the sequence

$\{f(x_n)\}$ converges to M. By applying once again the Bolzano-Weierstrass theorem for sequences, we construct a subsequence $\{x_{n_k}\}, k = 1,2,...$ which converges to $\xi \in I$. The continuity of $f(x)$ at ξ implies $\lim_{k\to\infty} f(x_{n_k}) = f(\xi)$, but since $\lim_{k\to\infty} f(x_{n_k}) = M$ as well, we get $M = f(\xi)$ which concludes the proof.

Similarly, a continuous function over a closed interval attains its infimum, i.e. has a minimum.

Theorem 4.3.5. Let $f(x)$ denote a continuous function over the interval $I = [a,b]$ with maximum and minimum M, m respectively, and consider an arbitrary number $c : m \le c \le M$. Then, there exists at least one $\xi : a \le \xi \le b$ such that $f(\xi) = c$.

The proof of Theorem 4.3.5 follows directly from the following lemma.

Lemma 4.3.1. Let $f(x)$ denote a continuous function over the interval $I = [a,b]$ which satisfies $f(a) < 0$, $f(b) > 0$. Then, there exists at least one point $\xi : a \le \xi \le b$ such that $f\xi) = 0$.

Proof.

Let $c = (a+b)/2$. If $f(c) = 0$ we choose $\xi = c$. If $f(c) < 0$ we define $a_1 = c$, $b_1 = b$ and if $f(c) > 0$ we take $a_1 = a$, $b_1 = c$. We continue the process with the interval $[a_1, b_1]$ where $f(a_1) < 0$, $f(b_1) > 0$ and $(b_1 - a_1) = (b-a)/2$. Prior to the $(n+1)$-th step we have an interval $[a_n, b_n]$ for which $f(a_n) < 0$, $f(b_n) > 0$ and $(b_n - a_n) = (b-a)/2^n$. Define $c = (a_n + b_n)/2$. If $f(c) = 0$ the process stops. If not, we have two cases: If $f(c) < 0$ we define $a_{n+1} = c$, $b_{n+1} = b_n$ and if $f(c) > 0$ the new interval for further search is $a_{n+1} = a_n$, $b_{n+1} = c$. In either case we proceed to the $(n+2)$-th step with $f(a_{n+1}) < 0$, $f(b_{n+1}) > 0$ and $(b_{n+1} - a_{n+1}) = (b-a)/2^{n+1}$.

Now, by virtue of Bolzano-Weierstrass theorem for sequences, we can find a subsequence $\{a_{n_k}\}$ of $\{a_n\}$ which converges to some $\alpha \in I$. Since $\lim_{k\to\infty}(b_{n_k} - a_{n_k}) = 0$ (why?) we get $\lim_{k\to\infty} b_{n_k} = \alpha$ as well. The continuity of $f(x)$ implies

$$\lim_{k\to\infty} f(a_{n_k}) = \lim_{k\to\infty} f(b_{n_k}) = f(\alpha)$$

but since $f(a_{n_k}) < 0$, $f(b_{n_k}) > 0$ for all k we obtain $f(\alpha) = 0$ (why?) which completes the proof of the lemma.

The proof of Theorem 4.3.5 follows immediately: Let $f(x_1)=M$ and $f(x_2)=m$. If $M=m$ the function is constant and the theorem's claim is trivial. Otherwise, we may assume $m<c<M$. The function $g(x)=f(x)-c$ is also continuous over $I=[a,b]$ and satisfies $g(x_1)=M-c>0$, $g(x_2)=m-c<0$. Therefore, by virtue of the previous lemma, there exists ξ between x_1 and x_2 (therefore within $[a,b]$) such that $g(\xi)=f(\xi)-c=0$, or $f(\xi)=c$.

The result of Theorem 4.3.5 implies that a continuous function $f(x)$ over a closed interval $I=[a,b]$ defines a range $R_f=[m,M]$ where M,m are the maximum and minimum of $f(x)$ respectively.

PROBLEMS

1. Let the function $f(x)$ be defined and continuous over the interval $[a,b]$ (At the endpoints we assume one-side continuity). Show that $g(x)=[f(x)]^n$ is continuous over the same interval.
2. Consider a continuous function $f(x)$ such that $f(x)\geq 0$ for all x.
   (a) Show that $g(x)=\sqrt{f(x)}$ is continuous at all $x=a$ such that $f(a)>0$
   (Hint: rewrite $\sqrt{f(x)}-\sqrt{a}=[f(x)-f(a)]/[\sqrt{f(x)}+\sqrt{a}]$).
   (b) Show that $g(x)=\sqrt{f(x)}$ is continuous even when $f(x)=0$.
3. Explain why the function $f(x)=\sin[x^3+\sin^2(x)]$ is continuous everywhere.
4. Find the domain of continuity for the functions: (a) $\sqrt{\sin(x)+\cos(x)}$ (b) $1/(x^4-1)$ (c) $\sqrt{2-\sin^2(x)}$ (d) $\sqrt{1/\sin(x)+1/\cos(x)}$.
5. Show that continuity of $f(x)$ guarantees the continuity of $|f(x)|$ but not the opposite.
6. Show that the function $f(x)=\sin(x)$ attains the value 0.55 somewhere within the interval $[\pi/6,\pi/2]$.

4.4 Continuity of Special Functions

In this section we present several results related to monotone and inverse functions.

Lemma 4.4.1. Let a continuous function $f(x)$ defined over a closed bounded interval $[a,b]$, satisfy $f(a)<f(b)$. Then, for every A, $f(a)<A<f(b)$ there exists $c: a<c<b$ (not necessarily unique) such that $f(c)=A$.

Proof.

The function $g(x) = f(x) - A$ is defined and continuous over $[a,b]$ and satisfies $g(a) = f(a) - A < 0$ and $g(b) = f(b) - A > 0$. By Lemma 4.3.1, there exists at least one $c : a < c < b$ such that $g(c) = f(c) - A = 0$, i.e., $f(c) = A$.

A repeated use of this lemma provides the next result.

Theorem 4.4.1. Let a continuous function $f(x)$ defined over a closed bounded interval $[a,b]$, obtain each value between $f(a)$ and $f(b)$ exactly once. Then, $f(x)$ is strictly monotone. In particular, if $f(a) < f(b)$, then $f(x)$ is strictly monotone increasing, and if $f(a) > f(b)$, $f(x)$ is strictly monotone decreasing.

Proof.

Let for example $f(a) < f(b)$ and assume $f(x_1) > f(x_2)$ for some $a \le x_1 < x_2 \le b$. If $f(x_1) > f(a)$ as well, then by Lemma 4.4.1 each value between $\max\{f(a), f(x_2)\}$ and $f(x_1)$ is attained at least twice, first at the interval $[a, x_1]$ and secondly at $[x_1, x_2]$. This contradicts the basic assumption of Theorem 4.4.1. A similar contradiction occurs if $f(x_1) > f(b)$.

Next, assume $f(a) > f(x_1)$. Since $f(a) < f(b)$ we get $f(x_1) < f(b)$ and consequently, by applying Lemma 4.4.1, each value between $f(x_2)$ and $\min\{f(b), f(x_1)\}$ is attained at least twice and the assumption of Theorem 4.4.1 is again evoked. Thus, $f(x_1) < f(x_2)$ (why not $f(x_1) = f(x_2)$?), i.e., $f(x)$ is strictly monotone increasing. The case $f(a) > f(b)$ in which $f(x)$ must be strictly monotone decreasing, is treated similarly. This completes the proof.

The next result relates to inverse functions.

Theorem 4.4.2. Let $f(x)$ denote a continuous strictly monotone function over a closed bounded interval $[a,b]$. Then, the inverse function $f^{-1}(x)$ exists and is continuous strictly monotone of the same type.

Proof.

Assume first that $f(x)$ is strictly monotone increasing. By Lemma 4.4.1, for each $y_0 : f(a) < y_0 < f(b)$ there is at least one x_0 such that $f(x_0) = y_0$. Since $f(x)$ is strictly monotone increasing x_0 is unique and inverse function of $f(x)$ denoted by $f^{-1}(y)$, which assigns x_0 to y_0 exists. Let y_1, y_2 satisfy $f(a) \le y_1 < y_2 \le f(b)$ and determine the unique x_1, x_2 for which

$f(x_1) = y_1$ and $f(x_2) = y_2$. Thus, $f^{-1}(y_1) = x_1$ and $f^{-1}(y_2) = x_2$. Clearly $x_1 \neq x_2$, otherwise $y_1 = y_2$ in contradiction with the previous assumption. If $x_1 > x_2$ then $y_1 > y_2$, which evokes the assumption that $f(x)$ is strictly monotone increasing. Consequently $x_1 < x_2$, i.e., $f^{-1}(y)$ is strictly monotone increasing.

The continuity of $f^{-1}(y)$ is shown as follows. Assume that $f^{-1}(y)$ is discontinuous at y_0. Then, there exists a sequence of distinct points y_n, $n = 1,2,\ldots$ within the interval $[f(a), f(b)]$, such that $\lim_{n\to\infty} y_n = y_0$ while the sequence $x_n = f^{-1}(y_n)$ does not converge to $x_0 = f^{-1}(y_0)$. Since $f^{-1}(y)$ is strictly monotone, all x_n are distinct and within $[a,b]$. Also, for some $\varepsilon > 0$ there exists an infinite subsequence $\{x_{n_i}\}$ such that $|x_{n_i} - x_0| \geq \varepsilon$ for all i. By Theorem 3.7.6 (Bolzano – Weierstrass theorem) an infinite subsequence of $\{x_{n_i}\}$ converges to some $X_0 \neq x_0$. Denote this final subsequence by $\{z_j\}$. Since $f(x_n) \to y_0$ as $n \to \infty$, $f(z_j) \to y_0$ as well and the continuity of $f(x)$ guarantees $f(X_0) = y_0$. This contradicts the relation $x_0 = f^{-1}(y_0)$ since $X_0 \neq x_0$. Thus $f^{-1}(y)$ is continuous everywhere which completes the proof. The case of a strictly monotone decreasing $f(x)$ is treated a similar manner.

Example 4.4.1. Let $f(x) = x^2$, $0 \leq x \leq 10$. The function is continuous and strictly monotone increasing. Therefore, its inverse $g(x) = \sqrt{x}$, $0 \leq x \leq 100$ is continuous strictly monotone increasing as well.

Example 4.4.2. Consider $f(x) = \sin(x)$, $0 \leq x \leq \pi/2$. The function is continuous strictly monotone increasing and consequently, so is its inverse denoted by $\sin^{-1}(x)$ (or frequently by $\arcsin(x)$) and defined over the interval $[0,1]$.

The exponential function $y = a^x$, $a > 0$ is quite frequently used and taken for granted as continuous. We will now introduce it properly and show its continuity. The process includes four stages: (1) Defining an integer power and showing basic rules (Eq. (4.4.1) below); (2) Extension to rational powers; (3) Extension to irrational powers; (4) Establishing continuity of the exponential function.

Definition 4.4.1. For arbitrary $a > 0$ and a positive integer n

$$a^0 = 1 \, , \, a^1 = 1 \, , \, a^{n+1} = a^n \cdot a \, , \, n \geq 1 \, ; \, a^{-n} = 1/a^n$$

Theorem 4.4.3. For arbitrary $a > 0$ and positive integers m, n we have

$$a^m a^n = a^{m+n}, (a^m)^n = a^{m \cdot n} \tag{4.4.1}$$

The proof, using induction, is straightforward, and can be easily extended for all integers. It is left as an exercise for the reader.

Our next result provides a proper definition to the $n-th$ root of a positive number.

Theorem 4.4.4. For arbitrary $a > 0$ and a positive integer n the equation $x^n = a$ has a unique solution denoted by $\sqrt[n]{a}$ or $a^{1/n}$.

Proof.

The continuous function $f(x) = x^n - a$ satisfies $f(0) < 0$, $f(1+a) > 0$ (the second result follows from Bernoulli's inequality). Therefore (by Lemma 4.3.1), there exists $x_1 : 0 < x_1 < 1+a$ such that $x_1^n = a$. If $x_2 > x_1$ $(x_2 < x_1)$ then $x_2^n > x_1^n$ $(x_2^n < x_1^n)$. Thus, x_1 is unique.

We now define powers with rational exponents.

Definition 4.4.2. For arbitrary $a > 0$ and positive integers m, n :

$$a^{m/n} = (a^{1/n})^m$$

$$a^{-(m/n)} = 1/a^{m/n}$$

The next result is an extension of Theorem 4.4.3.

An alternative definition for a rational power is $a^{m/n} = (a^m)^{1/n}$. This is a consequence of the following result.

Theorem 4.4.5. If a rational power is defined by Definition 4.4.2 then for arbitrary $a > 0$ and positive integers m, n :

$$a^{m/n} = (a^m)^{1/n}$$

Proof.

Let $x = (a^{1/n})^m$, $y = (a^m)^{1/n}$. Then, $x^{1/m} = a^{1/n}$ and $y^n = a^m$. The first relation provides $(x^{1/m})^n = a$. Thus, $(x^{1/m})^{nm} = a^m$ and by applying Theorem 4.4.3 we obtain $x^n = a^m$. Therefore $x^n = y^n$ and consequently $x = y$ which completes the proof.

Theorem 4.4.3 may now be extended to rational powers.

Theorem 4.4.6. For arbitrary rational numbers r_1, r_2 and $a > 0$ we have

$$a^{r_1} a^{r_2} = a^{r_1 + r_2} \,,\, (a^{r_1})^{r_2} = a^{r_1 r_2} \tag{4.4.2}$$

Proof.

Let $r_1 = p/q \,,\, r_2 = m/n$. Then $r_1 + r_2 = (pn + qm)/(qn)$. Therefore, by Theorem 4.4.3,

$$a^{r_1 + r_2} = a^{(pn+qm)/(qn)} = [a^{1/(qn)}]^{(pn+qm)} = a^{(pn)/(qn)} a^{(qm)/(qn)} = a^{r_1} a^{r_2}$$

The proof of the second claim is a bit more complicated. We have to show that

$$(a^{p/q})^{m/n} = a^{(pm)/(qn)}$$

Clearly, $(a^{1/q})^{1/n} = a^{1/(qn)}$ (why?). Therefore, by Theorems 4.4.3 and 4.4.5

$$(a^{p/q})^{m/n} = \{[(a^{1/q})^p]^{1/n}\}^m = \{[(a^{1/q})^{1/n}]^p\}^m = [(a^{1/(qn)})^p]^m = a^{(pm)/(qn)}$$

and the proof is completed.

We are now ready to define an irrational power of an arbitrary positive number a. We first confine ourselves to $a > 1$.

Definition 4.4.3. Let $a > 1$ and consider an arbitrary irrational number x. The power a^x is defined as

$$a^x = \lim_{n \to \infty} a^{r_n} \tag{4.4.3}$$

where $\{r_n, n = 1,2,\ldots\}$ is a decreasing sequence of rational numbers which converges to x.

Clearly, this definition is meaningful only if the sequence $\{a^{r_n}, n = 1,2,\ldots\}$ converges and if the limit is independent of the particular sequence $\{r_n, n = 1,2,\ldots\}$. The first request is satisfied since the sequence $\{a^{r_n}, n = 1,2,\ldots\}$ is monotonic decreasing and bounded below (for example by 0). Let $\{s_n, n = 1,2,\ldots\}$ be another decreasing sequence that converges to x. Denote $a^{r_n} \to A$, $a^{s_n} \to B$ as $n \to \infty$. For an arbitrary fixed m there exists $n_0(m)$ such that $s_n < r_m$, $n > n_0(m)$. Since $a > 1$ this yields $a^{s_n} < a^{r_m}$, $n > n_0(m)$ as well (why?). Consequently, $B \le a^{r_m}$ for all m. Thus, $B \le A$. Similarly $A \le B$ which yields $A = B$. Needless to say that by defining a^x via a monotonic increasing sequence that converges to x, one gets the same value (this is left as an exercise for the reader). Thus, the power a^x is *well defined* for all real numbers x and $a > 1$. For $0 < a < 1$ we define $a^x = 1/a^x$ and also $1^x = 1$.

We now establish the result regarding the continuity of the exponential function. This is done in two steps.

Theorem 4.4.7. The function a^x , $a > 1$ is strictly monotone increasing for all x .

The proof can be designed by arguments similar to those used in the discussion added to Definition 4.4.3, and is left for the reader.

Theorem 4.4.8. The function a^x , $a > 1$ is continuous for all x .

Proof.

(a) $x = 0$. We will show $a^y \to a^0$ as $y \to 0$. For arbitrary $\varepsilon > 0$ we must show the existence of $\delta > 0$ such that $|y| < \delta$ yields $|a^y - 1| < \varepsilon$. This is done as follows. Since $a^{1/n} \to 1, a^{-1/n} \to 1$ as $n \to \infty$, we can find n_0 sufficiently large such that

$$|a^{1/n} - 1| < \varepsilon, |a^{-1/n} - 1| < \varepsilon \quad , \quad n > n_0$$

However, since the exponential function is monotonic, for each $y : -1/n < y < 1/n$ we obtain $|a^y - 1| < \varepsilon$ which concludes the continuity at $x = 0$.

(b) $x \neq 0$. In this case $a^{x+\delta} - a^x = a^x a^\delta - a^x = a^x(a^\delta - 1)$ and since $\lim_{\delta \to 0} |a^\delta - 1| = 0$ as shown in part (a), we obtain $\lim_{\delta \to 0} |a^{x+\delta} - a^x| = 0$ which concludes the continuity at $x \neq 0$.

PROBLEMS

1. Let $f(x) = 2x^2 - x + 1, 1 \le x \le 2$. Prove that the function is strictly monotone increasing by showing that it uniquely attains every arbitrary value between $f(1)$ and $f(2)$.
2. Obtain the inverse function of $f(x)$ of problem 1 if it exists. Is it continuous?
3. Find the inverse of $y = \sqrt{1 + x^2}$ over $x : 0 < x < \infty$.
4. Show $\lim_{x \to \infty} a^x = \infty$ for all arbitrary $a > 1$.
5. Show that $y = \cos(x), 0 \le x \le \frac{\pi}{2}$ has an inverse. Does the function $y = \cos(x), -\frac{\pi}{2} \le x \le \frac{\pi}{2}$ have one?

4.5 Uniform Continuity

We now introduce the concept of *uniform continuity*. A function $f(x)$, $x \in D \subseteq \Re$ is said to be continuous at $a \in D$ if for every $\varepsilon > 0$, a number $\delta > 0$ such that $|x - a| < \delta$, $x \in D$ imply $|f(x) - f(a)| < \varepsilon$, can be found. Usually, δ depends on both ε and a. If, however, a continuous function $f(x)$ is such that for an arbitrary $\varepsilon > 0$, the number δ is independent of a, $f(x)$ is uniformly continuous. An equivalent definition is given next.

Definition 4.5.1. A continuous function $f(x)$, $x \in D \subseteq \Re$ such that given an arbitrary $\varepsilon > 0$, a number $\delta(\varepsilon) > 0$ which for all $x, y \in D$ satisfies

$$|x - y| < \delta \Rightarrow |f(x) - f(y)| < \varepsilon \tag{4.5.1}$$

is said to be *uniformly continuous.*

The main feature of uniform continuity is δ's independency of x, y.

Example 4.5.1. Consider the function

$$f(x) = \frac{1}{x}, 0 < x \le 1$$

It is clearly continuous throughout its domain of definition, but is not uniformly continuous. Indeed

$$|f(x) - f(y)| = \left|\frac{1}{x} - \frac{1}{y}\right| = \frac{|x - y|}{|xy|} \tag{4.5.2}$$

and given an arbitrary $\varepsilon > 0$, the inequality $|f(x) - f(y)| < \varepsilon$ is equivalent to $|x - y| < \varepsilon|xy|$. Assume the existence of δ, independent of x, y for some given $\varepsilon > 0$ and let $y = \delta$, $0 < x < \delta$. Then, $|x - \delta| < \varepsilon|x\delta|$ which leads to contradiction when $x \to 0$. Consequently, $f(x)$ is not uniformly continuous.

Example 4.5.2. The function $f(x) = x^3$, $0 < x < 1$ is uniformly continuous. Indeed, let $\varepsilon > 0$, $0 < x, y < 1$. Then

$$|x^3 - y^3| = |x - y||x^2 + xy + y^2| \le 3|x - y| \tag{4.5.3}$$

Consequently, the choice $\delta = \frac{\varepsilon}{3}$ guarantees

$$|x - y| < \delta \Rightarrow |f(x) - f(y)| < \varepsilon$$

i.e., x^3 is uniformly continuous over $(0,1)$.

Another, somewhat less trivial example is given next.

Example 4.5.3. The function $f(x)=\sqrt{x}\,,0\le x\le 1$ is uniformly continuous. Let $\varepsilon>0\,,0\le x,y\le 1$. Since $|x-y|=\left(\sqrt{x}+\sqrt{y}\right)\left|\sqrt{x}-\sqrt{y}\right|$, the relation $|f(x)-f(y)|=\left|\sqrt{x}-\sqrt{y}\right|<\varepsilon$ holds if and only if the inequality

$$|x-y|<\varepsilon\left(\sqrt{x}+\sqrt{y}\right)$$

holds as well. However, since $\left|\sqrt{x}-\sqrt{y}\right|\le\sqrt{x}+\sqrt{y}\,,0\le x,y\le 1$, the choice $\delta=\varepsilon^2$, *independent* of x,y, implies

$$|x-y|<\delta\Rightarrow\left|\sqrt{x}-\sqrt{y}\right|^2\le\left|\sqrt{x}-\sqrt{y}\right|\left(\sqrt{x}+\sqrt{y}\right)=|x-y|<\delta=\varepsilon^2 \tag{4.5.4}$$

i.e., $\left|\sqrt{x}-\sqrt{y}\right|<\varepsilon$.

The following result provides a sufficient condition for uniform continuity of an arbitrary continuous function.

Theorem 4.5.1. A function $f(x)$ continuous over a closed interval $[a,b]$ is uniformly continuous over the interval.

Proof.

If $f(x)$ is not uniformly continuous over the interval, an $\varepsilon>0$ can be found such that for any arbitrary $\delta>0$ a pair of points $x,y;a\le x,y\le b$ for which $|f(x)-f(y)|\ge\varepsilon$, exists. For each $\delta_n=(b-a)/2^n\,,n=0,1,2,\ldots$ let x_n,y_n denote a pair such that

$$|x_n-y_n|<\delta_n\,,|f(x_n)-f(y_n)|\ge\varepsilon \tag{4.5.5}$$

Since all the pairs are within the bounded interval $[a,b]$, converging subsequences $\{x_{n_k}\},\{y_{n_k}\},k=0,1,2,\ldots$ such that

$$\lim_{k\to\infty}x_{n_k}=\lim_{k\to\infty}y_{n_k}=A\,,0\le A\le B \tag{4.5.6}$$

exist. However, the continuity of $f(x)$ at the particular point A implies

$$\lim_{k\to\infty}f(x_{n_k})=\lim_{k\to\infty}f(y_{n_k})=f(A) \tag{4.5.7}$$

and consequently $\lim_{k\to\infty}[f(x_{n_k})-f(y_{n_k})]=0$ which contradicts Eq. (4.5.5). Thus, $f(x)$ is uniformly continuous as stated.

Theorem 4.5.1 can be easily extended to include continuous functions defined over an arbitrary closed set. This is left as an exercise for the reader.

PROBLEMS

1. Is the function $f(x)=x^3, 0\le x<1$ continuous? Is it uniformly continuous?
2. Let $f(x)=\dfrac{1}{x-1}, 0\le x<1$.
   (a) Is the function continuous?
   (b) Is the function uniformly continuous?
   (c) Explain the difference between this function and the function of problem 1.
3. Show that a function for which a Lipschitz condition holds, is uniformly continuous.
4. Which of the following functions are uniformly continuous:
   (a) $f(x)=\sqrt[3]{x}\,, -\infty<x<10$ (b) $f(x)=\sqrt[3]{x}\,, -10<x<1000$
   (c) $f(x)=\dfrac{1}{x}, 0.001<x<1$ (d) $f(x)=\sin\left(\dfrac{1}{\sqrt{x}}\right), 0<x<\dfrac{3\pi}{4}$
5. Show that if $f(x)$ is uniformly continuous over $[a,b]$ and over $[b,c]$ it is also uniformly continuous over $[a,c]$.
6. Show that if $f(x)$ is uniformly continuous over (a,b) and (b,c) and is continuous at $x=b$, it is also uniformly continuous over (a,c).
7. Which of the following functions satisfy a Lipschitz condition:
   (a) $f(x)=x^3, -1<x<3$ (b) $f(x)=\sqrt[4]{x}, 0\le x\le 3$
   (c) $f(x)=\sin\left(\dfrac{1}{x}\right), 0<x<\dfrac{\pi}{2}$ (d) $f(x)=\sqrt{x}\,, 0\le x\le 1$

5 Differentiable Functions

In the previous chapter we investigated a special class of functions, C, the continuous functions. In this chapter we investigate a small subset of C, called *differentiable functions* or functions that possess *derivatives*. The concept of derivative, introduced by Newton and Leibniz, marked the beginning of modern calculus. Once familiar with the limit notation of Chapter 3, we are ready to define the concept of the derivative of a given function.

5.1 A Derivative of a Function

Consider a function $y = f(x)$, $a \le x \le b$ (Fig. 5.1.1) and two arbitrary points $x_0, x_0 + \Delta x$ such that $a < x_0, x_0 + \Delta x < b$. Clearly, the number Δx measures the change in x while moving from x_0 to $x_0 + \Delta x$. Let Δy denote the change in y (the function's value), i.e. $\Delta y = f(x_0 + \Delta x) - f(x)$. Then, the quantity $\Delta y / \Delta x$ represents the average change (increase or decrease – depending on the sign of this number) in y over the interval $[x_0, x_0 + \Delta x]$.

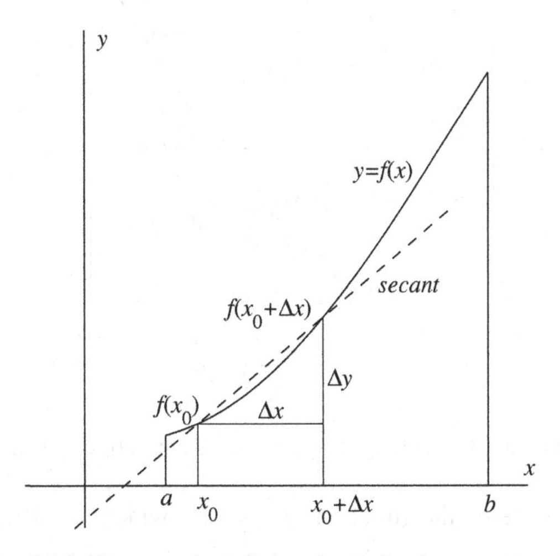

Fig. 5.1.1. First step in defining the derivative: average change.

M. Friedman and A. Kandel: Calculus Light, ISRL 9, pp. 107–146.
springerlink.com

The average change is also the slope of the secant defined by the points $(x_0, f(x_0))$ and $(x_0+\Delta x, f(x_0+\Delta x))$. The behavior of the slope as $\Delta x \to 0$ is one of the most basic concepts of calculus and is treated by the next definition.

Definition 5.1.1. Let $f(x)$ be defined at some neighborhood of x_0. The *derivative* of $f(x)$ at x_0 is

$$f'(x_0) = \lim_{\Delta x \to 0} \frac{f(x_0+\Delta x) - f(x_0)}{\Delta x} \tag{5.1.1}$$

provided that this limit exists and is finite.

Note that the requitement presented by Eq. (5.1.1) means that for *any* sequence $\{\Delta x_n\}$ which satisfies $\Delta x_n \to 0$ the right-hand side of Eq. (5.1.1) must converge to the *same* number $f'(x_0)$. Thus, we are not restricted for example to positive Δx. If by taking two different sequences $\{\Delta x'_n\}$, $\Delta x'_n \to 0$ and $\{\Delta x''_n\}$, $\Delta x''_n \to 0$ we obtain different limits, the function $f(x)$ does not possess a derivative at x_0.

The most significant part of Definition 5.1.1 is the word 'exists'. In many cases this limit does not exist, and if it does it is not necessarily finite. In terms of 'slope' and 'secant', the existence of a derivative means that the secant approaches a line called *tangent* which includes the point $(x_0, f(x_0))$ and whose slope is the limit of $\Delta y / \Delta x$ as $\Delta x \to 0$ (Fig. 5.1.2). While the slope of the particular secant in Fig. 5.1.2 is DB/BA, the derivative $f'(x_0)$ is given by CB/BA.

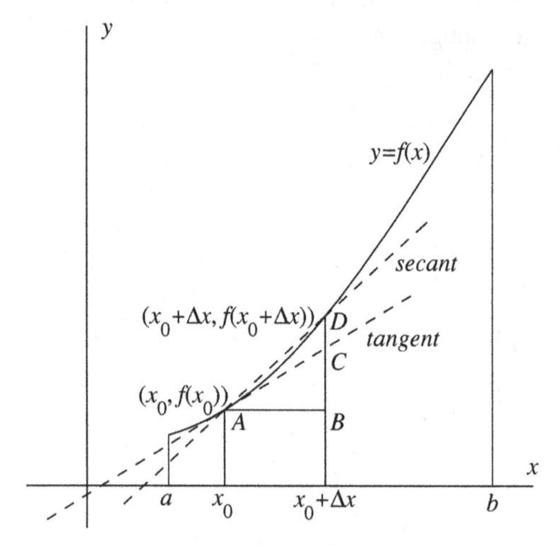

Fig. 5.1.2. The tangent's slope is the derivative $f'(x_0)$.

Example 5.1.1. Consider the function $f(x) = x^2$ which is defined everywhere. For arbitrary x_0 we get

$$\frac{f(x_0+\Delta x)-f(x_0)}{\Delta x}=\frac{(x_0+\Delta x)^2-x_0^2}{\Delta x}=2x_0+\Delta x$$

As $\Delta x \to 0$ the right-hand side converges to $2x_0$ which implies $f'(x_0)=2x_0$.

If a function $f(x)$ possesses a derivative everywhere, i.e. over the function's whole domain of definition D (for example the interval $[a,b]$ in Figs. 5.1.1-2) we define the derivative of $f(x)$ as the function

$$f'(x)=f'(x_0)\ ,\ x_0 \in D \tag{5.1.2}$$

and say that $f(x)$ *differentiable* everywhere. Thus, $(x^2)'=2x$. The geometrical implication is given in Fig. 5.1.3. The original function $y=x^2$ has the derivative function $y'=2x$ everywhere. At an arbitrary point x_0, the derivative is $2x_0$ and equals the slope of the tangent at (x_0,x_0^2), i.e.

$$y'(x_0)=2x_0=\tan(\alpha) \tag{5.1.3}$$

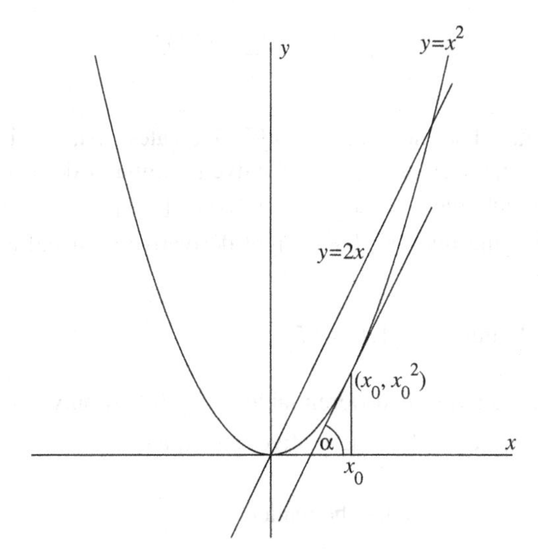

Fig. 5.1.3. The function $y=x^2$ and its derivative $y'=2x$.

For the sake of simplicity we usually replace the symbol Δx which denotes a *small increment* to x by h which represents a *small number*.

Example 5.1.2. Consider the function $y=x^n$ where n is an arbitrary positive integer. By Newton's binomial formula we have

$$(x+h)^n = x^n + \binom{n}{1}x^{n-1}h + \binom{n}{2}x^{n-2}h^2 + \ldots + \binom{n}{n-2}x^2h^{n-2} + \binom{n}{n-1}xh^{n-1} + h^n$$

which implies that

$$\frac{(x+h)^n - x^n}{h} = nx^{n-1} + \binom{n}{2}x^{n-2}h + \ldots + \binom{n}{n-1}xh^{n-2} + h^{n-1}$$

Except for the first term, all the elements at the right-hand side are multiples of powers of h and hence converge to zero as $h \to 0$. Consequently, $y' = (x^n)' = nx^{n-1}$ for all x. The case $n = 0$ can be easily included in this example. Here the function is 1 and its derivative 0. If $f(x) = c$ where c is an arbitrary constant we also have $f'(x) = 0$ for all x (the proof is trivial and is left for the reader).

Definition 5.1.2. A function $f(x)$ has a *right derivative* at x_0 if it is defined at a right neighborhood of x_0 and if the limit

$$f'(x_{0^+}) = \lim_{h \to 0^+} \frac{f(x_0 + h) - f(x_0)}{h} \tag{5.1.4}$$

exists and is finite. The notation $h \to 0^+$ indicates that h is positive and approaches zero from the right. A left derivative is similarly defined and by saying that $f(x)$ is differentiable over a closed interval $[a,b]$, we mean that $f(x)$ is differentiable inside the interval, has a right derivative at a and a left derivative at b.

A direct result of Definitions 5.1.1 and 5.1.2 is

Corollary 5.1.1. A function is differentiable at x_0 if and only if it has both right and left derivatives at x_0 and the derivatives are equal.

The proof is left as an exercise for the reader.

Example 5.1.3. The function $y = |x|$ is differentiable over the interval $[0,1]$. In particular, it has a right derivative at $x = 0$: $y'(0^+) = 1$. However, it is not differentiable over $[-0.1,1]$ since now it does not possess a derivative at $x = 0$ but rather a right derivative 1 and a left derivative -1.

As previously stated, the set of functions that possess derivative is a subset of the set C of all the continuous functions. The next theorem confirms this claim.

Theorem 5.1.1. A function $f(x)$ which possesses a derivative at x_0 is also continuous there.

Proof.

Given the existence of the limit

$$f'(x_0) = \lim_{h \to 0} \frac{f(x_0 + h) - f(x_0)}{h}$$

we obtain

$$\lim_{h \to 0}[f(x_0 + h) - f(x_0)] = \lim_{h \to 0} h \cdot \frac{f(x_0 + h) - f(x_0)}{h} = 0 \cdot f'(x_0) = 0$$

which implies the continuity of $f(x)$ at x_0.

The opposite is not true, i.e. continuity does not imply differentiability. This is demonstrated by the following example.

Example 5.1.4. Let

$$f(x) = \begin{cases} x\sin(1/x), & x \neq 0 \\ 0, & x = 0 \end{cases}$$

The continuity of $f(x)$ was previously established for all x. Consider now the particular point $x_0 = 0$. By definition

$$\frac{f(x_0 + h) - f(x_0)}{h} = \frac{h\sin(1/h)}{h} = \sin(1/h)$$

Unfortunately, $\sin(1/h)$ does not converge as $h \to 0$. For example the choice $h_n = 2/(n\pi)$, $n = 1,2,3,\ldots$ provides the sequence $\sin(1/h_n) = \{1,0,-1,0,1,0,-1,0,\ldots\}$ which does not converge in spite of $h_n \to 0$. Hence, $f(x)$ is not differentiable at $x_0 = 0$. The function is plotted in Fig. 5.1.4 for $0.01 \le x \le 0.05$. One can easily see that while $f(h)$ converges to zero as $h \to 0$, the slope of $f(x)$ keeps vibrating (with increasing speed!) between -1 and 1. This alone does not prevent the existence of derivative at $x_0 = 0$. The derivative there does not exist *only* because $\sin(1/h)$ diverges.

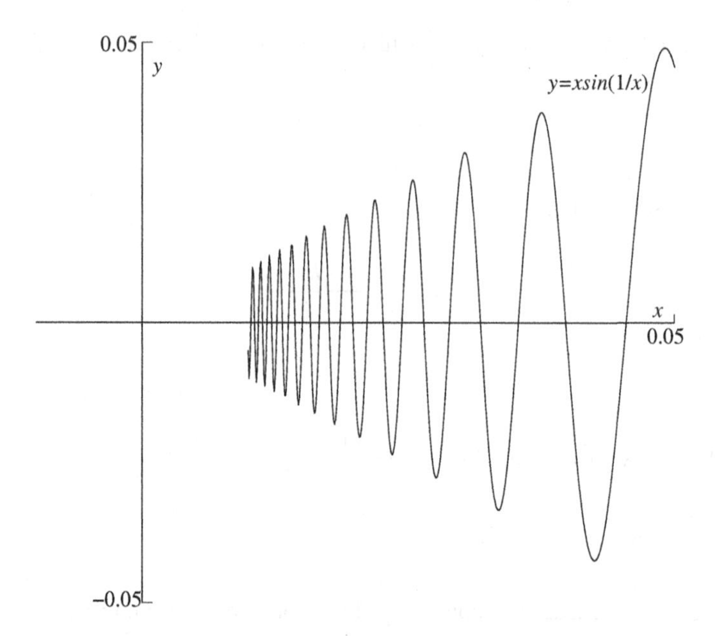

Fig. 5.1.4. A function – continuous but not differentiable at $x = 0$.

In the next example we treat a another vibrating function that due to its faster convergence rate at $x_0 = 0$, is both continuous and differentiable there.

Example 5.1.5. Consider the function

$$f(x) = \begin{cases} x^2 \sin(1/x) \, , & x \neq 0 \\ 0 \, , & x = 0 \end{cases}$$

The continuity of $f(x)$ at $x_0 = 0$ follows from

$$\lim_{h \to 0} [f(x_0 + h) - f(x_0)] = \lim_{h \to 0} h^2 \sin(1/h) = 0$$

since $\sin(1/h)$ is bounded and $h^2 \to 0$. The differentiability follows using similar reasoning, which leads to

$$\lim_{h \to 0} \frac{f(x_0 + h) - f(x_0)}{h} = \lim_{h \to 0} h \sin(1/h) = 0$$

While the slope (see Fig. 5.1.5) still vibrates in an increasing pace without converging (as $h \to 0$), the numbers $h \sin(1/h)$ converge to zero which guarantees the existence of a derivative at $x_0 = 0$. The non-converging vibration of the derivative near $x_0 = 0$ indicates that the derivative is *not continuous* there.

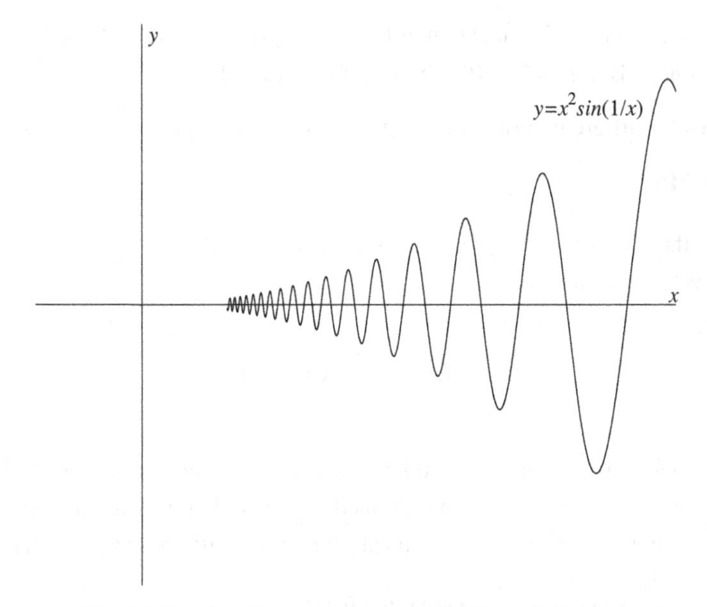

Fig. 5.1.5. A function – continuous and differentiable at $x = 0$.

The use of derivatives in sciences is enormous. A simple example taken from mechanics is defining a velocity at a given time. Let the time t be measured along the x - axis and let a particle travel along the y - axis. At an arbitrary time t the particle is located at $y = y(t)$. If the particle travels at the time interval $t_1 \le t \le t_2$ its *average velocity*, between $t = t_1$ and $t = t_2$ is defined as

$$v_{av}(t_1,t_2) = \frac{y(t_2) - y(t_1)}{t_2 - t_1} \tag{5.1.5}$$

The average velocity does not necessarily reflect on the velocity at all times between the specific time limits. If for example, the speed limit on the highway is $60\frac{miles}{hour}$ and $v_{av}(t_1,t_2) = 50\frac{miles}{hour}$ the driver may still get a ticket since at certain times between t_1 and t_2 he might have traveled at $70\frac{miles}{hour}$ while still maintaining an average speed of $50\frac{miles}{hour}$. However, if t_1 is fixed and if $t_2 - t_1$ decreases and eventually approaches 0 , the average velocity *may* converge to a specific limit called the *instantaneous velocity* (or simply *velocity*) *at time* t_1 . In other words

$$v(t_1) = \lim_{t_2 \to t_1} \frac{y(t_2) - y(t_1)}{t_2 - t_1} = y'(t_1) \tag{5.1.6}$$

For example, if a particle's location at time t is $y(t)=t^2+10t$, its velocity at an arbitrary time t is $v(t)=2t+10$. The proof is left for the student.

Basic rules for differentiation will be given in the next section.

PROBLEMS

1. Find the derivatives of $f(x)=x^3+x^2$ and $f(x)=8x+5$.
2. Show that the function

$$f(x)=\begin{cases} 0\,,\, x<0 \\ \sqrt{x}\,,\, 0\le x\le 1 \\ 1\,,\, x>1 \end{cases}$$

has a derivative except at $x=0,1$. What can be said about these points?

3. The function $f(x)=1/x$ is not defined at $x=0$. Is it possible to define it at this point so that the extended function will possess a derivative everywhere?
4. Use two approaches to show that the function

$$f(x)=\begin{cases} x^2\,,\, x<2 \\ 4.1\,,\, x\ge 2 \end{cases}$$

does not have a derivative at $x=2$. Does it possess a left derivative? A right derivative?

5. Consider the function

$$f(x)=\begin{cases} 1\,,\, x \text{ } rational \\ 0\,,\, x \text{ } irrational \end{cases}$$

At what points the function is continuous? Has a derivative?

6. Consider the function

$$f(x)=\begin{cases} 0\,,\, x<0 \\ x^2\,,\, 0\le x\le 1 \\ 2x-1\,,\, x>1 \end{cases}$$

Where does it have a derivative? Explain.

7. Let $f(x)=x^2$, $g(x)=x^2+2$. Show $f'(x)=g'(x)$. Any conclusion?
8. Let $f(x)$ be an even differentiable function. Compare between $f(a)$ and $f'(a)$ for arbitrary a within the domain of $f(x)$.
9. Repeat and solve problem 8 if $f(x)$ is an odd function.

10. Let $f(x)=\sqrt{x}\ ,\ x\geq 0$. Find where does the function possess a derivative and calculate it.
11. Show that the function

$$f(x)=\begin{cases} x^2 \ , x>0 \\ x^3 \ , x\leq 0 \end{cases}$$

has a derivative everywhere.
12. Does the function

$$f(x)=\begin{cases} x^2 \ , x>0 \\ x \ , x\leq 0 \end{cases}$$

have a derivative at $x=0$? Explain.

5.2 Basic Properties of Differentiable Functions

We start with evaluating the derivatives of $f(x)\pm g(x)$, $f(x)g(x)$, $f(x)/g(x)$ where $f(x)$ and $g(x)$ are two given differentiable functions defined over the same domain. This domain can be an arbitrary open set D within the real axis $-\infty<x<\infty$ although usually, throughout most of the book, we assume, for the sake of simplicity, a domain which is a single bounded interval, closed, open or semi-open.

Theorem 5.2.1. For arbitrary differentiable functions $f(x)$ and $g(x)$

1. $[f(x)\pm g(x)]'=f'(x)\pm g'(x)$
2. $[f(x)g(x)]'=f'(x)g(x)+f(x)g'(x)$ (5.2.1)
3. $[f(x)/g(x)]'=[f'(x)g(x)-f(x)g'(x)]/[g(x)]^2 \quad , \quad g(x)\neq 0$

Proof.

1. For arbitrary h we have

$$\frac{[f(x+h)+g(x+h)]-[f(x)+g(x)]}{h}=\frac{f(x+h)-f(x)}{h}+\frac{g(x+h)-g(x)}{h}$$

and since the right-hand side converges to $f'(x)+g'(x)$ as $h\to 0$ this value is also the limit of the left-hand side and part 1 of the proof is completed (the proofs for sum and difference of functions are identical).
2. The functions $f(x)$ and $g(x)$ are differentiable and therefore continuous as well. By subtracting and adding the same expression we may rewrite

$$\frac{f(x+h)g(x+h)-f(x)g(x)}{h}=\frac{f(x+h)g(x+h)-f(x)g(x+h)}{h}+\frac{f(x)g(x+h)-f(x)g(x)}{h}$$

The first term of the right-hand side is $g(x+h)\{[f(x+h)-f(x)]/h\}$. The continuity of $g(x)$ guarantees that $\lim_{h\to 0} g(x+h)=g(x)$ and consequently

$$\lim_{h\to 0} g(x+h)\frac{f(x+h)-f(x)}{h}=g(x)f'(x)$$

Similarly, the second term converges to $f(x)g'(x)$ as $h\to 0$ which completes the proof of part 2.
3. Let $g(x)\neq 0$ at some given x and over a small open interval which includes x as well. Then

$$\frac{f(x+h)/g(x+h)-f(x)/g(x)}{h}=\frac{f(x+h)g(x)-f(x)g(x+h)}{hg(x)g(x+h)}$$

To complete the proof, the same method of subtracting and adding an identical term, should be implemented before letting $h\to 0$. It is left as an exercise for the reader.

Corollary 5.2.1. For arbitrary differentiable function $f(x)$ and constant c we have

$$[cf(x)]'=cf'(x) \tag{5.2.2}$$

The proof is straightforward and is left as an exercise for the reader.

Example 5.2.1. Let $f(x)=x^3+x^2$, $g(x)=5x-1$. Then $f'(x)=3x^2+2x$, $g'(x)=5$. Consequently

$$[f(x)g(x)]'=(3x^2+2x)(5x-1)+5(x^3+x^2)=20x^3+12x^2-2x$$

$$\left[\frac{f(x)}{g(x)}\right]'=\frac{(3x^2+2x)(5x-1)-5(x^3+x^2)}{(5x-1)^2}=\frac{2x(5x^2+x-1)}{(5x-1)^2}$$

Example 5.2.2. Consider the function $f(x)=1/x^n$ where n is a positive integer. Then, by using part 3 of Theorem 5.2.1, we obtain

$$f'(x)=\left(\frac{1}{x^n}\right)'=\frac{0\cdot x^n-nx^{n-1}\cdot 1}{x^{2n}}=-nx^{-n-1}$$

This result combined with that of Example 5.1.2 implies

$$(x^n)'=nx^{n-1} \tag{5.2.3}$$

for all integers.

We now introduce a new symbol for the derivative which is useful tool in obtaining numerous theoretical results. Let $y = f(x)$ denote an arbitrary differentiable function. Then $dy/dx \equiv f'(x)$ is another symbol, introduced by Leibniz, which represents the derivative. Since $f'(x)$ is the limit of the fractions $\Delta x/\Delta y$ as $\Delta x \to 0$ one can wrongly visualize it as a 'final' quotient where dy and dx are 'infinitesimally small' numbers. From a pure mathematical point of view this is unacceptable. The derivative is a well defined limit of a sequence and an 'infinitesimally small' number does not exist. However, the beauty of Leibniz idea is first that it helps suspicious students get familiar with the new concept in several steps. Also, many theoretical results can be obtained in a most natural by treating dy/dx as a regular quotient and dx, dy as regular numbers. What important is the bottom line and since the final results are correct, these processes are acceptable.

The next result, is very helpful in calculating derivatives of complex functions.

Theorem 5.2.2. Let $y = f(x)$, $u = g(t)$ denote differentiable functions defined over open intervals. Consider a point x_0 such that $y_0 = f(x_0)$ is within the domain of $g(t)$, i.e. $g(y_0)$ exists. Then, the *composite function* $g \circ f \equiv g(f(x))$ is differentiable at x_0 and

$$(g \circ f)' = g'(y_0)f'(x_0) \tag{5.2.4}$$

Proof.

The proof is quite simple but should be carried with care. Since $f(x)$ is continuous, then for sufficiently small Δx, the point $f(x_0 + \Delta x) = y_0 + \Delta y$ is also the domain of $g(t)$. The quantity $\Delta g = (g \circ f)(x_0 + \Delta x) - (g \circ f)(x_0)$ is therefore well defined. We can clearly write

$$\frac{\Delta g}{\Delta x} = \frac{\Delta g}{\Delta y}\frac{\Delta y}{\Delta x} \tag{5.2.5}$$

provided that $\Delta y \neq 0$. There are two cases to be considered:

Case (a) Assume $\Delta y \neq 0$ for all Δx (already taken small enough!). If $\Delta x \to 0$ then $\Delta y \to 0$ as well (continuity of $f(x)$). Since $f(x)$ and $g(t)$ are simultaneously differentiable, we get

$$\lim_{\Delta x \to 0} \frac{\Delta g}{\Delta y} = g'(f(x_0)) \quad , \quad \lim_{\Delta x \to 0} \frac{\Delta y}{\Delta x} = f'(x_0)$$

Thus, $\lim\limits_{\Delta x \to 0} \frac{\Delta g}{\Delta x}$ exists and equals to $g'(f(x_0))f'(x_0)$ as stated.

Case (b) Assume that there is a sequence $x_n = x_0 + \Delta x_n$ where $\Delta x_n \to 0$ such that $\Delta y_n = f(x_n) - f(x_0) = 0$. Note that if such sequence does not exist, then $\Delta y \neq 0$ for all sufficiently small Δx and we are back to Case 1. Now, since $\Delta y_n = 0$, we have $f'(x_0) = 0$. Note that the differentiability of $f(x)$ yields that using a *single* specific sequence of $x_n = x_0 + \Delta x_n$ is enough to calculate the derivative, *provided* that $\Delta x_n \to 0$. Thus, $g'(f(x_0))f'(x_0) = 0$. We still must show that $\Delta g / \Delta x$ converges to zero as $\Delta x \to 0$.

Let indeed $\Delta x \to 0$ and consequently $\Delta y \to 0$. If $\Delta y = 0$ then $\Delta g / \Delta x = 0$ (since $\Delta g = 0$). However, if $\Delta y \neq 0$, Eq. (5.2.5) holds and $\Delta g / \Delta y$ approaches $g'(y_0)$ while $\Delta y / \Delta x$ approaches $f'(x_0) = 0$. Thus, no matter which route Δy takes, we obtain

$$\lim_{\Delta x \to 0} \frac{\Delta g}{\Delta x} = 0 = g'(f(x_0))f'(x_0)$$

and thus the proof is completed.

Note that by using Leibniz notation, we just proved

$$\frac{du}{dx} = \frac{du}{dy}\frac{dy}{dx}$$

a relation which is trivially correct if du, dy, dx are real numbers and $dy \neq 0$, $dx \neq 0$. This is a good example for the advantage of this notation in performing appropriate shortcuts.

Example 5.2.3. Let $y = f(x) = x^2 + 2x$ and $u = g(t) = t^3 + t$. The two functions are defined and possess derivatives everywhere in R. The composite function is $g \circ f = (x^2 + 2x)^3 + (x^2 + 2x)$, i.e. $g \circ f = x^6 + 6x^5 + 12x^4 + 8x^3 + x^2 + 2x$. A direct calculation provides

$$(g \circ f)' = 6x^5 + 30x^4 + 48x^3 + 24x^2 + 2x + 2$$

Using the chain theorem we get

$$(g \circ f)' = (3y^2 + 1)(2x + 2) = [3(x^2 + 2x)^2 + 1](2x + 2)$$

and the reader can easily verify that the results are identical.

Example 5.2.4. Consider $y = f(x) = \dfrac{x^2 + 1}{x + 1}$, $u = g(t) = \dfrac{1}{t^3}$. Let $F(x) = g \circ f$.

Then

$$\frac{dF}{dx} = g'(y)f'(x) = -\frac{3}{y^4}\frac{2x(x+1)-1\cdot(x^2+1)}{(x+1)^2} = -\frac{3(x+1)^2(x^2+2x-1)}{(x^2+1)^4}$$

Example 5.2.5. Let $y = x^{m/n}$ where m,n are arbitrary integers. Then $y^n = x^m$ and by Theorem 5.2.2 $ny^{n-1}y' = mx^{m-1}$. Consequently

$$y' = \frac{m}{n}\frac{x^{m-1}}{y^{n-1}} = \frac{m}{n}x^{(m/n)-1}$$

Corollary 5.2.2. If r is rational, the power function $y = x^r$ yields $y' = rx^{r-1}$.

The general case of differentiating a power function is treated next.

Theorem 5.2.3. For arbitrary real number α, the derivative of $y = x^\alpha$ is $y' = \alpha x^{\alpha-1}$.

The proof of this theorem is beyond the scope of this book.

Let $y = f(x)$ possess a derivative and consider the inverse function $y = f^{-1}(x)$. Clearly it may not exist and even if it does, it may not possess a derivative. For example, the inverse of $y = x^2$ is not uniquely defined since $x = \pm\sqrt{y}$. On the other hand the function $y = x^3$ possesses an inverse $x = \sqrt[3]{y}$ or $y = \sqrt[3]{x}$, should we prefer x, y to keep their usual roles, but the inverse is not differentiable at $x = 0$. In fact $(\sqrt[3]{x})' = -(1/3)x^{-(2/3)} \to \infty$ as $x \to 0$ and a direct computation shows that the derivative does not exist at $x = 0$.

The following theorem yields sufficient conditions for the existence of $[f^{-1}(x)]'$ and provides its value.

Theorem 5.2.4. Let $y = f(x)$, defined over an interval I, denote a differentiable strictly monotone function such that $f'(x) \neq 0$ everywhere. Then, $f^{-1}(x)$ possess a derivative and

$$[f^{-1}(x)]' = \frac{1}{f'(x)} \tag{5.2.6}$$

Note that to use Eq. (5.2.6), we first compute the inverse $x = g(y)$ and then substitute $g'(y)$ as the right-hand side of this equation.

Proof.

Let $x_0 = g(y_0)$ and $y \neq y_0$. Since $f(x)$ is strictly monotone, so is its inverse, i.e. $x \neq x_0$ as well. Therefore

$$\frac{g(y)-g(y_0)}{y-y_0} = \frac{x-x_0}{y-y_0} = \frac{1}{[(y-y_0)/(x-x_0)]}$$

Being differentiable, $f(x)$ is also continuous and consequently so is $x = g(y)$. Thus, $y \to y_0$ implies $x \to x_0$, and since $f'(x_0) \neq 0$ we get

$$\lim_{y\to y_0} \frac{g(y)-g(y_0)}{y-y_0} = \lim_{y\to y_0} \frac{1}{[(y-y_0)/(x-x_0)]} = \frac{1}{f'(x_0)}$$

which completes the proof.

We will later show that the condition $f'(x) \neq 0$ already implies that $f(x)$ is strictly monotone.

A relation such as $f(x,y) = 0$ is called an *implicit relation*. It includes the set of all the pairs (u,v) which satisfies $f(u,v) = 0$. Every function $y = f(x)$ yields the implicit relation $g(x,y) = f(x) - y = 0$ but the opposite is certainly not true. For example, the relation $x^2 - y^2 = 0$ does not produce a unique function. However, if an implicit relation produces a unique function we may not have to get the explicit expression for y in order to obtain y'. We simply apply the differentiation rule for composite functions.

Example 5.2.6. Let $y = 1/x$. Clearly, $xy - 1 = 0$ and by using the rule for composite functions we get $y + xy' = 0$, i.e., $y' = -y/x = -1/x^2$. Direct differentiation produces the same result. The implicit relation $xy^3 + y + 1 = 0$, $y > 1$ (i.e. the relation is restricted for pairs (u,v) where $v > 3$) is less trivial and if a unique function can be defined, then $y^3 + 3xy^2y' + y' = 0$, i.e., $y' = -y^3/(1+3xy^2)$.

Definition 5.2.1. Consider a function $f(x)$, defined over a domain D. The point $x_0 \in D$ is called a local maximum (minimum) point of $f(x)$, if for some $\delta > 0$, there exists an interval $I = (x_0 - \delta, x_0 + \delta) \in D$ such that $f(x_0) \geq f(x)$ ($f(x_0) \leq f(x)$) for all $x \in I$.

Example 5.2.7. The function in Fig. 5.2.1 has a local maximum at x_1 and a local minimum at x_2.

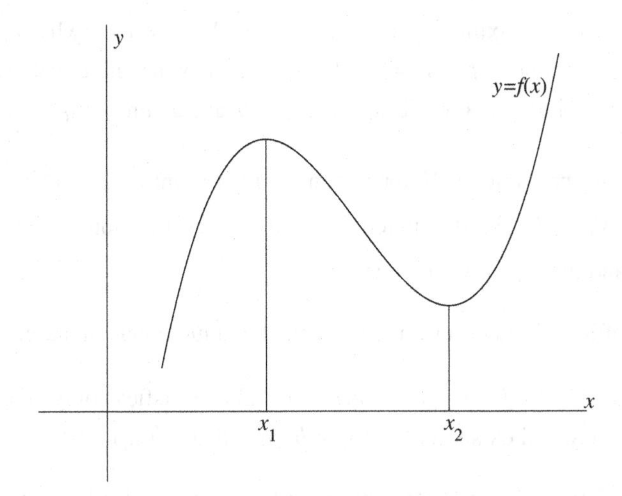

Fig. 5.2.1. Local extrema of $f(x)$.

An important property of a local extremum (i.e. maximum or minimum), which is visually clear from from Fig. 5.2.1 is stated next.

Theorem 5.2.5. If x_0 is a local extremum of a differentiable function $f(x)$, then $f'(x_0) = 0$.

Proof.

Let x_0 be a local maximum. Then, there exists an open interval $I = (x_0 - \delta, x_0 + \delta)$ such that $f(x) \le f(x_0)$ for all $x \in I$. Construct (how?) a sequence $\{x_n\}$, $n = 1,2,\ldots$ such that $x_n > x_0$ and $\lim_{x_n \to \infty} x_n = x_0$. Since $f(x_n) \le f(x_0)$ we obtain $f'(x_0) \le 0$. Similarly, by taking a sequence on the left of x_0, we get $f'(x_0) \ge 0$. Hence $f'(x_0) = 0$.

A direct result of Theorem 5.2.5 is the famous Rolle's theorem.

Theorem 5.2.6 (Rolle's theorem). Let $f(x)$ be a differentiable function over a closed interval $[a,b]$, and let $f(a) = f(b) = 0$. Then, there exists $x_0 : a < x_0 < b$ such that $f'(x_0) = 0$.

Proof.

If $f(x) = 0$ everywhere then $f'(x) = 0$ for all x and the theorem's claim is trivial. If the function is not identically zero, assume for example that it attains positive values. Since $f(x)$ is differentiable, it is continuous as well and possesses a global maximum at some x_0 within (a,b) (why?). This, by Theorem 5.2.5, implies $f'(x_0)=0$. If $f(x)$ attains negative values, it must possess a global minimum at some $x_0 : a < x_0 < b$ and again $f'(x_0)=0$.

Note that x_0 is not unique. If for example $f(x)=\sin(x)$, $a=0$, $b=2\pi$, then at $x_0^{(1)}=\pi/2$, $x_0^{(2)}=3\pi/2$, the function attains local maximum and minimum respectively and its derivative there is zero.

The validity of Rolle's theorem can be extended to a more general case.

Corollary 5.2.3. If $f(x)$ of Theorem 5.2.6 satisfies only the equality $f(a)=f(b)$ there still exists $x_0 : a < x_0 < b$ such that $f'(x_0)=0$.

The reader will find the proof straightforward by applying Rolle's theorem to the function $g(x)=f(x)-f(a)$ which satisfy $g(a)=g(b)=0$ and $g'(x)=f'(x)$.

The next result which follows almost directly from Rolle's theorem, is a powerful tool in the development of calculus – particularly the presentation of a function as a power series.

Theorem 5.2.7 (First mean-value theorem). Let $f(x)$ denote a differentiable function over a closed interval $[a,b]$. Then there exists $\xi : a < \xi < b$ such that

$$f(b)-f(a)=f'(\xi)(b-a) \tag{5.2.7}$$

Note that ξ, as in Rolle's theorem is not necessarily unique.

Proof.

Define

$$g(x)=f(x)-\frac{f(b)-f(a)}{b-a}(x-a)$$

The function $g(x)$ is differentiable over $[a,b]$ just like $f(x)$ (why?) and since it yields $g(a)=g(b)=f(a)$ we can implement Corollary 5.2.3 and conclude the existence of ξ for which $g'(\xi)=0$. Consequently

$$f'(\xi)-\frac{f(b)-f(a)}{b-a}=0$$

which implies Eq. (5.2.7).

The mean-value theorem actually states that the average change of a function between two points a and b, although usually different from the derivatives at these points, must be equal to the function's slope at some middle point. The idea presented at the proof of Theorem 5.2.7 is to rotate the function $f(x)$ until its values at the endpoints are the same (Fig. 5.2.2).

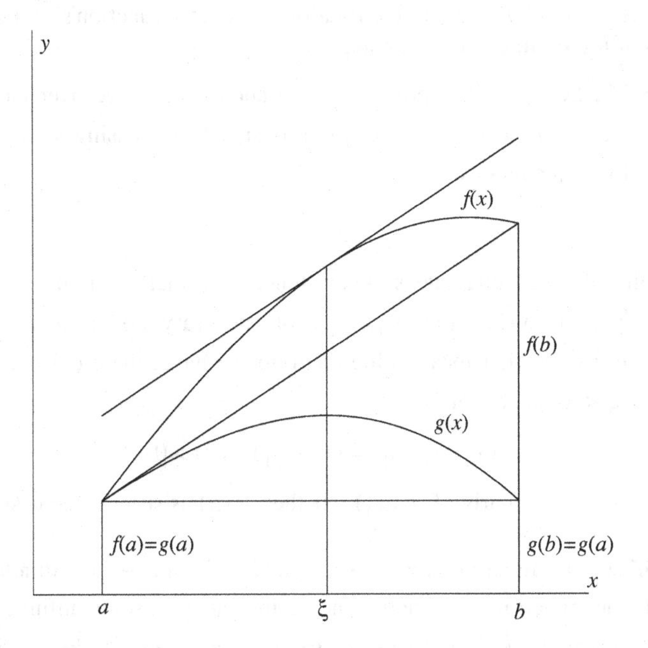

Fig. 5.2.2. Getting the mean-value theorem by rotating $f(x)$.

The next result is a straightforward conclusion from Theorem 5.2.7.

Theorem 5.2.8 (Second mean-value theorem). Let $f(x)$ and $g(x)$ be differentiable over the interval $[a,b]$ such that $g(b) \neq g(a)$ and $g'(x) \neq 0$ everywhere. Then, there exists an interim point $\xi : a < \xi < b$ for which

$$\frac{f'(\xi)}{g'(\xi)} = \frac{f(b) - f(a)}{g(b) - g(a)} \tag{5.2.8}$$

Proof.

Define

$$F(x) = f(x)[g(b) - g(a)] - g(x)[f(b) - f(a)] \tag{5.2.9}$$

which implies

$$F(a) = F(b) = f(a)g(b) - g(a)f(b)$$

and consequently, by the first mean-value theorem, there exists a $\xi : a < \xi < b$ such that $F'(\xi) = 0$. The rest is straightforward.

While the first mean-value theorem, by virtue of Fig. 5.2.2, seems visually logical, a geometric interpretation of Theorem 5.2.8 is not so clear since we have to observe the graph of $F(x)$ (Eq. (5.2.9)) rather than those given of $f(x)$ and $g(x)$.

We close this section by showing the relation between a function's derivative and its domains of increasing and decreasing.

Theorem 5.2.9. Let $y = f(x)$ possess a continuous derivative over an arbitrary domain D and let $f'(x_0) > 0$. Then $f(x)$ is strictly increasing at x_0, i.e. at a small interval which contains x_0.

Proof.

The continuity of $f'(x)$ guarantees the existence of a small interval I around x_0 such that $f'(x) > 0, x \in I$. Let x_1, x_2 denote arbitrary points at I such that $x_2 > x_1$. Then, by the first mean value theorem we have, there exists an interim point $\xi : x_1 < \xi < x_2$ such that

$$f(x_2) - f(x_1) = (x_2 - x_1) f'(\xi) > 0$$

i.e., $f(x_1) < f(x_2)$. Similarly, if $f'(x_0) < 0$ then $f(x)$ is strictly decreasing at x_0.

Example 5.2.8. The function $y = x^4 - x^2$ yields $y' = 4x^3 - 2x$. Simple algebra shows that the function is strictly increasing at the semi infinite interval $(\sqrt{2}/2, \infty)$ and at $(-\sqrt{2}/2, 0)$ and is strictly decreasing at the semi infinite interval $(-\infty, -\sqrt{2}/2)$ and at $(0, \sqrt{2}/2)$.

PROBLEMS

1. Find the derivatives of (a) $f(x) = x^8 - x^3 + x + 1$ (b) $f(x) = 100x^{10} + (\sqrt{x})^{\sqrt{2}}$ (c) $f(x) = x^{-3}\sqrt{x}$ (d) $f(x) = (x^3 + x)/(x^5 - 3\sqrt{x})$.
2. Find the derivatives of (a) $f(x) = (x^2 + x)^3$ (b) $f(x) = [(x + \sqrt{x})/(x^2 + 1/x)]^2$
3. For the following functions find the inverses and their derivatives: (a) $f(x) = 2x^5$ (b) $f(x) = 1/(\sqrt{x} + 1)$.
4. Show that the inverse of $f(x) = x + x^3$ exists and find its derivative.
5. Assume that a differentiable function y satisfies the equation $x^3 y^2 + yx = 0$ and calculate y'. As previously stated this implicit relation by itself does not guarantee that a unique function $y = f(x)$ exists.
6. Let $f(x, y) = y^2 x + x^2 y^3 - 2 = 0$ and assume that $y(x)$ is a differentiable function which solves this equation. (a) Calculate y'. (b) Show $f(1,1) = 0$. Does that mean $y(1) = 1$? (c) Assume $y(1) = 1$ and calculate $y'(1)$

7. Show that the derivative of $f(x) = x^3 + x^2$ attains the value 10 within $[1,2]$.
8. (a) Show that for some x_0 within $[1,2]$, the functions $f(x) = x^3 + x^2$ and $g(x) = x^2 + x$ satisfy $f'(x_0)/g'(x_0) = 2.5$.

   (b) Calculate x_0.
9. Let $f(x) = 3x^3$, $g(x) = x^2$ over the interval $[2,3]$. Find the unique point ξ inside the interval, at which the second mean-value theorem holds for $f(x)$ and $g(x)$.
10. How can you simplify the proof of Theorem 5.2.8 in the case $f(a) = f(b)$?

5.3 Derivatives of Special Functions

We are now ready to calculate the derivatives of the trigonometric and logarithmic functions and their inverses. Like most traditional authors we start this section with the function $y = \sin(x)$.

Theorem 5.3.1. Let x denote an arbitrary angle measured in radians. Then

$$\lim_{x \to 0} \frac{\sin(x)}{x} = 1 \tag{5.3.1}$$

Proof.

Consider a sector of the unit circle with arc (angle) x (Fig. 5.3.1) measured in radians. A basic statement from elementary plane geometry is that the area S of this sector is its arc multiplied by half the radius, i.e. $S = (x \cdot 1)/2$. This area is bounded between the triangle OAB with area S_1 (from below) and the triangle OAC with area S_2 (from above).

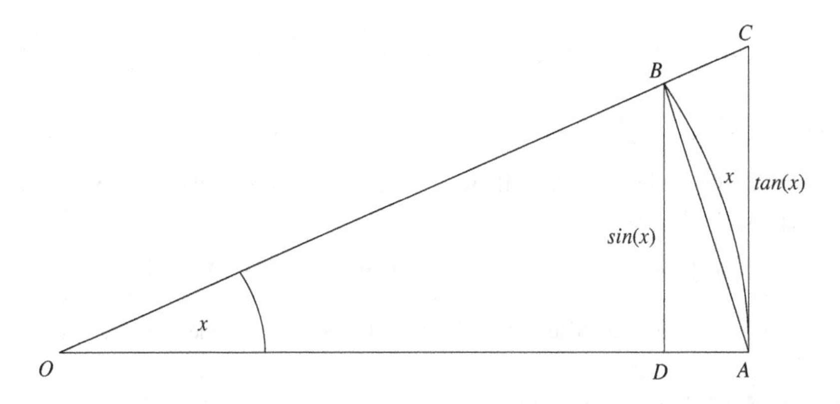

Fig. 5.3.1. Obtaining the inequality of Eq. (5.3.2)

Hence

$$S_1 = \frac{AO \cdot BD}{2} = \frac{1 \cdot \sin(x)}{2} < S = \frac{1 \cdot x}{2} < S_2 = \frac{OA \cdot AC}{2} = \frac{1 \cdot \tan(x)}{2}$$

which leads to (how?)

$$1 < \frac{x}{\sin(x)} < \frac{1}{\cos(x)} \tag{5.3.2}$$

The first part of this inequality, i.e., $0 < \sin(x) < x$ clearly guarantees $\lim_{x \to 0} \sin(x) = 0$. By virtue of the identity $\sin^2(x) + \cos^2(x) = 1$, we therefore also have $\lim_{x \to 0} \cos(x) = 1$. Thus, the whole inequality of Eq. (5.3.2) yields $\lim_{x \to 0} \frac{x}{\sin(x)} = 1$ which completes the proof.

The next result is a direct result of Theorem 5.3.1.

Theorem 5.3.2. $(\sin(x))' = \cos(x)$.

Proof.

Using a basic trigonometric identity we obtain

$$\frac{\sin(x+h) - \sin(x)}{h} = \frac{2\sin(h/2)\cos(x+h/2)}{h} = \frac{\sin(h/2)}{(h/2)}\cos(x+h/2)$$

By applying Theorem 5.3.1 and basic results from the theory of limits we therefore have

$$\lim_{h \to 0} \frac{\sin(x+h) - \sin(x)}{h} = \lim_{h \to 0} \frac{\sin(h/2)}{(h/2)} \lim_{h \to 0} \cos(x+h/2) = \cos(x)$$

Corollary 5.3.1. $(\cos(x))' = -\sin(x)$.

Proof.

Rewrite $\cos(x) = \sin(\pi/2 - x)$. If we define $z = \pi/2 - x$ then by virtue of Theorem 5.2.2 we get

$$(\cos(x))' = (\sin(z))'_z z'_x = \cos(z) \cdot (-1) = -\cos(\pi/2 - x) = -\sin(x)$$

The notation z'_x denotes differentiation of the function z by the variable x.

Example 5.3.1. Let $f(x) = \tan(x)$. Then

$$(\tan(x))' = \left(\frac{\sin(x)}{\cos(x)}\right)' = \frac{\cos(x)\cdot\cos(x) - (-\sin(x)\cdot\sin(x)}{\cos^2(x)} = \frac{1}{\cos^2(x)}$$

and similarly $(\cot(x))' = -1/\sin^2(x)$.

Next we will differentiate the logarithmic function with base e (see Example 3.7.2), called the *natural logarithmic function* and denoted by $\ln(x)$.

Theorem 5.3.3. The function $y = \ln(x)$, $x > 0$ yields $y' = 1/x$.

Proof.

For arbitrary $x > 0$ we have

$$\frac{\ln(x+h) - \ln(x)}{h} = \frac{1}{x}\left(\frac{x}{h}\right)\ln\left(\frac{x+h}{h}\right) = \frac{1}{x}\ln\left(\frac{x+h}{h}\right)^{x/h} = \frac{1}{x}\ln\left(1+\frac{x}{h}\right)^{x/h}$$

Since $x/h \to \infty$ as $h \to 0$ we get

$$\lim_{h\to 0}\left(1+\frac{x}{h}\right)^{x/h} = e$$

and the continuity of $\ln(x)$ at all x, particularly at $x = e$, yields

$$\lim_{h\to 0}\frac{\ln(x+h) - \ln(x)}{h} = \frac{1}{x}$$

Corollary 5.3.2. The function $y = e^x$ yields $y' = y$.

Proof.

The inverse function of the exponential function is $x = \ln(y)$. Using the differentiation rule for a composite function we get

$$1 = (1/y)y'_x$$

Hence $y' = y$.

Example 5.3.2. Consider the function $y = x\ln[\sin(x)]$. Its domain of definition is, $\sin(x) > 0$ i.e., $2k\pi < x < (2k+1)\pi$ for arbitrary integer k. The derivative (again we must use Theorem 5.2.2)

$$y' = \ln[\sin(x)] + x\cot(x)$$

is defined over the same domain. However, finding the zeroes of y, y' or y'' is certainly not an easy task.

The inverse functions of the trigonometric and logarithmic functions are just as important and used in numerous applications. We will define them and calculate their derivatives using, as we did for the exponential function (Corollary 5.3.2) the rules for composite functions.

Definition 5.3.1. The inverse function of $y = \sin(x)$ is defined as $x = g(y)$, $-1 \le y \le 1$ where for each y the value x is taken so that $y = \sin(x)$. Since $\sin(x)$ is periodic x is not uniquely defined. We usually choose the *main branch* of the inverse function for which $-\pi/2 \le x \le \pi/2$ and after switching back the roles of x and y, write $y = \arcsin(x)$ or $y = \sin^{-1}(x)$ where $-1 \le x \le 1$ and $-\pi/2 \le y \le \pi/2$ (Fig. 5.3.2).

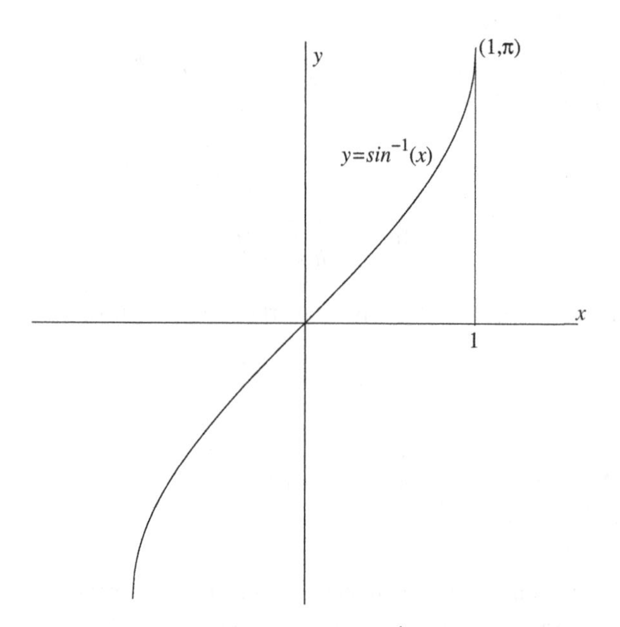

Fig. 5.3.2. The function $y = \sin^{-1}(x)$ - main branch.

The inverse of $y = \cos(x)$ is also not uniquely defined and to overcome this obstacle we define $y = \cos^{-1}(x)$ over $-1 \le x \le 1$ such that for a given x the value of $\cos^{-1}(x)$ is the number $y : 0 \le y \le \pi$ for which $x = \cos(y)$.

Definition 5.3.2. The inverse function of $y = \tan(x)$ (main branch) is $y = \arctan(x)$ or $y = \tan^{-1}(x)$ (Fig. 5.3.3) where for each $x : -\infty < x < \infty$, the value y is the unique number within the interval $[-\pi/2, \pi/2]$ for which $x = \tan(y)$.

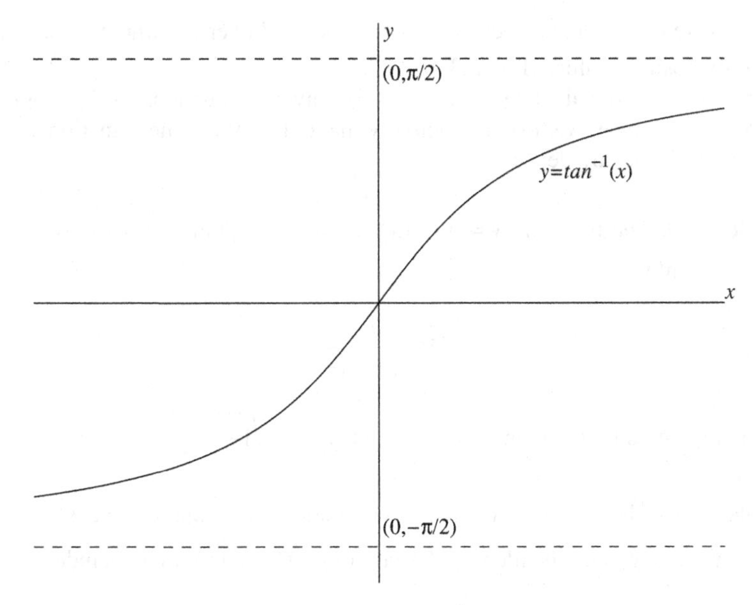

Fig. 5.3.3. The function $y = \tan^{-1}(x)$ - main branch.

The inverse of $y = \cot(x)$ is similarly defined except that for each $x: -\infty < x < \infty$, we choose the unique $y: 0 \le y \le \pi$ which satisfies $x = \cot(y)$.

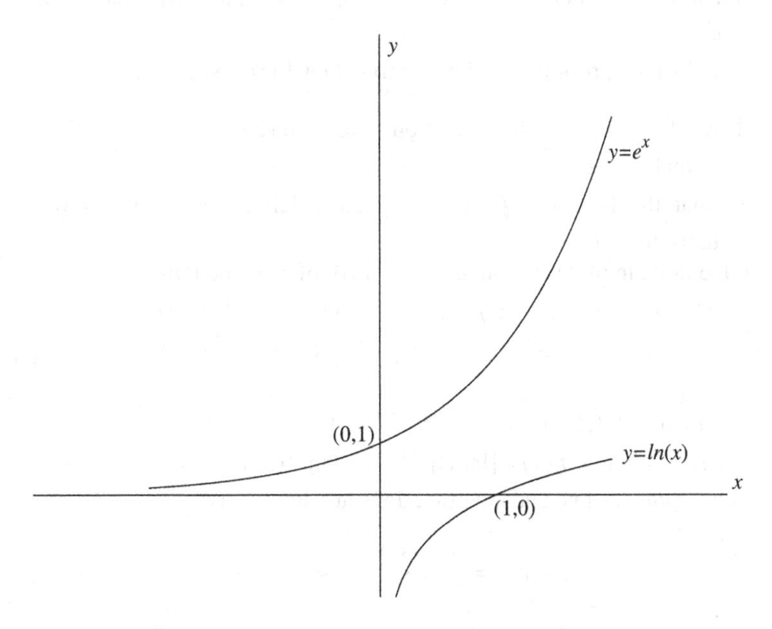

Fig. 5.3.4. The logarithmic and exponential functions.

The inverse of $y = \ln(x)$ is clearly $y = e^x$, defined over the whole x - axis and attains only positive values (Fig. 5.3.4).

The best way to calculate derivatives of inverse functions is by applying Theorem 5.2.4 since we do not actually have to invert the function before calculating the inverse's derivative.

Example 5.3.3. The function $y = \sin^{-1}(x)$ yields $y' = 1/(\sin(y))' = 1/\cos(y)$. The relation $x = \sin(y)$ yields

$$y' = \frac{1}{\pm\sqrt{1-x^2}}$$

but being restricted to the main branch we get $y' = 1\big/\sqrt{1-x^2}$.

Example 5.3.4. The function $y = \tan^{-1}(x)$ yields $y' = 1\big/(\tan(y))' = \cos^2(y)$. By virtue of $x = \tan(y)$ and the identity $1 + \tan^2(y) = 1\big/\cos^2(y)$ we conclude

$$y' = \frac{1}{1+x^2}$$

PROBLEMS

1. Differentiate the identity $\sin^2(x) + \cos^2(x) = 1$ and show that both sides provide 0.
2. Write a detailed proof to the claim $\lim_{h\to 0} \cos(x+h) = \cos(x)$.
3. Find y' from the implicit relation $x^2 e^y + \ln(2 + x^2 + y^2) = 0$. Does this exercise make sense?
4. Show that the function $f(x) = x^3 \sin(1/x)$, defined as 0 at $x = 0$, has a continuous derivative everywhere.
5. Find the domain of definition and derivative of the functions:
   (a) $f(x) = \ln(e^x \sin(x))$ (b) $f(x) = \tan^{-1}(\ln(x))$ (c) $f(x) = e^{\sin(\ln(x))}$.
6. Find the derivative of $y = x^x$ (Hint: apply the logarithmic function on both sides).
7. Find domains of definition and derivatives for:
   (a) $f(x) = x^{\ln(x)}$ (b) $f(x) = [\ln(x)]^{\ln(x)}$ (c) $\tan^{-1}(e^x)$.
8. The *hyperbolic* cosine and sine functions are defined as

$$\cosh(x) = \frac{e^x + e^{-x}}{2}, \; -\infty < x < \infty$$

$$\sinh(x) = \frac{e^x - e^{-x}}{2}, -\infty < x < \infty$$

Show: (a) $\cosh^2(x) - \sinh^2(x) = 1$ (b) $\cosh'(x) = \sinh(x)$ (c) $\sinh'(x) = \cosh(x)$.

9. Calculate the derivatives of $\tanh(x) = \sinh(x)/\cosh(x)$, $\coth(x) = \cosh(x)/\sinh(x)$.
10. Find the extremum points of $\cosh(x)$, $\sinh(x)$, draw the functions and compare with your results.
11. Find the domains where the following functions are defined: (a) $\ln[\cosh(x)]$ (b) $\sqrt{\tanh(x) - \cosh(x)}$ (c) $\ln[\cos(x) - 0.5\cosh(x)]$.
12. Calculate the derivatives of the following functions: (a) $x^{(x^x)}$ (b) $(x^x)^x$

5.4 Higher Order Derivatives; Taylor's Theorem

In this section we introduce higher order derivatives, obtain Taylor's theorem – one of the most useful tools in analysis, and present rules for studying the behavior of an arbitrary differentiable function.

Definition 5.4.1. Let $f(x)$ denote a differentiable function defined over an interval $[a,b]$ and let x_0 be an arbitrary point within the interval. The second derivative of $f(x)$ at $x = x_0$ is the number

$$f''(x_0) = \lim_{h \to 0} \frac{f'(x_0 + h) - f'(x_0)}{h} \tag{5.4.1}$$

provided that the limit at the right-hand side exists. If the limit exists for all $x_0 \in [a,b]$ then $f(x)$ possess a second derivative $f''(x)$ (written also as $f^{(2)}(x)$ or as $\frac{d^2y}{dx^2}$) over $[a,b]$ (The derivatives at the endpoints are right derivative at a and left derivative at b).

Example 5.4.1. The function $y = x^3 + \sin(x)$ satisfies $y' = 3x^2 + \cos(x)$ and its second derivative exists everywhere and equals to $y'' = 6x - \sin(x)$.

Note that in order for $f''(x_0)$ to exist, the first derivative $f'(x)$ must exist at least in a small neighborhood of x_0, otherwise the number $f''(x_0)$ is simply not defined. However, this is only a necessary condition since the expressions at the right-hand side of Eq. (5.4.1) must not only exist but also converge to a finite number.

Example 5.4.2. The function

$$f(x) = \begin{cases} x^3 \sin(1/x)\,, & x \neq 0 \\ 0\,, & x = 0 \end{cases}$$

is differentiable everywhere and satisfies $f'(0) = 0$ as can be easily seen. We also have $f''(x) = 3x^2 \sin(1/x) - x\cos(1/x)$ for all $x \neq 0$. However, the expression

$$\frac{f'(h) - f'(0)}{h} = \frac{3h^2 \sin(1/h) - h\cos(1/h)}{h} = 3h\sin(1/h) - \cos(1/h)$$

although bounded, does not converge but rather oscillates as $h \to 0$. Consequently, the second derivative does not exist at $x = 0$.

We previously defined the instantaneous velocity of an object as the derivative $x'(t) \equiv \frac{dx}{dt}$ where $x(t)$ denotes the location of the object at time t. Consider the velocities $x'(t_1)$, $x'(t_2)$ at times t_1, t_2 respectively. The quantity

$$a_{av}(t_1, t_2) = \frac{x'(t_2) - x'(t_1)}{t_2 - t_1} \tag{5.4.2}$$

is defined as the average acceleration of the object between t_1 and t_2. If this quantity converges as t_2 approaches t_1, the limit is called the instantaneous acceleration of the object at time t_1 and denoted by $a(t_1) = x''(t_1) \equiv \frac{d^2x}{dt^2}(t_1)$. If a force F is acting on an object with mass m and acceleration a, then by Newton's second law of motion

$$F = ma = m\frac{d^2x}{dt^2} \tag{5.4.3}$$

The next definition completes the definition of higher order derivatives.

Definition 5.4.2. Let the $n-th$ ($n \geq 2$) derivative of $f(x)$ exist in a small neighborhood of x_0. If the limit

$$A = \lim_{h \to 0} \frac{f^{(n)}(x_0 + h) - f^{(n)}(x_0)}{h}$$

exists, it is called the $(n+1)-th$ derivative of $f(x)$ at x_0 and denoted by $f^{(n+1)}(x_0)$ (or by $\frac{d^n y}{dx^n}$).

If $f'(x), f''(x), \ldots, f^{(n)}(x)$ exist and are continuous over the interval $[a,b]$, we write $f(x) \in C^n[a,b]$ (the continuity of $f^{(k)}(x)$, $1 \leq k \leq n-1$ is guaranteed by the existence of $f^{(n)}(x)$ - why?), i.e. $f(x)$ belongs to the set of all functions which possess n continuous derivatives over the closed interval $[a,b]$. We similarly define the function sets $C^n(a,b)$, $C^n[a,b)$ and $C^n(a,b]$.

Example 5.4.3. The function $f(x)=1/\cos(x)$ satisfies $f'(x)=\sin(x)/\cos^2(x)$ and $f''(x)=1/\cos(x)+2\sin^2(x)/\cos^3(x)$ everywhere except at the discrete points $x_n=(n+1/2)\pi$ where n is an arbitrary integer. Such a point is called *singular* and since $\cos(x_n)=0$, the function and all its derivatives are not defined at these points (Fig. 5.4.1).

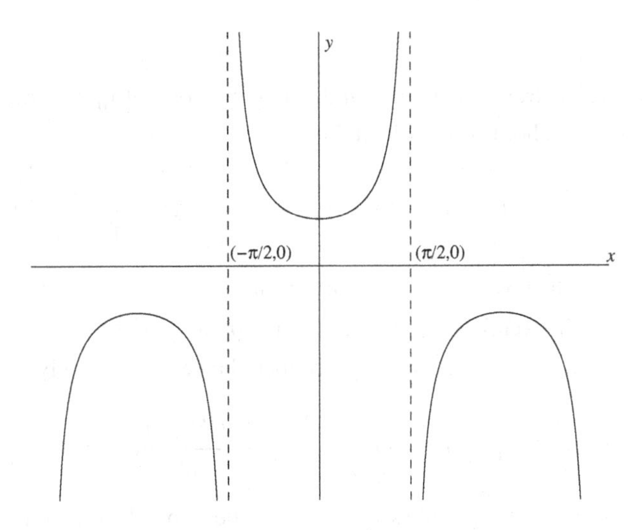

Fig. 5.4.1. Domain of definition of $f(x)=1/\cos(x)$: $-\infty<x<\infty$, $x\neq(n+1/2)\pi$

The next result is an extension to Theorem 5.2.7 and one of the most significant results in calculus. It enables us to approximate a function, occasionally quite complex, by a polynomial of arbitrary degree, provided that several derivatives of the function exist and can be calculated at a single point.

Theorem 5.4.1 (Taylor's theorem). Let $f(x)$ possess $n+1$ derivatives in some neighborhood of $x=x_0$. Then

$$f(x)=f(x_0)+f'(x_0)(x-x_0)+\frac{f''(x_0)}{2!}(x-x_0)^2+\ldots+\frac{f^{(n)}(x_0)}{n!}(x-x_0)^n+R_n$$

where R_n, called the *remainder*, is given by

$$R_n = \frac{f^{(n+1)}(\xi)}{(n+1)!}(x - x_0)^{n+1} \tag{5.4.4}$$

where ξ is some interim point between x_0 and x.

Thus if R_n is known to be relatively small, the function $f(x)$ can be approximated by Taylor's polynomial

$$p_n(x) = f(x_0) + f'(x_0)(x - x_0) + \frac{f''(x_0)}{2!}(x - x_0)^2 + \ldots + \frac{f^{(n)}(x_0)}{n!}(x - x_0)^n$$

A polynomial approximation has clear advantages as its properties are known and it is also easily manipulated.

Proof.

Let b denote an arbitrary fixed point in the neighborhood of x_0. In order to show that Taylor's formula holds for $x = b$, define

$$F(x) = f(x) + f'(x)(b - x) + \ldots + \frac{f^{(n)}(x)}{n!}(b - x)^n + \frac{A(b - x)^{n+1}}{(b - x_0)^{n+1}}$$

where A is a constant yet to be determined and x varies between x_0 and b. Clearly, $F(x)$ is differentiable between x_0 and b, and $F(b) = f(b)$. However, we may force $F(x_0) = f(b)$ as well, by a proper choice of A, namely,

$$A = F(x_0) - f(x_0) - f'(x_0)(b - x_0) - \ldots - \frac{f^{(n)}(x_0)}{n!}(b - x_0)^n \tag{5.4.5}$$

Since $F(x_0) = F(b) = f(b)$, Rolle's theorem can be applied to obtain $F'(\xi) = 0$ for some interim point ξ between x_0 and b. Simple algebra provides

$$F'(x) = \frac{f^{(n+1)}(x)}{n!}(b - x)^n - \frac{A(n+1)(b - x)^n}{(b - x_0)^{n+1}}$$

and by virtue of $F'(\xi) = 0$ we get that A, already determined by Eq. (5.4.5) must also satisfy

$$A = \frac{f^{(n+1)}(\xi)}{(n+1)!}(b - x_0)^{n+1}$$

The remaining of the proof, i.e., substituting in Eq. (5.4.5) and replacing $F(x_0)$ by $f(b)$ is straightforward and is left for the reader.

Note that by applying Taylor's theorem for $n = 0$ we obtain the familiar first mean-value theorem (Theorem 5.2.6).

Although Taylor's theorem is usually applied to approximate functions, we will first show how it is used to obtain simple rules for determining special points within a function's domain – the stationary points.

Definition 5.4.3. A point x_0 where $f'(x_0)=0$ is called a *stationary point.*

Definition 5.4.4. A function $f(x)$ is called *concave up* in an interval $I=(a,b)$, if $f'(x)$ is strictly increasing in I and *concave down* if $f'(x)$ is strictly decreasing in I.

Definition 5.4.5. Let $f(x)$ denote a differentiable function over an interval I. A point $x_0 \in I$ is called an *inflection* point if $f(x)$ is concave up at the left (right) of x_0 and concave down at the right (left) of x_0.

Example 5.4.4. The stationary points of $f(x)=x^4-3x^3+x^2$ are determined by $f'(x)=4x^3-9x^2+2x=0$, and are consequently $x_1=0$, $x_2=8$, $x_3=0.25$.

Example 5.4.5. Let $f(x)=x^3$, $-\infty<x<\infty$. The first derivative $f'(x)=3x^2$ is clearly strictly increasing for $x>0$ which implies that $f(x)$ is concave up there. For $x<0$, $f(x)$ is concave down (why?).

We usually distinguish between two classes of stationary points: (a) extremum points; and (b) other points which are usually inflection points if the function is twice continuously differentiable. The next result concerns the first class.

Theorem 5.4.2. Consider a function $f(x)\in C^2(a,b)$ such that $f'(x_0)=0$ and $f''(x_0)\neq 0$. Then x_0 is a local maximum of $f(x)$ if $f''(x_0)<0$ and a local minimum of $f(x)$ if $f''(x_0)>0$.

Proof.

By Taylor's theorem, we get for arbitrary $x\in(a,b)$

$$f(x)=f(x_0)+(x-x_0)f'(x_0)+\frac{(x-x_0)^2}{2!}f''(\xi)$$

where ξ (which depends on x) is between x_0 and x. Since $f'(x_0)=0$ this yields

$$f(x)=f(x_0)+\frac{(x-x_0)^2}{2!}f''(\xi)$$

The continuity of $f''(x)$ guarantees that $f''(x_0)$ and $f''(\xi)$ have the same sign provided that x is sufficiently close to x_0. Therefore, the sign of the remainder

$$R_2 = \frac{(x-x_0)^2}{2!} f''(\xi)$$

is the same on both sides of x_0. The rest of the proof is straightforward and is left for the reader.

Example 5.4.5. Define $y = x^3 - 2x^2 + x$ over the interval $[0,2]$ (Fig. 5.4.2). The derivative $y' = 3x^2 - 4x + 1$ vanishes at $x_1 = 1/3$ and $x_2 = 1$. The second derivative $y'' = 6x - 4$ satisfies $y''(1) = 2 > 0$ and $y''(1/3) = -2 < 0$, implying that x_1 is a local maximum and x_2 is a local minimum of y.

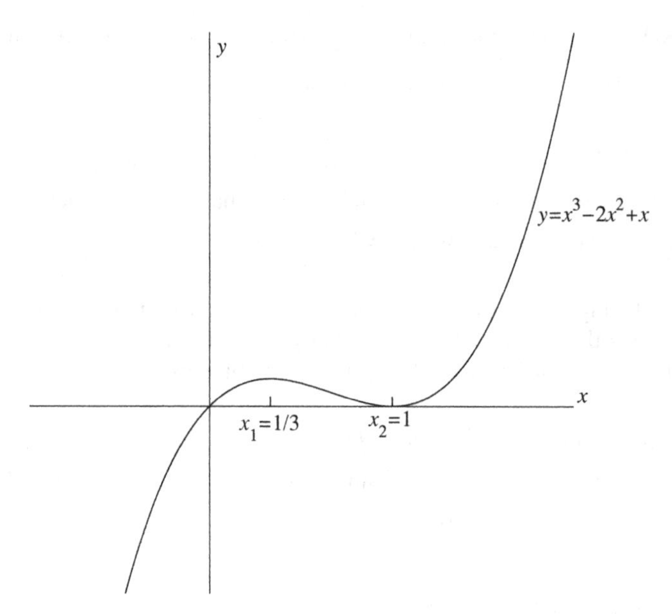

Fig. 5.4.2. Graph and extremum points of $y = x^3 - 2x^2 + x$.

Example 5.4.6. Consider the function $y = x^2 e^{-x}$ over the interval $[-1,3]$ (Fig. 5.4.3). The first and second derivatives are $y' = (2x - x^2)e^{-x}$ and $y'' = (2 - 4x + x^2)e^{-x}$ respectively. The solutions of $y' = 0$ are $x_1 = 0$, $x_2 = 2$ which are local minimum and maximum respectively, since $y''(0) = 2/e > 0$ and $y''(2) = -2/e^2 < 0$.

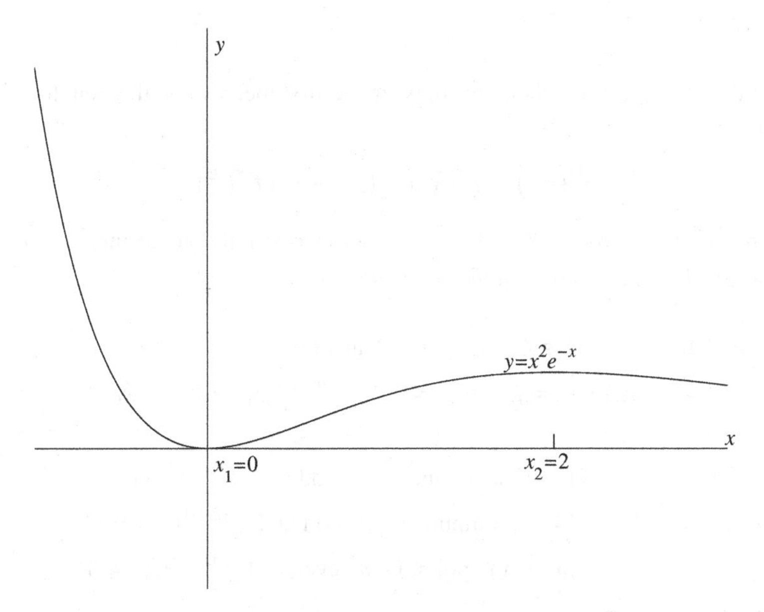

Fig. 5.4.3. Graph and extremum points of $y = x^2 e^{-x}$.

Clearly, the requirements of Theorem 5.4.2 present sufficient conditions for the existence of extremum points. They are certainly not necessary conditions. For example, the function $y = |x|$ is differentiable everywhere except at $x = 0$ where it attains a minimum. Another function defined as

$$f(x) = \begin{cases} 0\,,\ x \neq 3{,}5 \\ -1\,,\ x = 3 \\ 7\,,\ x = 5 \end{cases}$$

has a minimum at $x = 3$, maximum at $x = 5$ and is not differentiable or even continuous at these points.

By virtue of Theorem 5.4.2, it is clear that in the case of $f'(x_0) = f''(x_0) = 0$, one cannot, without further investigation, determine the nature of the point. Indeed, if $f'(x_0) = 0$, x_0 is usually either a local maximum, a local minimum, or an inflection point where the function moves from a concave up (down) zone to a concave down (up) zone.

The next result is an extension of Theorem 5.4.2 which enables to carry on the search for extremum points in the case of $f'(x_0) = f''(x_0) = 0$. We first prove the following lemma.

Lemma 5.4.1. If $f(x) \in C^2[a,b]$ and $f''(x) > 0$ $(f''(x) < 0)$, then $f(x)$ is concave up (concave down) over $[a,b]$.

Proof.

Let $f''(x) > 0$. If $x_1 < x_2$, then, by applying the first mean value theorem to $f'(x)$ we get

$$f'(x_2) - f'(x_1) = (x_2 - x_1) f''(\xi)$$

for some interim $\xi : a \le x_1 < x_2 \le b$. Thus $f'(x)$ is strictly increasing, i.e. $f(x)$ is concave up. The other case is similarly treated.

Theorem 5.4.3. Let $f(x) \in C^n[a,b]$, $n > 2$ and let

$$f'(x_0) = f''(x_0) = \ldots = f^{(m)}(x_0) = 0 \quad , \quad f^{(m+1)}(x_0) \ne 0 \quad , \quad m < n \qquad (5.4.6)$$

Then, x_0 is:

local maximum, if: m odd and $f^{(m+1)}(x_0) < 0$

local minimum, if: m odd and $f^{(m+1)}(x_0) > 0$

inflection point, if: m even and $f^{(m+1)}(x_0) \ne 0$

Proof.

By Taylor's theorem we get (for x close to x_0)

$$f(x) = f(x_0) + \frac{(x - x_0)^{(m+1)}}{(m+1)!} f^{(m+1)}(\xi) \qquad (5.4.7)$$

where ξ is an interim point between x_0 and x. If m is odd the remainder's sign is determined by the sign of $f^{(m+1)}(x_0)$ on *both sides* of x_0. If it is positive we have $f(x) > f(x_0)$ and a local minimum is obtained. If it is negative, we get a local maximum. If m is even (m is at least 2), we apply Taylor's theorem to the second derivative $f''(x)$ and obtain

$$f''(x) = \frac{(x - x_0)^{m-1}}{(m-1)!} f^{(m+1)}(\xi^*) \qquad (5.4.8)$$

where ξ^* is some interim point between x_0 and x (usually $\xi^* \ne \xi$). If $f^{(m+1)}(x_0) > 0$, then $f''(x) > 0$ at the right of x_0 provided that x is sufficiently close to x_0. This implies that $f(x)$ is concave up for $x > x_0$. Similarly, $f(x)$ is concave down for x at the left of x_0 and sufficiently close to it. The case $f^{(m+1)}(x_0) < 0$ provides a concave up zone at the left of x_0 and a concave down zone at the right.

Example 5.4.7. Let $f(x) = x\ln(x) - 2x$. The function domain of definition is $x \neq 0$. The stationary points are the solutions of $f'(x) = \ln(x) - 1 = 0$. We have a single solution $x_0 = e$ at which $f''(x_0) = 1/x_0 = 1/e > 0$, which implies that x_0 is a local minimum.

Example 5.4.8. Let $y = x + \sin(x)$. The stationary points are obtained by $y' = 1 + \cos(x) = 0$ which implies $x = x_k = (2k+1)\pi$ for arbitrary integer k. Since the second derivative $y'' = -\sin(x)$ yields $y''(x_k) = 0$, we must calculate the third derivative in order to determine the nature of the stationary points. Since $y''' = -\cos(x)$ we get $y'''(x_k) = 1$ and consequently all x_k are an inflection points (Fig. 5.4.4).

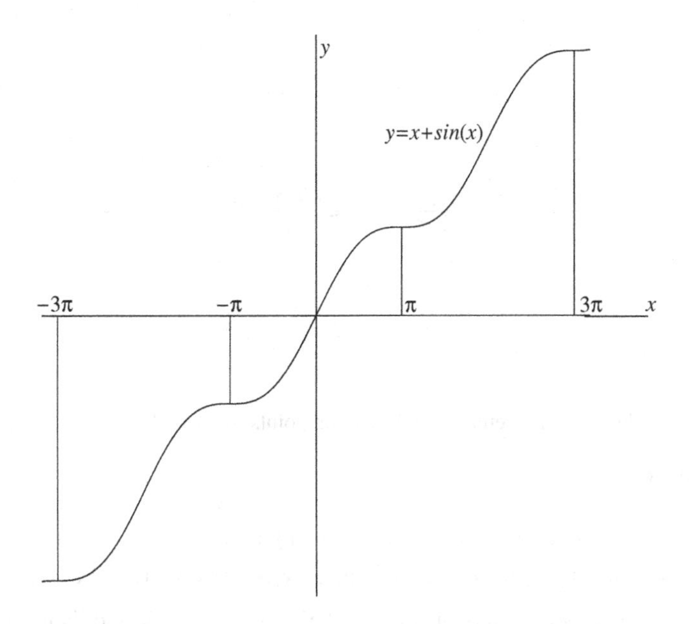

Fig. 5.4.4. Stationary points of $y = x + \sin(x)$: only inflection points.

Note that an inflection point of a function $f(x)$ is not necessarily a stationary point. Consider a function $f(x) \in C^3[a,b]$ for which x_0 is an inflection point such that $f'(x_0) \neq 0$. Assume that the function is concave up to the right and concave down to the left of this point.. If in addition $f''(x) > 0$ at the right and $f''(x) < 0$ at the left of x_0 (these are sufficient conditions for the concave up and down zones respectively) we must have $f''(x_0) = 0$. If now $f'''(x_0) \neq 0$, then x_0 is simultaneously an inflection point of $f(x)$ and an extremum point of $f'(x)$.

Example 5.4.9. The function $y = 2x^4 - x^2$ has three extremum points and two inflection points that are not stationary (Fig. 5.4.5). Indeed, $y' = 8x^3 - 2x = 0$ yields two local minima at $x_{1,2} = \pm 0.5$ and a single maximum at $x_3 = 0$. The solutions of $y'' = 24x^2 - 2 = 0$, i.e. $x_{4,5} = \pm 1/2\sqrt{3}$, are inflection points that are not stationary points of y but rather extremum points of y'. Indeed, the second derivative of y' which is $y''' = 48x$ provides $y'''(x_4) > 0$ (minimum) and $y'''(x_5) < 0$ (maximum).

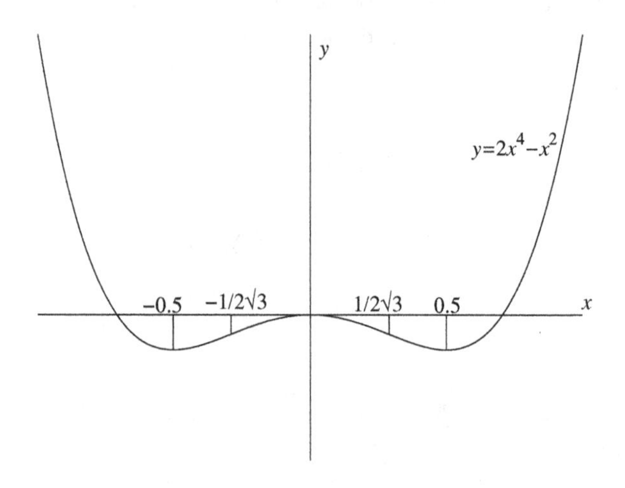

Fig. 5.4.5. Extremum and inflection points for $y = 2x^4 - x^2$.

PROBLEMS

1. Find the second derivatives of the following functions:
   (a) $y = \sin(e^x)$ (b) $y = e^x$ (c) $y = \ln[\sin(x) + \tan(x)]$ (d) $y = e^x/(x + e^x)$
2. Find the stationary points of $f(x) = x^n e^{-x}$, $-\infty < x < \infty$ and their types.
3. Find domain of definition, derivative, stationary points and their types and domains of monotonic increase and decrease for the following functions:
   (a) $y = x^4 - x^3 + x^2$ (b) $y = x\ln(1 + x)$ (c) $y = \sin(x) + \cos(x)$
4. Let $f(x) = 2x^2 - 1/x$. Find the domain of definition, stationary points and their types, domains of monotonic increase and decrease and draw the function.
5. Find domain of definition and stationary points for $f(x) = a\sin(x) + b\cos(x)$. Assume $a^2 + b^2 > 0$.
6. Show that the function $f(x) = x^m(a - x)^n$, $0 \le x \le a$, where m,n are positive integers, has a global maximum inside the interval and find it.

7. Find the stationary points of $f(x) = x^4 - 2x^3 + x^2$ and their types.
8. Use the first mean-value theorem to show $e^x > 1 + x$ for arbitrary $x > 0$. Find the stationary points of $f(x) = e^x - 1 - x$ and their types.
9. Let $f(x) = \sin(x)$, $x_0 = 0$. Use Taylor's theorem to evaluate R_4 and bound it for $|x| \le 1$.
10. Let $f(x) = xe^x$, $x_0 = 0$. Use Taylor's theorem to evaluate R_3.

5.5 L'Hospital's Rules

Quite often it is necessary to calculate limits such as

$$\lim_{x \to a} \frac{f(x)}{g(x)}$$

where a is usually an endpoint of an interval where the functions are defined, and either $\lim_{x \to a} f(x) = \lim_{x \to a} g(x) = 0$ or $\lim_{x \to a} f(x) = \lim_{x \to a} g(x) = \infty$. In either case the substitution $x = a$ leads to nowhere. The technique used to obtain these limits – if they exist is based on the first mean-value theorem and is known as L'Hospital's rules which help treat expressions such as $0/0$ and ∞/∞.

Theorem 5.5.1 (L'Hospital's first rule). Let $f(x), g(x)$ be continuous over a finite interval $[a,b]$, differentiable over (a,b) such that $\lim_{x \to a^+} f(x) = \lim_{x \to a^+} g(x) = 0$ and let $g'(x) \ne 0$ in a right neighborhood of a (not including a itself). Then

(1) If $\lim_{x \to a^+} \frac{f'(x)}{g'(x)} = A$ (finite) then $\lim_{x \to a^+} \frac{f(x)}{g(x)} = A$.

(2) If $\lim_{x \to a^+} \frac{f'(x)}{g'(x)} = \infty$ then $\lim_{x \to a^+} \frac{f(x)}{g(x)} = \infty$.

(3) If $\lim_{x \to a^+} \frac{f'(x)}{g'(x)} = -\infty$ then $\lim_{x \to a^+} \frac{f(x)}{g(x)} = -\infty$.

Proof.

Case (1) By virtue of Theorem 5.2.7 we have

$$\frac{f(x)}{g(x)} = \frac{f(x) - f(a)}{g(x) - g(a)} = \frac{f'(\xi)}{g'(\xi)}$$

for some interim point $\xi : a < \xi < x < b$ provided that x is in the neighborhood where $g'(x) \ne 0$. Since the right-hand side converges to A as $x \to a^+$, so does the left-hand side. Obviously, the same argument basically holds for Cases (2) and (3) as well. The exact details which include the separate definition of converging to ∞, are left as an exercise for the reader

It should be noted that in Theorems 5.5.1-5.5.2 we can replace a with b and a^+ with b^- since the limits are obtained by approaching b from the left.

Example 5.5.1. Let $f(x) = x^2 - 1$, $g(x) = x^2 + x - 2$, $a = 1$. Here $f(1) = g(1) = 0$ and by L'Hospital's first rule

$$\lim_{x \to 1} \frac{f(x)}{g(x)} = \lim_{x \to 1} \frac{2x}{2x+1} = 3$$

Example 5.5.2. Let $f(x) = \sin(x)$, $g(x) = 1 - \cos(x)$, $a = 0$. All the requirements of Theorem 5.5.1 are satisfied and consequently

$$\lim_{x \to 0} \frac{\sin(x)}{1 - \cos(x)} = \lim_{x \to 0} \frac{\cos(x)}{\sin(x)} = \infty$$

If $\lim\limits_{x \to a^+} [f'(x)/g'(x)]$ is also of the form $0/0$ or ∞/∞ we may check whether f', g' satisfy the requirements of L'Hospital's first rule. If so we can continue and try to obtain $\lim\limits_{x \to a^+} [f''(x)/g''(x)]$ etc.

Example 5.5.3. Let $f(x) = x^3 - 4x^2$, $g(x) = x^2 + x^4$, $a = 0$. Here

$$\lim_{x \to 0^+} \frac{x^3 - 4x^2}{x^2 + x^4} = \lim_{x \to 0^+} \frac{3x^2 - 8x}{2x + 4x^3} = \lim_{x \to 0^+} \frac{6x - 8}{2 + 12x^2} = -4$$

In many applications both numerator and denominator approach infinity as $x \to a^+$. This problem is treated next.

Theorem 5.5.2 (L'Hospital's second rule). Let $f(x), g(x)$ be differentiable over (a,b) such that $\lim\limits_{x \to a^+} f(x) = \lim\limits_{x \to a^+} g(x) = \infty$ and let $g'(x) \neq 0$ in a right neighborhood of a (not including a itself). Then

(1) If $\lim\limits_{x \to a^+} \dfrac{f'(x)}{g'(x)} = A$ (finite) then $\lim\limits_{x \to a^+} \dfrac{f(x)}{g(x)} = A$.

(2) If $\lim\limits_{x \to a^+} \dfrac{f'(x)}{g'(x)} = \infty$ then $\lim\limits_{x \to a^+} \dfrac{f(x)}{g(x)} = \infty$.

(3) If $\lim\limits_{x \to a^+} \dfrac{f'(x)}{g'(x)} = -\infty$ then $\lim\limits_{x \to a^+} \dfrac{f(x)}{g(x)} = -\infty$.

Proof.

The proof is more complex than that of the previous result. We will give it in detail only for Case 1. Given an arbitrary $\varepsilon > 0$ one needs to show the existence of $\delta > 0$ such that if $a < x < a + \delta$ then $|f(x)/g(x) - A| < \varepsilon$. However, by virtue of $\lim_{x \to a^+} [f'(x)/g'(x)] = A$, given $\varepsilon_1 = \varepsilon/3$ there exists $\delta_1 > 0$ such that

$$\left| \frac{f'(x)}{g'(x)} - A \right| < \varepsilon_1 \, , a < x \le a + \delta_1 \tag{5.5.1}$$

By applying Theorem 5.2.7 we get

$$\frac{f(x) - f(a+\delta_1)}{g(x) - g(a+\delta_1)} = \frac{f'(\xi)}{g'(\xi)} \, , a < x \le a + \delta_1$$

where ξ is an interim point between a and $a + \delta_1$, assuming that δ_1 is sufficiently close to a so that both denominators of the last equality are not zero and $g(x) \ne 0$ as well. Hence

$$\frac{\dfrac{f(x)}{g(x)} - \dfrac{f(a+\delta_1)}{g(x)}}{1 - \dfrac{g(a+\delta_1)}{g(x)}} = \frac{f'(\xi)}{g'(\xi)} \quad , \quad a < x \le a + \delta_1$$

which leads to

$$\frac{f(x)}{g(x)} = \left(1 - \frac{g(a+\delta_1)}{g(x)} \right) \frac{f'(\xi)}{g'(\xi)} + \frac{f(a+\delta_1)}{g(x)} \tag{5.5.2}$$

By virtue of $\lim_{x \to a^+} g(x) = \infty$, given an arbitrary ε_2 (not yet determined!) we can find $\delta_2 : a < \delta_2 < \delta_1$ such that

$$\left| \frac{g(a+\delta_1)}{g(x)} \right| < \varepsilon_2 \quad , \quad \left| \frac{f(a+\delta_1)}{g(x)} \right| < \varepsilon_2 \quad ; \quad a < x < \delta_2 \tag{5.5.3}$$

Therefore if we rewrite Eq. (5.5.2) as

$$\frac{f(x)}{g(x)} - A = \frac{f'(\xi)}{g'(\xi)} - A - \frac{g(a+\delta_1)}{g(x)} \frac{f'(\xi)}{g'(\xi)} + \frac{f(a+\delta_1)}{g(x)} \quad , \quad a < x \le \delta_1$$

we get, using Eqs. (5.5.2-3) the following relation:

$$\left| \frac{f(x)}{g(x)} - A \right| \le \left| \frac{f'(\xi)}{g'(\xi)} - A \right| + \varepsilon_2 (A + \varepsilon_1) + \varepsilon_2 \quad , \quad a < x < \delta_2$$

and finally we obtain

$$\left|\frac{f(x)}{g(x)} - A\right| \le \varepsilon_1 + \varepsilon_2[1 + A + \varepsilon_1] \ , \ a < x \le \delta_2$$

We now choose $\varepsilon_2 = \varepsilon_1/(1 + A + \varepsilon_1)$ and get

$$\left|\frac{f(x)}{g(x)} - A\right| < 2\varepsilon_1 = \frac{2\varepsilon}{3} < \varepsilon \ , a < x < \delta$$

for the choice $\delta = \delta_2$. This completes the proof.

The proof in Cases (2) and (3) are left as challenging exercises for the interested reader.

The results presented in Theorems 5.5.1-5.5.2 can be easily extended to include the case $a = \infty$.

Theorem 5.5.3. The results in Theorems 5.5.1-5.5.2 hold for $a = \infty$.

Proof.

In order to show validity of the previous results for $a = \infty$, it is sufficient to show that each case with $a = \infty$ is equivalent to a case with a finite a - already covered by Theorems 5.5.1-5.5.2.

Let $f(x), g(x)$ be differentiable over (b, ∞) such that $g'(x) \ne 0$ for sufficiently large x and let $\lim_{x \to \infty}([f'(x)/g'(x)] = A$ (finite or infinite). Define $t = 1/x$, $x > 0$ and let $F(t) = f(1/t)$, $G(t) = g(1/t)$. Thus, F, G are defined and differentiable at a right neighborhood of $t = 0$. Moreover, $G'(t) = g'(x)(-1/t^2) \ne 0$ near $t = 0$ (why?) and

$$\frac{F'(t)}{G'(t)} = \frac{f'(x)(-1/t^2)}{g'(x)(-1/t^2)} = \frac{f'(x)}{g'(x)}$$

which yields $\lim_{t \to 0^+}([F'(t)/G'(t)] = A$. Since L'Hospital's rules hold for the pair F and G, they also hold for f and g as well.

Example 5.5.3. Let $f(x) = 1/x$, $g(x) = \ln(\sin(x))$, $a = 0$. The functions approach $+\infty$ and $-\infty$ respectively as $x \to 0^+$. Their derivatives satisfy

$$\frac{f'(x)}{g'(x)} = -\frac{(1/x^2)}{\cot(x)} = -\frac{\sin(x)}{x}\frac{1}{\cos(x)}\frac{1}{x}$$

and the whole expression approaches $-\infty$ since $\sin(x)/x$ and $1/\cos(x)$ each approaches 1 as $x \to 0^+$. Thus $\lim_{x \to 0^+} 1/[x \ln(\sin(x))] = -\infty$.

Example 5.5.4. Let $f(x)=\ln(1+x)$, $g(x)=x$; $x>0$. Here

$$\lim_{x\to 0^+}\frac{\ln(1+x)}{x}=\lim_{x\to 0^+}\frac{1/(x+1)}{1}=1$$

which implies that for small positive x the function $\ln(1+x)$ can be approximated by x (this is easily extended for negative small x) as shown in Fig. 5.5.1.

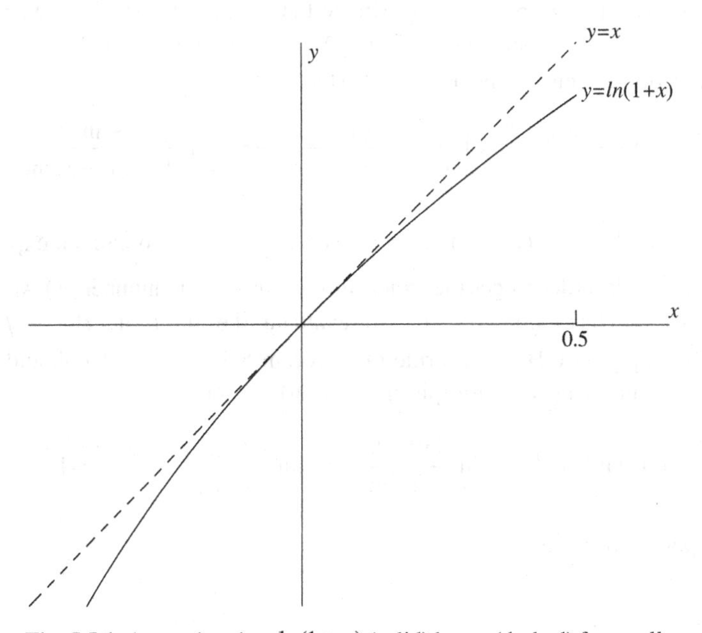

Fig. 5.5.1. Approximating $\ln(1+x)$ (solid) by x (dashed) for small x.

Example 5.5.5. The function $f(x)=e^x$ approach infinity as $x\to\infty$ and so is $g(x)=x^3$. Which one approaches faster? By repeatedly applying Theorem 5.5.3 we get

$$\lim_{x\to\infty}\frac{e^x}{x^3}=\lim_{x\to\infty}\frac{e^x}{3x^2}=\lim_{x\to\infty}\frac{e^x}{6x}=\lim_{x\to\infty}\frac{e^x}{6}=\infty$$

and this result is clearly valid for arbitrary power $g(x)=x^n$, $n>0$.

Sometimes it is needed to estimate a function $h(x)$ at points "near" a, when it is known to behave as one of the expressions $0\cdot\infty$, ∞^0, 1^∞, 0^0, $\infty-\infty$ when x approaches a. In such case we may try to express $h(x)=f(x)/g(x)$ where $f(x)$ and $g(x)$ satisfy the requirements needed for applying one of L'Hospital's rules.

Example 5.5.6. Let $h(x) = x\ln(x)$, $x > 0$. When $x \to 0^+$ we obtain a behavior of $0 \cdot \infty$, which can be treated by choosing $f(x) = \ln(x)$, $g(x) = 1/x$ thus obtaining

$$\lim_{x\to0^+} x\ln(x) = \lim_{x\to0^+} \frac{\ln(x)}{(1/x)} = \lim_{x\to0^+} \frac{(1/x)}{-(1/x^2)} = \lim_{x\to0^+} (-x) = 0$$

Example 5.5.7. Let $h(x) = 1/x - 1/\sin(x)$, $x \neq 0$. The function's behavior near $x = 0$ is of the type $\infty - \infty$. To approximate the function for small x we rewrite the function as $h(x) = (\sin(x) - x)/(x\sin(x))$. The nominator and denominator satisfy the requirements of Theorem 5.5.1. Hence

$$\lim_{x\to0^+} h(x) = \lim_{x\to0^+} \frac{\sin(x) - x}{x\sin(x)} = \lim_{x\to0^+} \frac{\cos(x) - 1}{\sin(x) + x\cos(x)} = \lim_{x\to0^+} \frac{-\sin(x)}{2\cos(x) - x\sin(x)} = 0$$

Example 5.5.8. Let $y = (1 + 1/x)^x$, $x > 0$. When $x \to \infty$ we obtain an expression of the type 1^∞ . In order to get the exact limit (which is the number e) we firstly rewrite $\ln(y) = x\ln(1 + 1/x)$ and try to evaluate the limit of $x\ln(1 + 1/x)$ as $x \to \infty$. By applying L'Hospital's rule of Theorem 5.5.2 (the reader should check that all the requirements hold for this application) we get

$$\lim_{x\to\infty} x\ln(1 + 1/x) = \lim_{x\to\infty} \frac{\ln(1 + 1/x)}{(1/x)} = \lim_{x\to\infty} \frac{-(1/x^2)/(1 + 1/x)}{-(1/x^2)} = 1$$

which implies $\lim_{x\to\infty} y = e$.

PROBLEMS

1. Calculate: (a) $\lim_{x\to1^-} \frac{x^3 - 1}{2x^3 - x^2 - 1}$ (b) $\lim_{x\to0^+} \frac{\sqrt{x} + \sin(x)}{x}$ (c) $\lim_{x\to0^+} \frac{x - \tan(x)}{x + \sin(x)}$.

2. Calculate: (a) $\lim_{x\to0} \frac{\ln[\cos(x)]}{x^2}$ (b) $\lim_{x\to0} \frac{\ln[e^{3x} - 5x]}{x}$.

3. Calculate: (a) $\lim_{x\to0} \frac{\sin(x) + \tan(x)}{\ln(1 + x)}$ (b) $\lim_{x\to0} \frac{\sin(x) - \tan(x)}{x\ln(1 + x)}$.

4. Calculate: (a) $\lim_{x\to\infty} \frac{\ln(x)}{x^\alpha}$, $\alpha > 0$ (b) $\lim_{x\to\infty} \frac{x^n}{e^x}$

5. Calculate $\lim_{x\to0} \frac{x^2 - \sin^2(x)}{x^2 \sin^2(x)}$ (Hint: use Taylor's theorem for $\sin(x)$).

6. Calculate $\lim_{x\to0^+} \frac{\ln(\cos(3x))}{\ln(\cos(2x))}$

6 Integration

The *integral* as represented in this chapter can be sometimes regarded as a mathematical modeling of the concept of *area*. However, once defined, the integral presents more than a tool for obtaining areas (or volumes), and calculation of integrals is quite popular, essential and helpful in all sciences, not just geometry. There are several definitions for the integral and in this chapter we follow most existing texts and introduce the most popular and applicable one: the *Riemann integral.*

6.1 The Riemann Integral

Throughout this section $f(x)$, unless otherwise specified, will denote a bounded function defined over a closed interval $[a,b]$ (i.e., $-\infty < a < b < \infty$). Prior to introducing the integral we define a *partition* (or *mesh*) of an interval.

Definition 6.1.1. A partition (or mesh) of a closed interval $[a,b]$ is a finite collection of distinct points $P = \{x_i\}_{i=0}^n$ such that $a = x_0 < x_1 < \cdots < x_{n-1} < x_n = b$. Each interval $[x_{i-1}, x_i]$ is called *subinterval* and the norm of P is $|P| = \max\limits_{1\le i\le n}\{|x_i - x_{i-1}|\}$. The partition's points are often referred to as *mesh points* and the norm is also called *mesh size.*

Example 6.1.1. Consider the interval $[1,7]$ with the partition $P = \{1,2,4.5,6.5,7\}$. In this case we have 3 subintervals and the partition's norm is 2.5.

For an arbitrary partition, the numbers

$$m_i = \inf[f(x)]\,,\, M_i = \sup[f(x)]\,;\, x_{i-1} \le x \le x_i \tag{6.1.1}$$

exist and we may therefore define the *lower* and *upper* sums

$$L(P,f) = \sum_{i=1}^{n} m_i \Delta x_i \,,\, U(P,f) = \sum_{i=1}^{n} M_i \Delta x_i \tag{6.1.2}$$

M. Friedman and A. Kandel: Calculus Light, ISRL 9, pp. 147–182.
springerlink.com

where $\Delta x_i = x_i - x_{i-1}$, $1 \le i \le n$. These sums shown in Figs. 6.1.1 and 6.1.2 are often called the Darboux sums and if $f(x) \ge 0$ they provide lower and upper bounds to the *area* bounded by $f(x)$, the $x-$axis and the straight lines $x = a$, $x = b$. Clearly, $L(f,P) \le U(f,P)$ for all $f(x)$ and P.

Example 6.1.2. Let $f(x) = \sin(x)$, $0 \le x \le \pi$ and $P = \{0, \pi/6, \pi/3, \pi/2, 3\pi/4, \pi\}$. The lower and upper sums are

$$L(P,f) = 0 \cdot \frac{\pi}{6} + \frac{1}{2} \cdot \frac{\pi}{6} + \frac{\sqrt{3}}{2} \cdot \frac{\pi}{6} + \frac{\sqrt{2}}{2} \cdot \frac{\pi}{2} + 0 \cdot \frac{\pi}{2} \approx 1.826$$

$$U(P,f) = \frac{1}{2} \cdot \frac{\pi}{6} + \frac{\sqrt{3}}{2} \cdot \frac{\pi}{6} + 1 \cdot \frac{\pi}{6} + 1 \cdot \frac{\pi}{2} + \frac{\sqrt{2}}{2} \cdot \frac{\pi}{2} \approx 3.920$$

For arbitrary partitions P, Q we may create the union $R = P \cup Q$ which is a partition which consists of all the mesh points of P and Q. We clearly get $P \subseteq R, Q \subseteq R$, i.e., all the mesh points of either P or Q belong to R as well. We therefore may refer to R as a *finer* partition than either P or Q. The next result is a major step in the process of defining the *integral*.

Lemma 6.1.1. Consider two partitions $P = \{x_i\}_{i=0}^{n}$ and Q which consists of all the mesh points of P and a single additional point $\xi : x_{k-1} < \xi < x_k$. Then

$$L(P,f) \le L(Q,f)\,,\, U(Q,f) \le U(P,f)$$

Proof.

By definition

$$L(P,f) = \sum_{i=1}^{n} m_i \Delta x_i = \sum_{i=1}^{k-1} m_i \Delta x_i + m_k \Delta x_k + \sum_{i=k+1}^{n} m_i \Delta x_i$$

$$L(Q,f) = \sum_{i=1}^{n} m_i \Delta x_i = \sum_{i=1}^{k-1} m_i \Delta x_i + m_{k,1}(\xi - x_{k-1}) + m_{k,2}(x_k - \xi) + \sum_{i=k+1}^{n} m_i \Delta x_i$$

where $m_{k,1} = \inf[f(x)]$, $x_{k-1} \le x \le \xi$ and $m_{k,2} = \inf[f(x)]$, $\xi \le x \le x_k$. Since $m_k \le m_{k,1}, m_{k,2}$ (why?) we obtain

$$m_k \Delta x_k = m_k[(\xi - x_{k-1}) + (x_k - \xi)] \le m_{k,1}(\xi - x_{k-1}) + m_{k,2}(x_k - \xi)$$

and consequently $L(P,f) \le L(Q,f)$. Similarly $U(Q,f) \le U(P,f)$ which completes the proof.

Corollary 6.1.1. If P,Q satisfy $P \subseteq Q$, then $L(P,f) \le L(Q,f)$, $U(Q,f) \le U(P,f)$, i.e., refinement of the mesh increases the lower sum and decreases the upper sum.

The proof by repeatedly applying Lemma 6.1.1 is left as an exercise for the reader.

Lemma 6.1.2. The relation $L(P,f) \le U(Q,f)$ holds for arbitrary P,Q.

Proof.

Let $R = P \cup Q$. Then $L(P,f) \le L(R,f) \le U(R,f) \le U(Q,f)$ which completes the proof. Thus an arbitrary lower sum is less than (or equal to) every upper sum.

Since the lower and upper sums are bounded, we may define *the definite lower and upper Darboux integrals* of a function $f(x)$, $a \le x \le b$ as

$$\underline{\int_a^b} f(x)dx = \sup_P [L(P,f)] \quad , \quad \overline{\int_a^b} f(x)dx = \inf_P [U(P,f)]$$

respectively, where the relation $\underline{\int_a^b} f(x)dx \le \overline{\int_a^b} f(x)dx$ always holds. If the two integrals are the same, the common value denoted by $(D)\int_a^b f(x)dx$ is called the *definite Darboux integral* over $[a,b]$, the function is said to be *Darboux integrable.*

Example 6.1.3. The function $f(x) = C$, $a \le x \le b$ satisfies $m_i = M_i = C$ for all possible partitions. Therefore, $L(f,P) = U(f,P) = C(b-a)$ for an arbitrary partition P which yields

$$\underline{\int_a^b} C\,dx = \overline{\int_a^b} C dx = C(b-a)$$

Intuitively, one would expect the lower and upper Darboux sums to converge to the lower and upper Darboux integrals respectively, when the norm $|P|=\max_{1\le i\le n}\{|x_i - x_{i-1}|\}$ approaches 0. This is established by the following results.

Theorem 6.1.1. For arbitrary $\varepsilon > 0$ a $\delta > 0$ can be found such that

$$|P|<\delta \Rightarrow L(P,f) > \underline{\int_a^b} f(x)\,dx - \varepsilon$$

Proof.

By definition, we can find an n- point partition $Q=\{x_i\}_{i=0}^n$ such that $L(Q,f) > \underline{\int_a^b} f(x)dx - \varepsilon/2$. Consider an arbitrary m- point partition $P=\{y_i\}_{i=0}^m$ with norm less than some given $\delta > 0$. The union partition $R = P \cup Q$ which has no more than $m+n-2$ mesh points satisfies

$$L(P,f), L(Q,f) \le L(R,f)$$

and consequently

$$L(R,f) > \underline{\int_a^b} f(x)dx - \varepsilon/2 \tag{6.1.3}$$

Given two consecutive mesh points y_{i-1}, y_i of P there are two possibilities: (a) The subinterval $[y_{i-1}, y_i]$ has no points of Q. In this case the term $m_i(y_i - y_{i-1})$ appears in both $L(P,f)$ and $L(R,f)$. (b) There are k points $z_1, z_2, \ldots, z_k$ $(k \ge 1)$ between y_{i-1} and y_i which belong to Q and to R but not to P, i.e., $y_{i-1} < z_1 < z_2 < \cdots < z_k < y_i$. Denote $z_0 = y_{i-1}, z_{k+1} = y_i$. Then, the contribution of $[y_{i-1}, y_i]$ to $L(R,f)$ is

$$A_i = m_{i,1}(z_1 - z_0) + m_{i,2}(z_2 - z_1) + \ldots + m_{i,k+1}(z_{k+1} - z_k)$$

where $m_{i,j} = \inf[f(x)],\ z_{i-1} \le x \le z_i$ satisfies $m_i \le m_{i,j}\ , 1 \le j \le k+1$. The difference between A_i and the term $m_i(y_i - y_{i-1})$ is

$$B_i = (m_{i,1} - m_i)(z_1 - z_0) + \ldots + (m_{i,k+1} - m_i)(z_{k+1} - z_k)$$

and consequently $|B_i| \le 2M(y_i - y_{i-1}) \le 2M\delta$ where M denotes any bound of $|f(x)|$ over $[a,b]$. Since the number of such subintervals in P (with interim points in Q) is no more than m (why?) we conclude that

$$L(P,f) \ge L(R,f) - 2mM\delta \tag{6.1.4}$$

The proof is completed by choosing $\delta = \varepsilon/(4mM)$, and applying Eq. (6.1.3).

The next result can be similarly proved.

Theorem 6.1.2. For arbitrary $\varepsilon > 0$ a $\delta > 0$ can be found such that

$$|P| < \delta \Rightarrow U(P,f) < \int_a^b f(x)\,dx + \varepsilon$$

Thus, the lower and upper Darboux sums indeed converge to the lower and upper Darboux integrals as $\max_{1 \le i \le n}\{|x_i - x_{i-1}|\} \to 0$.

Consider an arbitrary bounded function $f(x)$ and a partition $f(x)$ of $[a,b]$. Let ξ_i denote an arbitrary point within $[x_{i-1}, x_i]$ and generate the *Riemann sum*

$$S_n = \sum_{i=1}^{n} f(\xi_i)\Delta x_i \tag{6.1.5}$$

which depends on the particular P.

Definition 6.1.2. If $\lim S_n = S$ as $\max_{1 \prec i \le n}(x_i - x_{i-1}) \to 0$ the function $f(x)$ is said to be *Riemann integrable*, having the *definite Riemann integral* S which is often denoted by $(R)\int_a^b f(x)\,dx$. In this expression $f(x)$ is called the *integrand* while a,b are the *lower and upper limits* of the integral respectively. The symbol dx is a reminder of Δx and of the fact that the *integration* is with respect to the variable x.

Another form of Definition 6.1.2 is: If for arbitrary $\varepsilon > 0$, a $\delta > 0$ can be found such that $|J - S_n| < \varepsilon$ for *all partitions* with norm less than δ and *all choices* of interim points $\xi_i : x_{i-1} \le \xi_i \le x_i$, $1 \le i \le n$, $f(x)$ is Riemann integrable and its Riemann integral is J.

Since $L(P,f) < S_n < U(P,f)$ for arbitrary $f(x)$, we clearly get

$$\underline{\int_{a}^{b}} f(x)dx \le S_n \le \overline{\int_{a}^{b}} f(x)dx$$

Therefore, if the Riemann integral exists, it must satisfy

$$\underline{\int_{a}^{b}} f(x)dx \le (R)\int_{a}^{b} f(x)dx \le \overline{\int_{a}^{b}} f(x)dx \tag{6.1.6}$$

We thus concluded two different definitions for the *integral*, but apparently, if they exist they provide the same value, as stated and shown next.

Theorem 6.1.3. Let $f(x)$ denote a bounded function over a bounded closed interval $[a,b]$. Then, its Darboux integral exists if and only its Riemann integral exist as well. Furthermore, if the two integrals exist, then

$$(R)\int_{a}^{b} f(x)dx = (D)\int_{a}^{b} f(x)dx \tag{6.1.7}$$

Proof.

(a) Assume $f(x)$ is Darboux integrable and let $\varepsilon > 0$. Denote $I = (D)\int_{a}^{b} f(x)dx$.

Then, by virtue of Theorem 6.1.1, a $\delta_1 > 0$ exists such that

$$\sum_{i=1}^{n} m_i \Delta x_i > I - \frac{\varepsilon}{2}$$

for any partition $P = \{x_i\}_{i=0}^{n}$ with norm less than δ_1. Similarly, by Theorem 6.1.2, a $\delta_2 > 0$ exists such that

$$\sum_{i=1}^{n} M_i \Delta x_i < I + \frac{\varepsilon}{2}$$

provided that $|P| < \delta_2$. Consequently,

$$\left| \sum_{i=1}^{n} M_i \Delta x_i - \sum_{i=1}^{n} m_i \Delta x_i \right| < \varepsilon$$

for arbitrary P with norm less than $\delta = \min(\delta_1, \delta_2)$. However, for such partition and for any particular choice of interim points $\xi_i : x_{i-1} \le \xi_i \le x_i \,, 1 \le i \le n$, the Riemann sum of Eq. (6.1.5) satisfies

$$\sum_{i=1}^{n} m_i \Delta x_i \le S_n \le \sum_{i=1}^{n} M_i \Delta x_i$$

and therefore also yields $|S_n - I| < \varepsilon$, i.e., $f(x)$ is Riemann integrable and its Riemann integral equals its Darboux integral I.

(b) Let $f(x)$ be Riemann integrable over $[a,b]$, denote its Riemann integral by J and let $\varepsilon > 0$. By Definition 6.1.2, a $\delta > 0$ can be found, such that for any partition with norm less than δ and arbitrary choice of $\xi_i : x_{i-1} \le \xi_i \le x_i \,, 1 \le i \le n$, the inequality

$$J - \varepsilon < S_n < J + \varepsilon \tag{6.1.8}$$

where S_n is the Riemann sum of Eq. (6.1.5), holds. Let us choose, for such partition P, a set of interim points $K_1 = \{\xi_i, 1 \le i \le n\}$ such that

$$f(\xi_i) \le m_i + \frac{\varepsilon}{(b-a)} \,, 1 \le i \le n$$

This yields

$$S_n < L(P, f) + \varepsilon$$

and by applying Eq. (6.1.8) we get $J - \varepsilon < S_n < L(P, f) + \varepsilon$ which leads to $J - 2\varepsilon < L(P, f)$. Similarly, by choosing another set of interim points, we can show $J + 2\varepsilon > U(P, f)$. Therefore,

$$U(P, f) < J + 2\varepsilon < (L(P, f) + 2\varepsilon + 2\varepsilon) = L(P, f) + 4\varepsilon$$

which yields $\inf_P [U(P, f)] < \sup_P [L(P, f)] + 4\varepsilon$. Since this holds for all $\varepsilon > 0$ we get $\inf_P [U(P, f)] \le \sup_P [L(P, f)]$. However, we always have $\inf_P [U(P, f)] \ge \sup_P [L(P, f)]$ which yields $\inf_P [U(P, f)] = \sup_P [L(P, f)]$, i.e., the Darboux integral exists and by Eq. (6.1.6) it must equal the Riemann integral of $f(x)$.

Example 6.1.3. Let $f(x)=x^2, 0 \le x \le 2$. Consider an evenly spaced partition with n subintervals. This lower and upper Darboux sums are

$$L(P,f)=h[0^2+h^2+\ldots+(2-h)^2] \quad , \quad U(P,f)=h[h^2+\ldots+(2-h)^2+2^2]$$

where $h=2/n$ and $U(P,f)-L(P,f)=4h=8/n$. Therefore, $f(x)$ is integrable (by both Darboux and Riemann) (why?) and its integral is (for example)

$$\int_0^2 x^2 dx = \lim_{h\to 0} L(P,f) = \lim_{n\to\infty}\left\{\frac{2}{n}\left(\frac{2}{n}\right)^2\left(1^2+2^2+\ldots+(n-1)^2\right)\right\}=\frac{8}{3}$$

Deriving the last limit is a simple algebraic result and is left for the reader.

Example 6.1.4. Define $f(x)$ as follows:

$$f(x)=\begin{cases} 0\,, x \textit{ rational}\,, 0\le x\le 1 \\ 1\,, x \textit{ irrational}\,, 0\le x\le 1 \end{cases}$$

Since at every subinterval of $[0,1]$ one can find infinite rational and irrational numbers, we clearly get

$$\underline{\int_0^1} f(x)dx = 0 \quad , \quad \overline{\int_0^1} f(x)dx = 1$$

Thus, both the Darboux and the Riemann integrals of $f(x)$ do not exist. The lower Darboux integral is 0, while the upper Darboux integral equals to 1.

We close this section with a result that can be applied as an *integrability test* and whose proof is left as an exercise for the reader.

Theorem 6.1.4. A function $f(x)$ is integrable if and only if for an arbitrary $\varepsilon>0$ a partition P of the interval $[a,b]$ can be found such that $U(P,f)-L(P,f)<\varepsilon$.

PROBLEMS

1. (a) Show that the Riemann integral of the step function

$$f(x)=\begin{cases} 1\,, 0\le x\le 1 \\ 2\,, 1< x\le 3 \end{cases}$$

exists and calculate its value.

(b) Redefine $f(1)=2,3,10$. Does the integral still exist? Does its value change?

2. Let $f(x)=1/x\,,1\le x\le 2$ and consider an evenly spaced partition $P=\{x_i\}_{i=0}^{n}$ of the interval $[1,2]$ with $n=10$. Calculate $L(P,f)\,,U(P,f)$.
3. In the previous example calculate $U(P,f)-L(P,f)$ for arbitrary n. Does the Darboux integral exist? What about the Riemann integral of $f(x)$?
4. In Example 6.1.4 redefine

$$f(x)=\begin{cases}1\,,\ x\ rational\,,\ 0\le x\le 1\\ 0\,,\ x\ irrational\,,\ 0\le x\le 1\end{cases}$$

Does the integral of $f(x)$ exist?

5. Show that

$$f(x)=\begin{cases}\sin(1/x)\,,\ 0< x\le 1\\ 0\,,\ x=0\end{cases}$$

is not integrable over $[0,1]$.

6. Let $a=x_0<x_1<\ldots<x_n=b$ and consider the function

$$f(x)=c_i\,,\ x_{i-1}<x<x_i\ ;0\le i\le n$$

Show that $f(x)$ is Riemann integrable, calculate $\int_a^b f(x)dx$ and show that it does not depend on the function's values at the nodes $x_i\,,0\le i\le n$.

6.2 Integrable Functions

The set of all integrable functions over a given bounded interval $I=[a,b]$ does not include all bounded functions defined over I as seen from Example 6.1.4. However, this set is apparently far larger than other familiar sets such as the set of all differentiable or continuous functions. This section provides basic results which establish several *sufficient conditions for integrability*.

Theorem 6.2.1. A continuous function $f(x)$ over a bounded interval $I=[a,b]$ is Riemann integrable.

Proof.

Since $f(x)$ is also uniformly continuous, given $\varepsilon > 0$ one can find a $\delta > 0$ such that

$$|t_1 - t_2| < \delta, a \le t_1, t_2 \le b \Rightarrow |f(t_1) - f(t_2)| < \varepsilon/(b-a)$$

Now, consider any partition P with norm less than δ. The continuity of $f(x)$ guarantees the existence of y_i, z_i within any subinterval $[x_{i-1}, x_i]$ of P, such that

$$m_i = f(y_i)\,,\, M_i = f(z_i)$$

which leads to

$$U(P,f) - L(P,f) = \sum_{i=1}^{n} [f(z_i) - f(y_i)]\Delta x_i < \frac{\varepsilon}{(b-a)} \sum_{i=1}^{n} \Delta x_i = \varepsilon$$

and since this true for all $\varepsilon > 0$, $f(x)$ is Riemann integrable which completes the proof.

Example 6.2.1. The functions x^4, e^x, $\sin(x)$ and $1/(1+x^2)$ are all Riemann integrable. So is the function

$$f(x) = \begin{cases} x\sin(1/x)\,,\, 0 < x \le 3 \\ 0\,,\, x = 0 \end{cases}$$

which is continuous over the closed interval $[0,3]$.

Note that discontinuity of $f(x)$ at a single point, a finite number of points or sometimes even at an infinite number of points, does not rule out the possibility that $f(x)$ is integrable. For example, let x_0 denote a positive integer between 0 and 1 and define define

$$f(x) = \begin{cases} 1\,,\, 0 \le x \le 1\,,\, x \ne x_0 \\ 0\,,\, x = x_0 \end{cases}$$

This function has a single discontinuity at $x = x_0$, yet it is integrable. The proof is left for the student.

Corollary 6.2.1. An arbitrary differentiable function over a bounded interval is also integrable.

The proof is simple: a differentiable function is also continuous and hence integrable.

Theorem 6.2.2. Let $f(x)$ denote a monotone bounded function over a bounded interval $I = [a,b]$. Then $f(x)$ is integrable over I.

Proof.

Consider a monotone increasing $f(x)$ and let $\varepsilon > 0$. If $\max\limits_{1 \le i \le n} \Delta x_i < \delta$ then

$$\sum_{i=1}^{n}(M_i - m_i)(x_i - x_{i-1}) = \sum_{i=1}^{n}[f(x_i) - f(x_{i-1})]\Delta x_i \le \delta \sum_{i=1}^{n}[f(x_i) - f(x_{i-1})]$$

$$\le \delta[f(b) - f(a)]$$

and the choice $\delta < \varepsilon / [f(b) - f(a)]$ guarantees $U(P,f) - L(P,f) < \varepsilon$, i.e., $f(x)$ is integrable. The proof for a monotone decreasing function is similar and is left for the reader.

Example 6.2.2. The function

$$f(x) = \begin{cases} x \,, 0 \le x \le 1 \\ \sqrt{x+1} \,, 1 < x \le 3 \end{cases}$$

is monotone increasing over the interval $[0,3]$ and therefore integrable. Thus, the jump at $x = 1$ does not affect the function's integrability.

Example 6.2.3. The integral of

$$f(x) = \begin{cases} \sqrt[4]{x}\cos(1/x) \,, 0 < x \le 1 \\ 0 \,, x = 0 \end{cases}$$

exists, since the function is continuous everywhere, including at $x = 0$.

PROBLEMS

1. Why are the following functions Riemann integrable:

a. $f(x) = \dfrac{1+\sin(x)}{0.1+x^2}, -1 \le x \le 2$ (b) $f(x) = |x| + x, -2 \le x \le 5$

b. $f(x)=\begin{cases} x^2\,,0\le x\le 1 \\ 2\,,1<x\le 3 \\ 2+\sqrt{x}\,,3<x\le 7 \end{cases}$

2. Show that

$$f(x)=\begin{cases} 0\,,0\le x\le 1\,,\,x\ne 1/2,1/3 \\ 1\,,\,x=1/2,1/3 \end{cases}$$

is Riemann integrable.

3. Define

$$f(x)=\begin{cases} 1/2^n\,,1/2^{n+1}<x\le 1/2^n\;;n=0,1,2,\ldots \\ 0\,,\,x=0 \end{cases}$$

and show the integrability of this function.

4. Let $f(x)$ be a bounded integrable function over a bounded interval $[a,b]$. Show that by changing the function's definition at a single point, keeping it bounded, the integral still exists and maintains the same value. Can you extend the result for any finite set of changes.
5. Is the function

$$f(x)=\begin{cases} \sin\left(\dfrac{1}{x}\right),0<x\le\pi \\ \\ 0\,,\,x=0 \end{cases}$$

Riemann integrable? Why?

6.3 Basic Properties of the Riemann Integral

Theorem 6.3.1. Let $f(x),g(x)$ denote bounded Riemann integrable functions over a bounded interval $[a,b]$. Then

$$\int_a^b [f(x)\pm g(x)]dx=\int_a^b f(x)dx\pm\int_a^b g(x)dx \tag{6.3.1}$$

Proof.

Let $\varepsilon > 0$. Since $f(x)$ is Riemann integrable, there exists a $\delta_1 > 0$ such that

$$\left|\sum_{i=1}^{n} f(\xi_i)\Delta x_i - \int_a^b f(x)dx\right| < \frac{\varepsilon}{2}$$

for arbitrary partition $P = \{x_i\}_{i=0}^{n}$ and provided that $\max_{1 \le i \le n} \Delta x_i < \delta_1$. Similarly, a $\delta_2 > 0$ can be found such that

$$\left|\sum_{i=1}^{n} g(\xi_i)\Delta x_i - \int_a^b g(x)dx\right| < \frac{\varepsilon}{2}$$

for any partition and arbitrary interim points which satisfy $\max_{1 \le i \le n} \Delta x_i < \delta_2$. Thus, if we confine ourselves to partitions $P = \{x_i\}_{i=0}^{n}$ with $\max_{1 \le i \le n} \Delta x_i < \delta = \min(\delta_1, \delta_2)$, we get for arbitrary choice of $\xi_i : x_{i-1} \le \xi_i \le x_i$, $1 \le i \le n$,

$$\left|\sum_{i=1}^{n} [f(\xi_i) + g(\xi_i)]\Delta x_i - \int_a^b f(x)dx - \int_a^b g(x)dx\right| \le \left|\sum_{i=1}^{n} f(\xi_i)\Delta x_i - \int_a^b f(x)dx\right|$$

$$+\left|\sum_{i=1}^{n} g(\xi_i)\Delta x_i - \int_a^b g(x)dx\right| < \varepsilon$$

Therefore, the Riemann integral of $f(x) + g(x)$ exists and equals the sum of the single integrals as stated. The second part, i.e., showing the existence of the integral of $f(x) - g(x)$ and calculating it, is similarly done.

Theorem 6.3.2. Let $f(x)$ denote bounded Riemann integrable functions over a bounded interval $[a,b]$. Then, for arbitrary constant c

$$\int_a^b [cf(x)]\,dx = c\int_a^b f(x)\,dx \tag{6.3.2}$$

Proof.

For any partition $P=\{x_i\}_{i=0}^n$ with arbitrary interim points $\xi_i : x_{i-1} \le \xi_i \le x_i$, $1 \le i \le n$ we have

$$\sum_{i=1}^{n} cf(\xi_i)\Delta x_i = c\sum_{i=1}^{n} f(\xi_i)\Delta x_i$$

Consider an arbitrary $\varepsilon > 0$. Since $f(x)$ is integrable, a $\delta > 0$ can be found such that

$$\left| \sum_{i=1}^{n} f(\xi_i)\Delta x_i - \int_a^b f(x)\,dx < \frac{\varepsilon}{c} \right|$$

provided that $\max\limits_{1\le i\le n} \Delta x_i < \delta$. Consequently,

$$\left| \sum_{i=1}^{n} cf(\xi_i)\Delta x_i - c\int_a^b f(x)\,dx < \varepsilon \right| \quad , \quad \max_{1\le i\le n} \Delta x_i < \delta$$

which yields that the integral of $cf(x)$ *exists* and equals to $c\int_a^b f(x)\,dx$.

Theorem 6.3.3. Let $f(x)$, $g(x)$ denote bounded Riemann integrable functions over a bounded interval $[a,b]$ such that $f(x) \le g(x)$, $a \le x \le b$. Then

$$\int_a^b f(x)dx \le \int_a^b g(x)dx \tag{6.3.3}$$

Proof.

For an arbitrary partition $P=\{x_i\}_{i=0}^n$ let m_i, m_i^* denote $\inf[f(x)]$, $x_{i-1} \le x \le x_i$ and $\inf[g(x)]$, $x_{i-1} \le x \le x_i$ respectively. Assume $m_i > m_i^*$, i.e., $m_i = m_i^* + \varepsilon_0$ for some $\varepsilon_0 > 0$. Let $\xi : x_{i-1} \le \xi \le x_i$ be such that $g(\xi) < m_i^* + \varepsilon_0$. This leads to contradiction since $f(\xi) \le g(\xi) < m_i^* + \varepsilon_0 = m_i$. Consequently, $m_i \le m_i^*$ which clearly yields

$$\overline{\int_a^b} f(x)dx \le \overline{\int_a^b} g(x)dx$$

Since both functions are Riemann integrable, we finally get

$$\int_a^b f(x)dx \le \int_a^b g(x)dx$$

which completes the proof.

Theorem 6.3.4. For arbitrary integrable function $f(x)$ the integral of $|f(x)|$ exists and

$$\left|\int_a^b f(x)dx\right| \le \int_a^b |f(x)|dx \tag{6.3.4}$$

Proof.

First we must show the integrability of $|f(x)|$. Let

$$f_+(x) = \begin{cases} f(x)\,,\ f(x) \ge 0 \\ 0\,,\ f(x) < 0 \end{cases}\,,\quad f_-(x) = \begin{cases} 0\,,\ f(x) > 0 \\ f(x)\,,\ f(x) \le 0 \end{cases}$$

Clearly, $f(x) = f_+(x) + f_-(x)$ and $|f(x)| = f_+(x) - f_-(x)$. Consider a partition $P = \{x_i\}_{i=0}^n$ and define

$$m_i^{\pm} = \inf[f_{\pm}(x)]\,,\ M_i^{\pm} = \sup[f_{\pm}(x)]\,;\ x_{i-1} \le x \le x_i$$

The integrability of $f_+(x)$ is established as follows. If $f_+(\xi) > 0$ for some $\xi \in I_n$ then $M_i^+ = M_i$. Obviously $m_i^+ \ge m_i$ which yields $M_i^+ - m_i^+ \le M_i - m_i$. However, this inequality holds even if such $\xi \in I_n$ does not exist, since then $f_+(x) = 0$, $x_{i-1} \le x \le x_i$ and consequently $m_i^+ = M_i^+ = 0$ which yields $0 = M_i^+ - m_i^+ \le M_i - m_i$. Generating the lower and upper sums we obtain

$$U(P, f_+) - L(P, f_+) \le U(P, f) - L(P, f)$$

and since $f(x)$ is integrable, so is $f_+(x)$. The relation $f_-(x) = f(x) - f_+(x)$ and Theorem 6.3.1 yield that $f_-(x)$ is integrable as well and the relation $|f(x)| = f_+(x) - f_-(x)$ combined with Theorem 6.3.1 establishes the integrability of $|f(x)|$. Since $\pm f(x) \le |f(x)|$, we may apply Theorem 6.3.3 and get the inequalities

$$\pm \int_a^b f(x)dx \le \int_a^b |f(x)|dx$$

which complete the proof.

Theorem 6.3.5. Let $f(x)$, $g(x)$ denote bounded Riemann integrable functions over a bounded interval $[a,b]$. Then, the Riemann integral of $h(x) = f(x)g(x)$ exists.

Proof.

Unlike in the Theorems 6.3.1 and 6.3.2, we do not have a simple specific value for the integral of $h(x)$, which is solely based on the integrals of $f(x)$ and $g(x)$. Therefore, the only reasonable way to show its existence is to compare between the lower and upper sums of $h(x)$ for an arbitrary partition $P = \{x_i\}_{i=0}^n$. Let

$$m_i = \inf[f(x)], M_i = \sup[f(x)]; x_{i-1} \le x \le x_i$$

$$m_i^* = \inf[g(x)], M_i^* = \sup[g(x)]; x_{i-1} \le x \le x_i \tag{6.3.5}$$

$$m_i^{**} = \inf[h(x)], M_i^{**} = \sup[h(x)]; x_{i-1} \le x \le x_i$$

We distinguish between two cases:

(a) $f(x) \ge 0$, $g(x) \ge 0$. Denote $M = \sup[f(x)]$, $M^* = \sup[g(x)]; a \le x \le b$. Then

$$L(P,h) = \sum_{i=1}^n m_i^{**} \Delta x_i \ge \sum_{i=1}^n m_i m_i^* \Delta x_i$$

$$U(P,h) = \sum_{i=1}^n M_i^{**} \Delta x_i \le \sum_{i=1}^n M_i M_i^* \Delta x_i$$

leading to

$$U(P,h)-L(P,h)\le\sum_{i=1}^{n}(M_iM_i^*-m_im_i^*)\Delta x_i$$

$$=\sum_{i=1}^{n}M_i(M_i^*-m_i^*)\Delta x_i+\sum_{i=1}^{n}m_i^*(M_i-m_i)\Delta x_i$$

$$\le M\cdot[U(P,g)-L(P,g)]+M^*\cdot[U(P,f)-L(P,f)]$$

Given an $\varepsilon>0$ we can find a $\delta_1>0$ and $\delta_2>0$ such that

$$U(P,g)-L(P,g)<\varepsilon/(2M)\,,\ |P|<\delta_1$$

$$U(P,f)-L(P,f)<\varepsilon/(2M^*)\,,\ |P|<\delta_2$$

and thus $U(P,h)-L(P,h)<\varepsilon$ provided that $|P|<\delta=\min(\delta_1,\delta_2)$. This completes the proof of (a).

(b) General $f(x)$, $g(x)$. Find constants A,B such that $F(x)=f(x)+A\ge 0$ and $G(x)=g(x)+B\ge 0$. Since any constant is a Riemann integrable function, so are the functions $F(x)$ and $G(x)$ (why?). Also

$$f(x)g(x)=[F(x)-A][G(x)-B]=F(x)G(x)-AG(x)-BF(x)+AB$$

Thus, the left-hand side is obtained by adding and subtracting four Riemann integrable functions and must therefore be itself Riemann integrable.

Theorem 6.3.6. Let $f(x)$, $g(x)$ denote bounded Riemann integrable functions over a bounded interval $[a,b]$ and let $\inf_{a\le x\le b}[g(x)]>0$. Then, the function $h(x)=f(x)/g(x)$ is also Riemann integrable over $[a,b]$.

Proof.

The proof is left as an exercise for the reader.

Theorem 6.3.7. If $f(x)$ is integrable over $[a,b]$, then it is integrable over any subinterval of $[a,b]$.

Proof.

Let $a \le c < d \le b$. For clarity let the lower and upper sums defined by Eq. (6.1.2) be re-denoted by $L(P,f,\alpha,\beta)$, $U(P,f,\alpha,\beta)$ respectively, where $[\alpha,\beta]$ is any subinterval of $[a,b]$. We need to show that given an arbitrary $\varepsilon > 0$, a $\delta > 0$ can be found such that for any partition $P = \{x_i\}_{i=0}^{n}$ with $|P| < \delta$,

$$U(P,f;c,d) - L(P,f;c,d) < \varepsilon$$

Since $f(x)$ is integrable over $[a,b]$, we can find a $\delta_1 > 0$ such that for any partition $Q = \{y_i\}_{i=0}^{m}$ of $[a,b]$ with $|Q| < \delta_1$, the inequality

$$U(Q,f;a,b) - L(Q,f;a,b) < \varepsilon$$

holds. Let $P = \{x_i\}_{i=0}^{n}$ denote a partition of $[c,d]$ with $|P| < \delta_1$ and extend it to any partition P' over $[a,b]$, provided that $|P'| < \delta_1$. The special choice of δ_1 guarantees

$$U(P',f;a,b) - L(P',f;a,b) < \varepsilon$$

Also,

$$U(P,f;c,d) - L(P,f;c,d) \le U(P',f;a,b) - L(P',f;a,b)$$

since P is included in P'. Therefore $U(P,f;c,d) - L(P,f;c,d) < \varepsilon$ for any partition with norm less than δ_1 and the proof is completed.

Theorem 6.3.8. If $f(x)$ is integrable over $[a,b]$ and over $[b,c]$, then it is integrable over $[a,c]$ and

$$\int_a^c f(x)dx = \int_a^b f(x)dx + \int_b^c f(x)dx \tag{6.3.6}$$

Proof.

By Theorem 6.1.4 there is a partition P of $[a,b]$ such that

$$U(P,f;a,b) - L(P,f;a,b) < \frac{\varepsilon}{2}$$

and another partition Q of $[b,c]$ such that

$$U(Q,f;b,c)-L(Q,f;b,c)<\frac{\varepsilon}{2}$$

The partition $R=P\cup Q$ is clearly a partition of $[a,c]$ such that

$$L(R,f;a,c)=L(P,f;a,b)+L(Q,f;b,c)$$
$$U(R,f;a,c)=U(P,f;a,b)+U(Q,f;b,c)$$

which consequently yields

$$U(R,f;a,c)-L(R,f;a,c)<\varepsilon$$

and by Theorem 6.1.4 the proof is concluded.

We have previously established that all continuous functions $C[a,b]$ and all monotone functions $M[a,b]$ are also integrable (Theorems 6.2.1 and 6.2.2), i.e., belong to $R[a,b]$ - the set of all Riemann integrable functions over $[a,b]$. Another important subset of $R[a,b]$ is the class $V[a,b]$ of all functions with bounded variation (see 4.5).

Theorem 6.3.9. Let $f(x)$ denote a function with bounded variation over the interval $[a,b]$. Then $f(x)$ is Riemann integrable.

Proof.

By virtue of Theorem 4.4.4 $f(x)$ can be represented as a difference between two monotone functions. We now apply Theorems 6.2.2 and 6.3.1 and conclude that $f(x)$ is Riemann integrable.

We close this section by extending the definition of the Riemann integral to include the case $b\le a$ and to simplify the proof of the fundamental theorem of calculus (see below).

Definition 6.3.1. Let $f(x)$ denote a Riemann integrable function over $[a,b]$ where $a<b$. We define

$$\int_b^a f(x)dx=-\int_a^b f(x)dx\,,\;\int_a^a f(x)dx=0 \tag{6.3.7}$$

Corollary 6.3.1. Theorem 6.3.8 holds for arbitrary a,b,c. The proof, based on Eq. (6.3.6), is left for the reader.

PROBLEMS

1. Let $f(x)$ denote an integrable function over the interval $[a,b]$ and let

$$a = x_0 < x_1 < \cdots < x_n = b$$

   Use induction to show

$$\int_a^b f(x)dx = \sum_{i=1}^{n} \int_{x_{i-1}}^{x_i} f(x)dx$$

2. Consider a continuous function $f(x)$ and an increasing monotone function $g(x)$ over $[a,b]$ and let $g(a) = 0.001$. Are the functions $f(x)+g(x)$, $f(x)g(x)$ and $f(x)/g(x)$ integrable?
3. Let $f_1(x), f_2(x), \ldots, f_n(x)$ be integrable over the interval $[a,b]$. Show that the function $F(x) = \prod_{i=1}^{n} f_i(x)$ is also integrable.
4. Show that the integral of $f(x) = \dfrac{e^x}{1-x+x^2}$, $-1 \le x \le 4$ exists.
5. Show that $\int_0^{\pi} \dfrac{\cos(x)dx}{10^{-6}+x}$ exists.

6.4 The Fundamental Theorem of Calculus

Let $f(x)$ be integrable over the interval $[a,b]$. By virtue of Theorem 6.3.5 it is integrable over any subinterval $[a,x]$, $a \le x \le b$. We define

$$F(x) = \int_a^x f(t)dt \tag{6.4.1}$$

as the Riemann integral of the function $f(t)$ over the interval $[a,x]$. This integral which is a function of x is called an *indefinite integral* of $f(t)$. Since we chose the letter x as the independent variable of the indefinite integral $F(x)$, we replaced the independent variable of the integrand in Eq. (6.4.1) by t. The basic features of the indefinite integral are given next.

Theorem 6.4.1. The function $F(x)$ defined by Eq. (6.4.1) is continuous over $[a,b]$.

Proof.

For any fixed $x_0 \in [a,b]$ and an arbitrary x, we may apply Corollary 6.3.1 and get

$$F(x) - F(x_0) = \int_{x_0}^{x} f(t)dt \tag{6.4.2}$$

Since $f(t)$ is bounded over $[a,b]$, there exists $M > 0$ such that $|f(t)| \le M$ for all t and consequently, by using the definition of the Riemann integral, we obtain

$$|F(x) - f(x_0)| \le M|x - x_0|$$

and the continuity of $F(x)$ is straightforward.

The differentiability of $F(x)$ is not guaranteed except in the case discussed next.

Theorem 6.4.2 (The fundamental theorem of calculus). If $f(t)$ is continuous at x_0 then $F(x)$ has a derivative at x_0 and

$$F'(x_0) = f(x_0) \tag{6.4.3}$$

Proof.

Using the standard terminology, we need to show

$$\lim_{h \to 0} = \frac{F(x_0 + h) - F(x_0)}{h} = f(x_0)$$

i.e., for arbitrary $\varepsilon > 0$ there is a $\delta > 0$ such that

$$\left| \frac{F(x_0 + h) - F(x_0)}{h} - f(x_0) \right| < \varepsilon$$

provided that $|h| < \delta$. By Example 6.1.3 $C = \frac{1}{b-a}\int_a^b C dt$ for an arbitrary constant C and hence, by virtue of Theorem 6.3.5,

$$\left|\frac{F(x_0+h)-F(x_0)}{h} - f(x_0)\right| = \left|\frac{1}{h}\int_{x_0}^{x_0+h} f(t)dt - f(x_0)\right| = \left|\frac{1}{h}\int_{x_0}^{x_0+h} [f(t)-f(x_0)]dt\right|$$

$$\le \frac{1}{h}\int_{x_0}^{x_0+h} | f(t)-f(x_0) | dt$$

The continuity of f at x_0 guarantees the existence of $\delta > 0$ such that

$$|t - x_0| < \delta \Rightarrow |f(t) - f(x_0)| < \frac{\varepsilon}{2}$$

Thus, by applying Theorem 6.3.4, if $|h| < \delta$ then

$$\left|\frac{F(x_0+h)-F(x_0)}{h} - f(x_0)\right| \le \frac{1}{h}\cdot h \cdot \frac{\varepsilon}{2} < \varepsilon$$

which completes the proof.

Theorem 6.4.2 provides us with a tool for calculating integrals. Recall the first mean-value theorem (Theorem 5.2.7) which stated that an arbitrary differentiable function $f(x), a \le x \le b$ satisfies $f(b) - f(a) = (b-a)f'(\xi)$ for some interim point $\xi : a < \xi < b$. An immediate consequence is that if $f'(x) = 0, a \le x \le b$, then $f(x)$ is constant over the whole interval (the proof is left as an exercise for the student). Now, let

$$F(x) = \int_a^x f(t)dt$$

where $f(t)$ is continuous everywhere, and assume the existence of some $G(x), a \le x \le b$, called a *primitive* of $f(x)$, which satisfies $G'(x) = f(x), a \le x \le b$. By Theorem 6.4.2 $F'(x) = f(x)$ and therefore $(F(x) - G(x))' = f(x) - f(x) = 0$. Consequently

$$F(x) - G(x) = C \ , a \le x \le b$$

where C is a constant. By substituting $x=a$ we get $C=-G(a)$ and since $F(a)=0$ obtain $F(x)=G(x)-G(a)$. Thus, in order to get the integral $F(x)$, it is sufficient to find *any* $G(x)$ with the property $G'(x)=f(x)$ and calculate $G(x)-G(a)$.

Example 6.4.1. Let

$$F(x)=\int_1^x t^n dt$$

Since the function $G(x)=x^{n+1}/(n+1)$ satisfies $G'(x)=x^n$ everywhere, we obtain

$$F(x)=G(x)-G(a)=\frac{x^{n+1}-1}{n+1}$$

Example 6.4.2. Let

$$F(x)=\int_0^x \cos(t)dt$$

The integrand is a derivative of $G(x)=\sin(x)$ and therefore $F(x)=\sin(x)$ since here $G(0)=0$.

If for some $f(x)$, a method to obtain $G(x)$ cannot be found, we have no choice except approximating $F(x)$ by taking a fine partition and calculating a lower and upper sums, such that the difference between them is below a prefixed degree of accuracy.

We usually denote an arbitrary primitive by $F(x)$ and express the definite integral as

$$\int_a^b f(x)dx=[F(x)]_a^b=F(x)\Big|_a^b=F(b)-F(a)$$

Since $F(x)+C$ is also primitive for an arbitrary constant C, we often write $\int f(x)dx=F(x)+C$ as a "definition" of the indefinite integral, when in fact it is simply the set of all primitive functions of $f(x)$.

The length of a curve

We previously introduced the definite integral as mathematical modeling of the concept of area. We close this section by showing that it can also be used for modeling another concept: the *length* of a curve. For the sake of simplicity we restrict ourselves to the curve created by a single differentiable function $y = f(x)$, $a \le x \le b$ such that $f'(x)$ is continuous.

Let $P: a = x_0 < x_1 < \ldots < x_n = b$ denote an arbitrary partition of the interval $[a,b]$. Let $\Delta x_k = x_k - x_{k-1}$, $\Delta y_k = y_k - y_{k-1}$; $1 \le k \le n$ where $y_k = f(x_k)$, $0 \le k \le n$. Clearly, by using the Pythagorean theorem, the sum

$$S_P = \sum_{k=1}^{n} \sqrt{\Delta x_k^2 + \Delta y_k^2} = \sum_{k=1}^{n} \sqrt{1 + \frac{\Delta y_k^2}{\Delta x_k^2}} \Delta x_k \tag{6.4.4}$$

can be easily accepted as an *approximation* to the "length" of the curve defined by the set of points $\{(x, f(x), a \le x \le b)\}$. The continuity of $f'(x)$ guarantees

$$\lim_{\max \Delta x_k, 1 \le k \le n} S_P = \int_a^b \sqrt{1 + [f'(x)]^2}\, dx \tag{6.4.5}$$

and the integral at the right-hand side of Eq. (6.4.5) is defined the length of the given curve.

Example 6.4.3. The length of $y = 2x$, $0 \le x \le 1$ is $\int_0^1 \sqrt{1 + 2^2}\, dx = \sqrt{5}$ while the length of $y = x^6$, $0 \le x \le 1$ is $\int_1^2 \sqrt{1 + 36x^{10}}\, dx$.

PROBLEMS

1. Find the following definite integrals:

(a) $\int_1^3 \frac{x^3}{2} dx$ (b) $\int_0^2 [x - \sin(x)] dx$ (c) $\int_1^2 \frac{1+x}{x} dx$

2. Find the following indefinite integrals:

(a) $\int_1^x \frac{1+t^2+t^4}{1+t^2}dt$ (b) $\int_1^x \frac{te^t+1}{t}dt$ (c) $\int_1^x [\sin(t+1)+\cos(t)+\frac{2}{t^2}]dt$

3. Let

$$f(x)=\begin{cases} x\,, 0\le x\le 1 \\ 2-x\,, 1<x\le 2 \end{cases}$$

Find a primitive function of $f(x)$ if there are any and calculate $\int_0^{3/2} f(x)dx$.

4. Find the definite integral $\int_0^{\pi} [\sin(5x)\cos(3x)]dx$.

5. Calculate (a) $\int_{-1}^{\sqrt{3}} \frac{dx}{1+x^2}$ (b) $\int_1^a (x+2)^7 dx$ (c) $\int_0^{\pi} x^3\cos(x^4)dx$

6. Find the primitive functions of (a) $xe^{(x^2)}$ (b) $\sin(x)\cos(x)$

6.5 The Mean-Value Theorems

Like in differential calculus there are several mean-value theorems for integrals which provide some information related to an integral's value without actually calculating it. The first result is a general mean-value theorem which is practically equivalent to the first mean-value theorem for derivatives (see Section 5.2).

Theorem 6.5.1. Let $f(x)$ denote a Riemann integrable function over the interval $[a,b]$ with $m=\inf_{a\le x\le b}[f(x)]\,, M=\sup_{a\le x\le b}[f(x)]$. Then, there exists $\lambda : m\le\lambda\le M$ such that

$$\int_a^b f(x)dx=\lambda(b-a) \tag{6.5.1}$$

Furthermore, if $f(x)$ is continuous then $\lambda = f(\xi)$ for some $\xi : a\le\xi\le b$.

Proof.

The double inequality

$$m(b-a) \le \int_a^b f(x)dx \le M(b-a)$$

clearly holds (why?) and therefore

$$m \le \frac{\int_a^b f(x)dx}{b-a} \le M$$

which provides the choice $\lambda = \int_a^b f(x)dx/(b-a)$. If $f(x)$ is continuous every value between m and M is obtained at least once by f, which completes the proof.

Theorem 6.5.2 (The first mean-value theorem for integrals). Consider two Riemann integrable functions $f(x), g(x); a \le x \le b$ such that $g(x) \ge 0$ everywhere and denote

$$m = \inf_{a \le x \le b} [f(x)], M = \sup_{a \le x \le b} [f(x)]$$

Then, there exists $\lambda : m \le \lambda \le M$ such that

$$\int_a^b f(x)g(x)dx = \lambda \int_a^b g(x)dx \tag{6.5.2}$$

Proof.

Quite clearly $mg(x) \le f(x)g(x) \le Mg(x)$, $a \le x \le b$ and by Theorem 6.3.3

$$m\int_a^b g(x)dx \le \int_a^b f(x)g(x)dx \le M\int_a^b g(x)dx$$

By dividing these inequalities by $\int_a^b g(x)dx$ we obtain

$$m \le \frac{\int_a^b f(x)g(x)dx}{\int_a^b g(x)dx} \le M$$

and the proof is completed by choosing

$$\lambda = \left(\int_a^b f(x)g(x)dx \right) \Big/ \left(\int_a^b g(x)dx \right)$$

The particular case $g(x) = 1$, $a \le x \le b$ provides the result of Theorem 6.5.1.

Corollary 6.5.1. If $f(x)$ is continuous, then λ of Eq. (6.5.1) can be replaced by $f(\xi)$ where ξ is some interim point within the interval.

Proof. The continuity of $f(x)$ guarantees the existence of $x_1, x_2 : a \le x_1, x_2 \le b$ such that $m = f(x_1)$, $M = f(x_2)$. Furthermore, for every $\alpha : m \le \alpha \le M$ there is a $\beta : x_1 \le \beta \le x_2$ such that $f(\beta) = \alpha$. In particular $f(\xi) = \lambda$ for some $\xi : x_1 \le \xi \le x_2$. This completes the proof.

Example 6.5.1. Let $f(x) = x^2$, $1 \le x \le 3$. Here

$$\int_1^3 x^2 dx = \frac{x^3}{3}\bigg|_1^3 = \frac{3^3}{3} - \frac{1^3}{3} = \frac{26}{3}$$

Since $m = 1$, $M = 9$ there exists $\lambda : 1 \le \lambda \le 9$ such that $(26/3) = \lambda(3-1)$. This yields $\lambda = 13/3$ which is obviously between 1 and 9. The continuity of $f(x)$ guarantees the existence of $\xi : 1 \le \xi \le 3$ such that $\xi^2 = 13/3$. This implies $\xi \approx 2.08$ which is indeed within the interval [1,3].

Example 6.5.2. Consider the integral $\int_0^2 (x-1)^2 x^3 dx$ where both $f(x) = (x-1)^2$ and $g(x) = x^3$ are integrable and $g(x) \ge 0$. The integral, using previous basic results, is easily calculated and equals to $28/15$ (this is left for the student to verify). Also

$$\int_0^2 x^3 dx = 4 \quad \text{(how?)}$$

and $m = \inf_{0 \le x \le 2} [(x-1)^2] = 0$, $M = \sup_{0 \le x \le 2} [(x-1)^2] = 1$. Therefore, by Theorem 6.5.2 there exists a $\lambda : 0 \le \lambda \le 1$ such that $28/15 = 4\lambda$. The last equation determines

$\lambda = 7/15$ which is indeed between 0 and 1. Also, since $f(x)$ is continuous, there exists a $\xi : 0 \le \xi \le 2$ such that $f(\xi) = 7/15$. This yields $\xi \approx 0.683$ which is between 0 and 2.

Theorem 6.5.3 (The first mean-value theorem for integrals). Let $f(x), g(x)$ denote two integrable functions over the interval $[a,b]$ such that $g(x) \ge 0$, $a \le x \le b$ and $f(x)$ is monotone. Then, there exists a $\xi : a \le \xi \le b$ which satisfies

$$\int_a^b f(x)g(x)dx = f(a)\int_a^{\xi} g(x)dx + f(b)\int_{\xi}^b g(x)dx \tag{6.5.3}$$

Proof.

Define

$$h(t) = f(a)\int_a^t g(x)dx + f(b)\int_t^b g(x)dx$$

By Theorem 6.4.1 both integrals at the right-hand side are continuous functions of t, i.e., $h(t)$, $a \le t \le b$ is continuous as well. Since $g(x) \ge 0$ everywhere, we may apply Theorem 6.5.2 and get $\int_a^b f(x)g(x)dx = \lambda \int_a^b g(x)dx$ for some λ between $m = \inf_{a \le x \le b} [f(x)]$ and $M = \sup_{a \le x \le b} [f(x)]$. However, $f(x)$ is monotone which implies that λ lies between $f(a)$ and $f(b)$. By substituting $t = a,b$ we obtain

$$h(a) = f(b)\int_a^b g(x)dx \, , \, h(b) = f(a)\int_a^b g(x)dx$$

and clearly the number $\lambda \int_a^b g(x)dx$ must lie between $h(a)$ and $h(b)$. Now, the continuity of $h(t)$ implies (see Theorem 4.3.5) that for an arbitrary value η between $h(a)$ and $h(b)$ there is a number $\xi : a \le \xi \le b$ such that $h(\xi) = \eta$. This is particularly true for $\eta = \lambda \int_a^b g(x)dx$ which yields

$$h(t) = f(a)\int_a^{\xi} g(x)dx + f(b)\int_{\xi}^{b} g(x)dx = \lambda\int_a^b g(x)dx = \int_a^b f(x)g(x)dx$$

and concludes the proof.

PROBLEMS

1. Consider a continuous function $f(x)$, $a \le x \le b$ which satisfies $\int_c^d f(x)dx = 0$ for all $a \le c < d \le b$. Show $f(x) = 0$, $a \le x \le b$.
2. Let $f(x) = e^x$, $0 \le x \le 2$. Apply Eq. (6.5.1), calculate λ and find x_0, $0 \le x_0 \le 2$ such that $e^{x_0} = \lambda$. Is x_0 unique?
3. Repeat and solve problem 2 for $f(x) = \sin(x)$, $0 \le x \le \pi$.
4. Apply Theorem 6.5.2 to $f(x) = \sin(x^2)$, $g(x) = x$, $0 \le x \le \pi$ and obtain λ.
5. Apply Theorem 6.5.3 to $f(x) = \sin(x)$, $g(x) = \cos(x)$, $0 \le x \le \pi$ and find ξ.
6. Repeat and solve problem 5 for $f(x) = \sin(x)$, $g(x) = \cos^2(x)$, $0 \le x \le \pi$.

6.6 Methods of Integration

Definition 6.6.1. An arbitrary function that can be obtained from polynomial, trigonometric, inverse trigonometric, exponential, and logarithm functions by a finite number of additions, subtractions, multiplications, divisions and compositions is called an *elementary function.*

Example 6.6.1. The functions $x^2\sin(x)$, $\dfrac{x+e^{-x}}{\tan(x)}$, $\sin(\tan(e^{x+5}))$ are all elementary functions. The third function is obtained by composition of trigonometric, exponential and polynomial functions.

It is clearly desirable to express an indefinite integral $\int_a^x f(t)dt$ as an *elementary function* $F(x)$, and obtain $\int_a^b f(t)dt = F(b) - F(a)$. Indeed, one then performs two simple calculations to get the *exact* result. Unfortunately, this is usually impossible. For example, it can be shown that the popular integral $\int_a^x e^{-t^2}dt$, which occurs in many physical problems cannot be presented as an elementary function. Consequently, in most cases, one must turn to *numerical methods* (see 10.6) which only *approximate* $\int_a^b f(t)dt$. Two methods that can sometimes be applied to obtain an indefinite integral as an elementary function are given below.

Integration by parts

Let $f(x)$, $g(x)$, $h(x)$ denote differentiable functions such that

$$h(x) = f(x)g'(x) \tag{6.6.1}$$

Since $(f(x)g(x))' = f'(x)g(x) + f(x)g'(x)$ we obtain

$$\int_a^x h(t)dt = \int_a^x (f(t)g(t))'dt - \int_a^x f'(t)g(t)dt$$

which leads to

$$\int_a^x h(t)dt = f(x)g(x) - f(a)g(a) - \int_a^x f'(t)g(t)dt \tag{6.6.2}$$

Thus, if the integral of $f'(x)g(x)$ is an elementary function, so is the integral of $h(x)$. The relation given by Eq. (6.6.2) is called *integration by parts* and is usually written as

$$\int f(x)g'(x)dx = f(x)g(x) - \int f'(x)g(x)dx + C \tag{6.6.3}$$

where C is an arbitrary constant.

Example 6.6.2. Consider the integral $\int xe^x dx$. By writing $f(x) = x$, $g'(x) = e^x$ and applying Eq. (6.6.3) one gets

$$\int xe^x dx = xe^x - \int e^x dx + C = xe^x - e^x + C$$

which is an elementary function.

Example 6.6.3. In order to express $\int x\ln(x)dx$ as an elementary function, denote $f(x)=\ln(x)$, $g'(x)=x$. By Eq. (6.6.3) we obtain

$$\int x\ln(x)dx=\frac{x^2}{2}\ln(x)-\int\frac{x^2}{2}\frac{1}{x}dx+C=\frac{x^2}{2}\ln(x)-\frac{x^2}{4}+C$$

Integration by substitution

Consider the indefinite integral $\int f(x)dx$ and let $x=g(u)$, i.e., $\frac{dx}{du}=g'(u)$ or formally $dx=g'(u)du$. Then under certain quite general conditions

$$\int f(x)dx=\int f(g(u))g'(u)du \tag{6.6.4}$$

$$\int_a^b f(x)dx=\int_c^d f(g(u))g'(u)du \tag{6.6.5}$$

here $a=g(c)$, $b=g(d)$.

While the left-hand side of Eq. (6.6.4) may be difficult to calculate as an elementary function, the left-hand side may be easily manipulated. This approach is called *integration by substitution.* Usually, the substitution is in the form of $h(x)=g(u)$, i.e., $h'(x)dx=g'(u)du$ - depending on the expression defining the integrand $f(x)$.

Example 6.6.4. Consider the integral $\int x\sin(x^2+1)dx$. By substituting $u=x^2+1$ we get $du=2xdx$. Consequently

$$\int x\sin(x^2+1)dx=\int x\sin(u)dx=\int\frac{\sin(u)}{2}du=-\frac{\cos(u)}{2}+C=-\frac{\cos(x^2+1)}{2}+C$$

Example 6.6.5. In the indefinite integral $\int\frac{2xdx}{x^2+3}$ substitute $u=x^2+3$. Thus, $du=2xdx$, i.e.,

$$\int\frac{2xdx}{x^2+3}=\int\frac{du}{u}=\ln(u)+C=\ln(x^2+3)+C \tag{6.6.6}$$

This example is a particular case of a more general integral $\int \frac{f'(x)}{f(x)} dx$ where $f(x)$ is an arbitrary function with continuous derivative. By substituting $u = f(x)$ we get $du = f'(x)dx$ which implies

$$\int \frac{f'(x)}{f(x)} dx = \int \frac{du}{u} = \ln(u) + C = \ln(f(x)) + C \tag{6.6.7}$$

PROBLEMS

1. Use integration by parts to calculate (a) $\int x\sin(x)dx$ (b) $\int xe^{x}dx$
2. 2. Use integration by parts to calculate (a) $\int x^{4}\ln(x)dx$ (b) $\int x^{3}\cos(x)dx$
3. Use integration by parts to calculate (a) $\int e^{x}\cos(x)dx$ (b) $\int \frac{xdx}{e^{-x}}$
4. Use substitution to calculate (a) $\int_{0}^{1} x\sqrt{1-x^{2}}dx$ (b) $\int \frac{x^{3}+2x+1}{x^{2}+x}dx$
5. Use substitution to calculate $\int \frac{\ln(x)dx}{x}$.
6. Calculate $\int x^{n}e^{x}dx$ and use the result to obtain $\int x^{3}e^{x}dx$.
7. Calculate the integrals (a) $\int \tan(x)dx$ (b) $\int \frac{dx}{x\ln(x)}$
8. Calculate the integral $\int \frac{dx}{\sin(x)}$ (Hint: replace $\sin(x)$ with $2\sin\left(\frac{x}{2}\right)\cos\left(\frac{x}{2}\right)$ and use the substitution method).
9. Use problem 8 to obtain $\int \frac{dx}{\cos(x)}$.
10. Calculate the integral $\int \frac{dx}{(x-a)(x-b)}$ for two cases: (a) $a \ne b$ (b) $a = b$.
11. Calculate the integral $\int \frac{dx}{(x-a)(x-b)(x-c)}$ where a,b,c are non-equal.
12. Let $I_n = \int \sin^{n}(x)dx$. Show $I_n = -\frac{\cos(x)\sin^{n-1}(x)}{n} + \frac{n-1}{n} I_{n-2}$ and use the formula to obtain I_2, I_3.
13. Similarly to problem 12, obtain an expression for $J_n = \int \cos^{n}(x)dx$.

6.7 Improper Integrals

So far we dealt with the class of "proper" integrals $\int_a^b f(x)dx$, where a,b are finite numbers and $f(x)$ defined and bounded for all $x: a \le x \le b$. We will now extend this class to include other integrals called "improper integrals".

Definition 6.7.1. The expression $\int_a^b f(x)dx$ is called an *improper integral of first kind* if at least one of the endpoints of the integration interval is infinite, i.e., if $a = -\infty$ or $b = \infty$ or both. If $f(x)$ is unbounded at one or more points, it is called an *improper integral of second kind.* If at least one endpoint is infinite and the integrand is not bounded simultaneously, the integral is called an *improper integral of third kind.*

Example 6.7.1. The expression $\int_0^\infty e^{-x}dx$ is an improper integral of the first kind, while $\int_0^2 \frac{dx}{\sqrt{x}}$ is an improper integral of second kind.

Example 6.7.2. The expression $\int_0^\infty \frac{e^{-x}dx}{\sqrt{x}}$ is an improper integral of third kind since the right endpoint is infinite while the integrand is not bounded at the left endpoint but converges there to ∞.

Defining the value of an improper integral is straightforward. We start with the first kind.

Definition 6.7.2. Consider a function $f(x)$, $a \le x < \infty$ such that a is finite and the integral $\int_a^b f(x)dx$ exists for all $b: a \le b < \infty$. If the integral

$$I(b) = \int_a^b f(x)dx$$

converges to a finite value I^* as $b \to \infty$, we write

$$\int_a^\infty f(x)dx = I^* = \lim_{b\to\infty} \int_a^b f(x)dx \tag{6.7.1}$$

We similarly define the case where b is finite and $a \to \infty$.

Example 6.7.3. Let $I = \int_1^\infty \frac{dx}{x}$. For all finite b we have $\int_1^b \frac{dx}{x} = \ln(x)\Big|_1^b = \ln(b)$ and clearly $\ln(b) \to \infty$ as $b \to \infty$. Hence, the given integral – an improper integral of first kind, does not converge. On the other hand the integral $\int_1^\infty \frac{dx}{x^2}$ exists since

$$\int_1^b \frac{dx}{x^2} = -\frac{1}{x}\Big|_1^b = 1 - \frac{1}{b} \to 1 \quad as\ b \to \infty \tag{6.7.2}$$

Calculating an improper integral of second kind is given next.

Definition 6.7.3. Consider a function $f(x)$, $a \le x \le b$ such that a, b are finite, $f(x)$ is bounded everywhere except at b and the integral $\int_a^{b'} f(x)dx$ exists for all $b' : a \le b' < b$. If the integral $\int_a^{b'} f(x)dx$ converges as $b' \to b$ we write

$$\int_a^b f(x)dx = \lim_{b'\to\infty} \int_a^{b'} f(x)dx \tag{6.7.3}$$

The case where the integrand is not bounded at the lower endpoint is similarly treated.

Example 6.7.4. The improper integral $\int_0^1 \frac{dx}{\sqrt{x}}$ is well defined (i.e., is finite) since

$$\int_\varepsilon^1 \frac{dx}{\sqrt{x}} = 2\sqrt{x}\Big|_\varepsilon^1 = 2 - 2\sqrt{\varepsilon} \to 2 \quad as\ \varepsilon \to 0 \tag{6.7.4}$$

We finally define the value of an improper integral of third kind, whenever it exists. There are four possibilities – one is discussed in detail. The other can be similarly obtained.

Definition 6.7.4. Consider a function $f(x)$, $-\infty < a < x < \infty$ such that $f(x)$ is not bounded at $x = a$. If the integral $\int_{a'}^{b} f(x)dx$ exists for all $a' > a$, $b < \infty$ and if

$$\exists I^* = \lim \int_{a'}^{b} f(x)dx\,,\ a' \to a\,,\ b \to \infty \tag{6.7.5}$$

we define $I^* = \int_{a}^{\infty} f(x)dx$.

It should be noted that in Eq. (6.7.5) $a' \to a$, $b \to \infty$ *independently*.

Example 6.7.5. Let

$$I = \int_0^\infty f(x)dx = \int_0^\infty \frac{e^{-\sqrt{x}}dx}{\sqrt{x}}$$

A specific primitive function of $f(x)$ is $F(x) = -2e^{-\sqrt{x}}$ and clearly

$$\lim_{x\to\infty} F(x) = 0$$
$$\lim_{x\to 0} F(x) = -2$$

which implies that I is an improper integral of third kind which exists and equals 2.

Example 6.7.6. Let

$$I = \int_0^\infty f(x)dx = \int_0^\infty \frac{e^{-\sqrt{x}}dx}{x}$$

This is an improper integral that does not converge at $x = 0$. Indeed, for sufficiently small x (say $x \le \varepsilon$) the integrand is larger than $\frac{1}{2x}$ (since $\lim_{x\to 0} e^{-\sqrt{x}} = 1$). However, since $\int_{\eta}^{\varepsilon} \frac{dx}{2x}$ diverges as $\eta \to 0$ (why?), so is I.

PROBLEMS

1. Show that the following integrals are convergent improper integrals of the first kind: (a) $\int_1^{\infty} \frac{dx}{x^3}$ (b) $\int_0^{\infty} e^{-tx} dx \,, t > 0$

2. Which of the following improper integrals of the first kind are divergent:

$$\text{(a)} \int_{-\infty}^{-0.01} \frac{dx}{x^2} \quad \text{(b)} \int_1^{\infty} \frac{dx}{x} \quad \text{(c)} \int_0^{\infty} \sin^2(x)\, dx$$

3. Find which of the following improper integrals of the second type are convergent and which are divergent: (a) $\int_0^1 \frac{\ln(x)dx}{x}$ (b) $\int_0^{\pi/2} \frac{dx}{\sin(x)}$ (c) $\int_1^2 \frac{dx}{\ln(x)}$

4. Determine the kind of following improper integrals and calculate their values whenever they exist: (a) $\int_2^{\infty} x^3 e^{-x} dx$ (b) $\int_{-1}^{\infty} 2xe^{-3x^2} dx$ (c) $\int_1^2 \frac{dx}{x^3 - 1}$

(d) $\int_0^{\infty} \frac{dx}{x\sqrt{1+x^4}}$

7 Infinite Series

This chapter presents the concept of infinite series and provides several useful tests for determining whether a given series is convergent.

7.1 Convergence

Consider a sequence $\{a_n\}_{n=1}^{\infty}$. For each n, the sum $S_n=\sum_{k=1}^{n} a_k$ is called the n - *th* partial sum of the sequence and the generated sequence $\{S_n\}_{n=1}^{\infty}$ defines an *infinite series* which is usually denoted by $\sum_{i=1}^{\infty} a_n$.

Definition 7.1.1. If $\lim_{n\to\infty} S_n = S$, where $-\infty < S < \infty$, the infinite series is *convergent*, it converges to S and the state of convergence is denoted by

$$S = \sum_{i=1}^{\infty} a_n \tag{7.1.1}$$

Otherwise, the infinite series *diverges*.

Example 7.1.1. The infinite series generated from the sequence $\{1, 1/2, 1/2^2, \ldots,\}$ converges to 2, while the infinite series of $\{1, -1, 1, -1, \ldots,\}$ diverges. The proof is left as an exercise for the student.

Theorem 7.1.1. If $\sum_{i=1}^{\infty} a_n$ is a convergent infinite series, then $\lim_{n\to 0} a_n = 0$.

Proof.

Since $a_n = S_n - S_{n-1}$ we get $\lim_{n\to\infty} a_n = \lim_{n\to\infty} S_n - \lim_{n\to\infty} S_{n-1} = S - S = 0$.

M. Friedman and A. Kandel: Calculus Light, ISRL 9, pp. 183–216.
springerlink.com

Thus, the relation $\lim_{n\to 0} a_n = 0$ is a necessary condition for the convergence of an infinite series. It is certainly not a sufficient condition as demonstrated by the next example.

Example 7.1.2. Consider the *harmonic series* $\sum_{i=1}^{\infty} \frac{1}{n}$. Obviously, $\lim_{n\to 0} \frac{1}{n} = 0$ yet since

$$\frac{1}{3} + \frac{1}{4} > \frac{1}{2} \quad , \quad \frac{1}{5} + \frac{1}{6} + \frac{1}{7} + \frac{1}{8} > \frac{1}{2} \quad etc.$$

we easily conclude (see Example 3.7.5) that the series does not converge but rather satisfies $\lim_{n\to\infty} S_n = \infty$, and therefore, by Definition 7.1.1, diverges.

A necessary and sufficient condition for the convergence of an infinite series was already presented by Theorem 3.7.9 (Section 3.7.3): Given an arbitrary sequence $\{a_n\}_{n=1}^{\infty}$, the associated infinite series is convergent, if and only if $\{S_n\}_{n=1}^{\infty}$ is a Cauchy sequence, i.e., if for any given $\varepsilon > 0$, an integer $n_0(\varepsilon)$ can be found such that $|S_n - S_m| < \varepsilon$ for all $n, m > n_0(\varepsilon)$. Thus, an infinite series converges if and only if, given $\varepsilon > 0$, an integer $n_0(\varepsilon)$ can be found such that for any n, m which satisfy $n_0 < n < m$, the inequality $|a_n + a_{n+1} + \ldots + a_m| < \varepsilon$ holds.

Example 7.1.3. Consider the infinite series $1 + 1/3 + 1/3^2 + \ldots + 1/3^n + \ldots$. Here

$$|S_m - S_n| = |1/3^{n+1} + 1/3^{n+2} + \ldots + 1/3^m| = (1/3^{n+1})(1 + 1/3 + \ldots + 1/3^{m-n-1})$$

which leads to

$$|S_m - S_n| = (1/3^{n+1})\frac{1 - 1/3^{m-n}}{1 - 1/3} < \frac{1}{2 \cdot 3^n}$$

Thus, given $\varepsilon > 0$ we may choose n_0 such that $\frac{1}{2 \cdot 3^{n_0}} < \varepsilon$, which implies

$$n_0 > \frac{\ln[1/(2\varepsilon)]}{\ln 3}$$

Example 7.1.4. The infinite series $a + aq + \ldots + aq^n + \ldots$ converges for arbitrary $q : -1 < q < 1$ and satisfies

$$\sum_{n=1}^{\infty} aq^{n-1} = \frac{a}{1-q} \quad , \quad -1 < q < 1 \qquad (7.1.2)$$

The proof, similar to that given in Example 7.1.3, is left as an exercise for the student.

Although we presented an infinite series as $\sum_{i=1}^{\infty} a_n$, quite often the sequence generating the series may start with a_m where m is an arbitrary fixed integer not equal necessarily to 1. Also, we may for some reason drop or ignore the first k terms of a series. This has usually *no effect* on the convergence or divergence of the series. In order to include these cases, we will usually denote a sequence by $\{a_n\}$ and an infinite series by $\sum a_n$.

Theorem 7.1.2. Consider an infinite series $\sum a_n$ such that $a_n \geq 0$ for all n. Then , the series converges if and only if its partial sums are bounded above.

Proof.

If the sequence $\{S_n\}$ is bounded above, it must converge by virtue of Theorem 3.7.1, as an increasing bounded monotone sequence, i.e., $\lim_{n \to \infty} S_n = S$. Conversely, if the partial sums converge, the sequence $\{S_n\}$ is obviously bounded.

The reader may notice that Theorem 7.1.2 holds even if $a_n \geq 0$ only for all $n \geq n_0$ for *some* fixed n_0 (why?).

Theorem 7.1.3. If $\sum a_n$ converges so is $\sum Ma_n$ for arbitrary constant M.

The proof can be found in Section 3.5 (In the proof of Theorem 3.5.1 choose $b_n = M$ for all n).

Theorem 7.1.4. If $\sum a_n$ and $\sum b_n$ converge, then $\sum (a_n \pm b_n)$ converge as well.

The proof, using Cauchy's criteria is left as an exercise for the reader.

PROBLEMS

1. Is the series generated by $\{1,0,1/2,0,0,1/4,0,0,0,1/8,\ldots\}$ convergent?
2. Is the series generated by $\{1,0,1/2,0,0,1/3,0,0,0,1/4,\ldots\}$ convergent?
3. An infinite series $\sum a_n$ is such that $a_n \ge 0$, $n \ge n_0$, its first 20 partial sums are bounded by 1000 and the rest of them are bounded by 5.
   (a) Does the series converge?
   (b) What can be said about $S = \sum a_n$?

4. Find whether $\sum a_n$ converges in the following cases:

   (a) $a_n = \dfrac{1}{3^n + 2^n}$ (b) $a_n = \dfrac{1}{3^n - 2^n}$ (c) $a_n = \dfrac{2^k}{3 \cdot 2^k + k}$ (d) $a_n = \dfrac{1}{\sqrt{n}}$

7.2 Tests for Convergence

In this section we present several useful tests for convergence of infinite series. We start with the *comparison test* which applies the already known convergence of one series for determining the convergence of another.

Theorem 7.2.1 (Comparison test). Let $\sum a_n$ and $\sum b_n$ denote two infinite series such that $0 \le a_n \le b_n$ for all n. Then, the convergence of $\sum b_n$ implies the convergence of $\sum a_n$ and the divergence of $\sum a_n$ implies the divergence of $\sum b_n$.

Proof.

Denote the partial sums of two series by $\{S_{a,n}\}$ and $\{S_{b,n}\}$ respectively. Then clearly $S_{a,n} \le S_{b.n}$ for all n. If $\lim_{n\to\infty} S_{b,n} = S_b$ then $\{S_{b,n}\}$ is bounded above and so is $\{S_{a,n}\}$. Therefore, by Theorem 7.1.2, the sequence $\{S_{a,n}\}$ is convergent. On the other hand, if $\{S_{a,n}\}$ diverges and $\{S_{b,n}\}$ converges, we obtain contradiction to this theorem's first part. Thus, $\{S_{b,n}\}$ must also diverge.

As in Theorem 7.1.2, the result holds even if $0 \le a_n \le b_n$ only for *almost all* n, i.e., for all $n \ge n_0$ where n_0 is any fixed positive integer. The proof is left for the reader.

Example 7.2.1. The infinite series $\sum_{n=1}^{\infty} \frac{1}{n!}$ converges. Indeed, $\frac{1}{n!} < \frac{1}{2^n}$, $n \ge 4$ as can be easily shown. The series $\sum (1/2^n)$ is a converging geometric series and by Theorem 7.2.1 this implies the convergence of $\sum (1/n!)$ as well.

Example 7.2.2. The series $\sum_{i=1}^{\infty} \frac{1}{\sqrt{n!}}$ converges. The proof, similar to that used in the previous example, is left for the student.

Another popular comparison test is given next.

Theorem 7.2.2 (Limit Comparison Test). Consider two infinite series $\sum a_n$ and $\sum b_n$ such that $a_n \ge 0$, $b_n > 0$ for all $n \ge n_0$ and let $\lim_{n \to \infty} \frac{a_n}{b_n}$ exist and be finite. Then, convergence of $\sum b_n$ implies the convergence of $\sum a_n$.

Proof.

The relation $\lim_{n \to \infty} \frac{a_n}{b_n} = M$ implies $\frac{a_n}{b_n} \le M + 1$ for sufficiently large $n \ge n_1$. Consequently, $a_n \le (M+1)b_n$, $n \ge n_1$. Since $\sum b_n$ converges, so is $\sum [(M+1) \cdot b_n]$ (Theorem 7.1.3) and by virtue of Theorem 7.2.1 so is $\sum a_n$.

Corollary 7.2.1. Consider the infinite series $\sum a_n$ and $\sum b_n$ such that $a_n > 0$, $b_n \ge 0$ for all $n \ge n_0$. If $\lim_{n \to \infty} \frac{b_n}{a_n}$ exists and is finite and if $\sum b_n$ diverges, then $\sum a_n$ diverges.

Proof.

If the opposite holds, i.e., if $\sum a_n$ converges, the series $\sum b_n$ must converge by virtue of Theorem 7.2.2. However, this contradicts the *a priori* given divergence of this series and therefore $\sum a_n$ must diverge.

Example 7.2.3. Let $a_n = \frac{1}{n+\sqrt{n}}$. The series defined by $b_n = \frac{1}{n}$ diverges and also yields

$$\lim_{n\to\infty} \frac{b_n}{a_n} = \lim_{n\to\infty} \left(1 + \frac{1}{\sqrt{n}}\right) = 1$$

which implies the divergence of $\sum a_n$.

Our next results is one of the most popular convergence tests.

Theorem 7.2.3 (Integral Test). Let $f(x), 1 \le x < \infty$ denote a nonnegative continuous monotone decreasing function such that $f(n) = a_n$, $n \ge 1$. Then, the infinite series $\sum a_n$ converges if and only if the integral $\int_1^\infty f(x)dx$ converges.

Proof.

Since $f(n+1) \le f(x) \le f(n)$, $n \le x \le n+1$ we may apply Theorem 6.3.3 and get

$$\int_n^{n+1} f(n+1)dx \le \int_n^{n+1} f(x)dx \le \int_n^{n+1} f(n)dx \quad , \quad n \le x \le n+1$$

which leads to

$$a_2 + a_3 + \ldots + a_{n+1} \le \int_1^{n+1} f(x)dx \le a_1 + a_2 + \ldots + a_n$$

Hence $S_{n+1} - a_1 \le \int_1^{n+1} f(x)dx \le S_n$, $n \ge 1$.

Let $\sum a_n$ converge. Then clearly $\int_1^{n+1} f(x)dx$ is bounded (why?) and since $\int_1^{n+1} f(x)dx$ also increases with n, the integral $\int_1^\infty f(x)dx$ converges. On the other hand, if $\int_1^\infty f(x)dx$ converges, the inequality $a_2 + a_3 + \ldots + a_{n+1} \le \int_1^{n+1} f(x)dx \le \int_1^\infty f(x)dx$ implies

$$S_{n+1} - a_1 \le \int_1^\infty f(x)dx$$

i.e., the sequence $\{S_n\}$ is bounded. Since it is also monotone increasing, it converges. This completes the proof.

Example 7.2.4. Let $a_n = \frac{1}{n^\alpha}$, $n \ge 1$ where α is an arbitrary fixed number greater than 1. We associate this sequence with the decreasing nonnegative continuous function $f(x) = \frac{1}{x^\alpha}$, $x \ge 1$ whose integral over $[0,\infty)$ exists and equals to

$$\int_1^\infty \frac{dx}{x^\alpha} = \frac{1}{1-\alpha} x^{1-\alpha} \Big|_1^\infty = \frac{1}{\alpha - 1}$$

Therefore, the series $\sum \frac{1}{n^\alpha}$ converges for all $\alpha > 1$.

Example 7.2.5. Since the series $\sum \frac{1}{n}$ diverges so is the integral $\int_1^\infty \frac{dx}{x}$. The student must though observe that all the requirements of Theorem 7.2.3 are fulfilled.

Theorem 7.2.4 (Cauchy Condensation Test). Let $\{a_n\}$ denote a nonnegative monotone decreasing sequence for $n \ge n_0$, i.e., $a_n \ge 0$, $a_n \ge a_{n+1}$ for all $n \ge n_0$. Then, $\sum_{n=1}^\infty a_n$ converges if and only if $\sum_{n=0}^\infty 2^n a_{2^n}$ converges.

Proof.

We may assume $n_0 = 1$ (showing that this assumption causes no loss of generality is left as an exercise for the student). Let $S_n = \sum_{k=1}^n a_k$ and $T_m = \sum_{k=0}^m 2^k a_{2^k}$ denote the partial sums of the two series respectively. If $n \le 2^m$ we have

$$\begin{aligned} S_n &= a_1 + a_2 + \ldots + a_n \le a_1 + (a_2 + a_3) + (a_4 + a_5 + a_6 + a_7) + \ldots \\ &+ (a_{2^m} + a_{2^m+1} + \ldots + a_{2^{m+1}-1}) \le a_1 + 2a_2 + 4a_4 + \ldots + 2^m a_{2^m} = T_m \end{aligned} \tag{7.2.1}$$

while for $n \ge 2^m$ we get

$$S_n = a_1 + a_2 + \ldots + a_n \ge a_1 + a_2 + (a_3 + a_4) + (a_5 + a_6 + a_7 + a_8) + \ldots$$
$$+ (a_{2^{m-1}+1} + a_{2^{m-1}+2} + \ldots + a_{2^m}) \ge \frac{a_1}{2} + a_2 + 2a_4 + 4a_8 + \ldots + 2^{m-1} a_{2^m} = \frac{T_m}{2} \quad (7.2.2)$$

Both sequences $\{S_n\},\{T_m\}$ are increasing. Therefore, if $\{S_n\}$ converges it is bounded above (Theorem 7.1.2) and since $\{T_m\}$ is bounded above by virtue of Eq. (7.2.2), it converges as well. Similarly, If $\{T_m\}$ converges, it is bounded above which implies (Eq. (7.2.1)) that $\{S_n\}$ is also bounded above and being an increasing sequence must also converge.

Example 7.2.6. We previously proved the convergence of $\sum \frac{1}{n^\alpha}$, $\alpha > 1$ by applying the integral test. Another method is Cauchy condensation test. Indeed, if $\alpha > 0$ the series $\sum \frac{1}{n^\alpha}$ whose terms are nonnegative and decreasing, converges if and only if the series $\sum 2^n \frac{1}{(2^n)^\alpha}$ converges. The new series can be rewritten as $\sum \left(\frac{1}{2^{\alpha-1}}\right)^n$ which is a geometric series that converges if and only if $\alpha - 1 > 0$, i.e., $\alpha > 1$. Not surprisingly, the result is identical to that of Example 7.2.4.

Example 7.2.7. Let $a_n = \frac{1}{n[\ln(n)]^\alpha}$, $n \ge 2$. The sequence has only nonnegative terms and is clearly decreasing for arbitrary $\alpha > 0$. Therefore, its infinite series converges if and only if the series $\sum 2^n \frac{1}{2^n (\ln 2^n)^\alpha} = \sum \frac{1}{n^\alpha (\ln 2)^\alpha}$ converges. By Example 7.2.6 this is true for $\alpha > 1$ but not for $\alpha = 1$. Thus, for example, the series $\sum \frac{1}{n \ln(n)}$ does not converge while $\sum \frac{1}{n[\ln(n)]^2}$ does.

Theorem 7.2.5 (Ratio Comparison Test). Let $\sum a_n$ and $\sum b_n$ be two infinite series such that $a_n, b_n > 0$ and $\frac{a_{n+1}}{a_n} \le \frac{b_{n+1}}{b_n}$ for all $n \ge n_0$. If $\sum b_n$ converges then $\sum a_n$ converges as well.

Proof. The inequality $\frac{a_{n+1}}{a_n} \le \frac{b_{n+1}}{b_n}$ yields $\frac{a_{n+1}}{b_{n+1}} \le \frac{b_n}{a_n}$, $n \ge n_0$. Thus, the sequence $\{c_n\}$ defined by $c_n = a_n / b_n$, $n \ge n_0$ is bounded above by $M = a_{n_0} / b_{n_0}$. This implies $a_n \le M b_n$, $n \ge n_0$. The convergence of $\sum b_n$ guarantees the convergence of $\sum M b_n$ (Theorem 7.1.3) and the comparison test (Theorem 7.2.1) implies the convergence of $\sum a_n$.

Example 7.2.8. Let $a_n = \frac{1}{n!}$, $b_n = \frac{1}{2^n}$. By Example 7.1.4 the series $\sum b_n$ converges. Also

$$\frac{a_{n+1}}{a_n} = \frac{1}{n+1} \le \frac{1}{2} = \frac{b_{n+1}}{b_n}, \ n \ge 1$$

and therefore $\sum_{n=1}^{\infty} \frac{1}{n!}$ converges.

Theorem 7.2.6 (D'Alembert's Ratio Test). Consider an infinite series $\sum a_n$ such that $a_n > 0$, $n \ge n_0$. Then

1. If $\frac{a_{n+1}}{a_n} \le q$, $n \ge n_0$ for some $q: 0 < q < 1$, $\sum a_n$ converges.
2. If $\frac{a_{n+1}}{a_n} \ge q$, $n \ge n_0$ for some $q \ge 1$, $\sum a_n$ diverges.

Proof.

In the first case we have $\frac{a_{n+1}}{a_n} \le q = \frac{q^{n+1}}{q^n}$ and since $\sum q^n$ converges, so does $\sum a_n$, by virtue of Theorem 7.2.5. In the second case we easily obtain $a_m \ge a_{n_0}$, $n \ge n_0$ and therefore the general term a_n does not converge to 0. Consequently, by Theorem 7.1.1, $\sum a_n$ diverges which completes the proof.

A direct consequence of D'Alembert's ratio test is:

Theorem 7.2.7 (Cauchy's Ratio Test). Consider an infinite series $\sum a_n$ such that $a_n > 0$, $n \ge n_0$. Then

1. If $\lim_{n\to\infty}\frac{a_{n+1}}{a_n}<1 \Rightarrow \sum a_n$ converges.

2. If $\lim_{n\to\infty}\frac{a_{n+1}}{a_n}>1 \Rightarrow \sum a_n$ divverges.

The proofs of Theorem 7.2.7 and of the next result are left as an exercise for the reader.

Theorem 7.2.8 (Root Test). Consider an infinite series $\sum a_n$ such that $a_n \ge 0$, $n \ge n_0$. Then

1. If $\sqrt[n]{a_n} \le \alpha$, $n \ge n_1$ for some $n_1 \ge n_0$ and $0 \le \alpha < 1$, then $\sum a_n$ converges.
2. If $\sqrt[n]{a_n} \ge 1$, $n \ge n_1$ for some $n_1 \ge n_0$, then $\sum a_n$ diverges.

The next result is the classic Cauchy's root test.

Theorem 7.2.9 (Cauchy's Root Test). Let $\sum a_n$ denote an infinite series such that $a_n \ge 0$ for all $n \ge n_0$. If

$$\overline{\lim_{n\to\infty}}\sqrt[n]{a_n} < 1 \tag{7.2.3}$$

the series converges and if

$$\overline{\lim_{n\to\infty}}\sqrt[n]{a_n} > 1 \tag{7.2.4}$$

the series diverges.

Proof.

The fact that several terms in the series (a finite number) may be negative should not disturb, since we ignore them and start from n_0 on. If the shortened series converges, it will still converge if we add the first n_0 terms, and if it diverges it will diverge with the initial n_0 terms as well.

In the first case, find a number q such that $\overline{\lim}\sqrt[n]{a_n} < q < 1$. There exists a positive integer n_1 $(\ge n_0)$ such that $\sqrt[n]{a_n} < q$, $n \ge n_1$. This implies $a_n < q^n$, $n \ge n_1$ and by Theorem 7.2.1 and Example 7.1.4 the series converges. To show the second part of the theorem we again apply Theorem 7.2.1 and the fact that $\sum q^n$ diverges for all $q > 1$ (This is left as an exercise for the student). If $\overline{\lim}\sqrt[n]{a_n} = 1$ the series may converge or diverge and further investigation is needed.

Example 7.2.9. Consider the infinite series $1 + \frac{1}{2} + 1 + \frac{1}{4} + 1 + \frac{1}{8} + \ldots + 1 + \frac{1}{2^n} + \ldots$. Here $\overline{\lim}\sqrt[n]{a_n} = 1$ and it is not clear whether the series converges or not. However, all the terms which generate the series are positive and the partial sums are not bounded above. Consequently, by virtue of Theorem 7.1.2, the series diverges.

Example 7.2.10. The infinite series $\frac{1}{3} + \frac{1}{5^2} + \frac{1}{3^3} + \frac{1}{5^4} + \ldots$ provides $\overline{\lim}\sqrt[n]{a_n} = \frac{1}{3} < 1$, i.e., it converges.

PROBLEMS

1. Let $a_n = \frac{1}{0.1n + 5\sqrt{n}}$, $n \ge 1$. Show that $\sum_{n=1}^{\infty} a_n$ diverges.
2. Use the integral test to show that $\sum_{n=1}^{\infty} \frac{1}{n\sqrt{n}}$ converges.
3. Use the integral test to show that $\sum_{n=2}^{\infty} \frac{1}{n\ln(n)}$ diverges and $\sum_{n=2}^{\infty} \frac{1}{n(\ln(n))^2}$ converges.
4. Does the series $\sum_{n=2}^{\infty} \frac{1}{n(\ln(n))^{\alpha}}$, $\alpha > 1$ converge?
5. Use Cauchy's ratio test to show the convergence of $\sum_{n=0}^{\infty}\left[\left(\frac{1}{2}\right)^n + \left(\frac{2}{3}\right)^n\right]$.
6. Can you use Cauchy's root test to show that $\sum_{n=1}^{\infty} \frac{1}{n^2}$ is convergent.

7.3 Conditional and Absolute Convergence

So far we have mainly discussed infinite series with nonnegative elements or with almost only nonnegative elements, i.e., $\sum a_n$ such that $a_n \ge 0$ for all $n \ge n_0$.

However, quite often this is not the case. For example, a very popular infinite series is an alternating series defined as follows.

Definition 7.3.1. Consider a sequence $\{a_n\}$ whose terms are all nonnegative. The series $\sum(-1)^{n+1}a_n$, $\sum(-1)^n a_n$ are called are called *alternating series.*

Example 7.3.1. The series $1-\frac{1}{2}+\frac{1}{3}-\frac{1}{4}+\ldots$ and $-\frac{1}{3}+\frac{1}{5}-\frac{1}{7}+\frac{1}{9}-\ldots$ are alternating series while $1+\frac{1}{2}-\frac{1}{3}-\frac{1}{4}+\frac{1}{5}+\frac{1}{6}-\ldots$ is not.

If by removing the first n_0 terms of a series, an alternating series is obtained, the original series can be treated as an alternating series for all convergence matters.

Theorem 7.3.1. (Leibniz's Alternating Series Test). Consider a sequence $\{a_n\}$ whose terms satisfy $0 \le a_{n+1} \le a_n$, $n \ge n_0$ and $\lim_{n\to\infty} a_n = 0$. Then, the series $\sum(-1)^{n+1}a_n$, $\sum(-1)^n a_n$ converge.

Proof.

(a) $\sum(-1)^{n+1}a_n$. Without loss of generality we may assume $n_0 = 1$. The series is therefore

$$a_1 - a_2 + a_3 - a_4 + a_5 - \ldots$$

and it clearly satisfies

$$S_2 \le S_4 \le S_6 \le \cdots \le S_5 \le S_3 \le S_1$$

The sequence $\{S_{2n}\}$ is increasing and bounded above by S_1. Consequently, it converges to some S^*. Also, the sequence $\{S_{2n-1}\}$ is decreasing and bounded below. It therefore converges to some S^{**}. However, $S_{2n+1} - S_{2n} = a_{2n+1}$ which yields

$$\lim_{n\to\infty} S_{2n+1} - \lim_{n\to\infty} S_{2n} = \lim_{n\to\infty} a_{2n+1} = 0$$

Thus $S^* = S^{**}$, i.e., $\{S_n\}$ converges as stated.

(b) $\sum(-1)^n a_n$. The series is $-a_1+a_2-a_3+a_4-a_5+\dots$, it satisfies

$$S_1 \le S_3 \le S_5 \le \dots \le S_6 \le S_4 \le S_2$$

and again, the odd and even partial sums must converge to the same value.

Example 7.3.2. The series $1-\frac{1}{2}+\frac{1}{3}-\frac{1}{4}+\dots$ converges since the sequence $\left\{\frac{1}{n}\right\}$ fulfills all the requirements of Leibniz's test. On the other hand, the sequence $\left\{\frac{1}{3},\frac{1}{2},1,\frac{1}{6},\frac{1}{5},\frac{1}{4},\frac{1}{9},\frac{1}{8},\frac{1}{7},\dots\right\}$ satisfies $\lim_{n\to\infty} a_n = 0$ but is not a decreasing sequence. The problem of determining whether the associated alternating series converges is left as an exercise for the student.

Example 7.3.3. Consider the sequence $\left\{1,\frac{1}{2},\frac{1}{3},\frac{1}{4},\frac{1}{3},\frac{1}{4},\frac{1}{5},\frac{1}{6},\frac{1}{5},\frac{1}{6},\frac{1}{5},\frac{1}{6},\dots\right\}$. It satisfies the requirement $\lim_{n\to\infty} a_n = 0$, it is not a decreasing sequence and its associated alternating series does not converge. This can be shown as follows. A subsequence of the generated alternating sequence is

$$S_2 = 1-\frac{1}{2}=\frac{1}{2} \quad , \quad S_6 = \frac{1}{2}+\frac{2}{3\cdot 4} \quad , \quad S_{12} = \frac{1}{2}+\frac{2}{3\cdot 4}+\frac{3}{5\cdot 6} \quad , \dots$$

which can be rewritten as

$$S_2 = \frac{1}{2}\cdot 1 \quad , \quad S_6 = \frac{1}{2}\cdot\left(1+\frac{1}{3}\right) \quad , \quad S_{12} = \frac{1}{2}\cdot\left(1+\frac{1}{3}+\frac{1}{5}\right) \quad , \dots$$

These partial sums are not bounded and this yields that the series diverges.

Definition 7.3.2. An infinite series $\sum a_n$ is *absolutely convergent* (or *converges absolutely*) if the series $\sum |a_n|$ converges.

Example 7.3.4. The series $1-\frac{1}{2^2}+\frac{1}{3^2}-\frac{1}{4^2}+\dots$ converges. Indeed, by Example 7.2.4, the series $\sum\frac{1}{n^\alpha}$ converges for all $\alpha > 1$, and in particular for $\alpha = 1$.

Example 7.3.5. The series $\sum \frac{\sin(n)}{n^2}$ is absolutely convergent. Indeed, $\left|\frac{\sin(n)}{n^2}\right| \le \frac{1}{n^2}$ and $\sum \frac{1}{n^2}$. What about $\sum \frac{\sin(n)}{n^2}$ which is not an alternating series of $\sum \left|\frac{\sin(n)}{n^2}\right|$? The next fundamental result provides the answer.

Theorem 7.3.2. A series which converges absolutely is also convergent, i.e., the convergence of $\sum |a_n|$ yields the convergence of $\sum a_n$ as well.

Proof.

If $\sum |a_n|$ converges, then, for an arbitrary $\varepsilon > 0$ there exists an integer $n_0 > 0$ such that $\big||a_n| + |a_{n+1}| + \ldots + |a_m|\big| < \varepsilon$ for all $n_0 < n < m$. But since

$$|a_n + a_{n+1} + \ldots + a_m| \le |a_n| + |a_{n+1}| + \ldots + |a_m| < \varepsilon$$

The partial sums of $\sum a_n$ form a Cauchy sequence and the series converges.

The opposite is clearly false. For example, the harmonic series diverges (Example 7.1.2) while its alternating series converges by Leibniz' test. Also, if $\sum a_n$ diverges, so is $\sum |a_n|$, otherwise this contradicts Theorem 7.3.2.

Example 7.3.6. Let $\sum a_n = \sum \frac{\cos(n)}{n^2}$. This infinite series is not a monotone decreasing alternating series and Leibniz's test for convergence cannot be applied. Yet, $\left|\frac{\cos(n)}{n^2}\right| \le \frac{1}{n^2}$ which, by virtue of Theorem 7.2.1 (Comparison test) and Example 7.2.4, implies the convergence of $\sum \left|\frac{\cos(n)}{n^2}\right|$. Hence, by Theorem 7.3.2, $\sum \frac{\cos(n)}{n^2}$ converges.

Consider now a general infinite series $\sum a_n$ and define the sequences $\{a_n^+\}, \{a_n^-\}$ as follows:

$$a_n^+ = \begin{cases} a_n \,,\, a_n \ge 0 \\ 0 \,,\, a_n < 0 \end{cases} \quad , \quad a_n^- = \begin{cases} a_n \,,\, a_n \le 0 \\ 0 \,,\, a_n > 0 \end{cases} \tag{7.3.1}$$

Thus, $\{a_n^+\}$ consists of all the nonnegative terms with 0's replacing all the negative terms and $\{a_n^-\}$ - of all the negative and zero terms with 0's replacing all the positive terms. Clearly,

$$a_n = a_n^+ + a_n^- \,,\, |a_n| = a_n^+ - a_n^- \tag{7.3.2}$$

for all n. Let S_n^+, S_n^-, S_n^* denote the partial sums of the associated infinite series of $\{a_n^+\}, \{a_n^-\}, \{|a_n|\}$ respectively. Then, Eq. (7.3.2) yields $S_n = S_n^+ + S_n^-$ and $S_n^* = S_n^+ - S_n^-$. The relation between the convergence of S^* and that of S_n^+, S_n^- is given by the next result.

Theorem 7.3.3. The series $\sum |a_n|$ converges if and only if both sequences $\{S_n^+\}, \{S_n^-\}$ converge.

Proof.

(a) If both $\{S_n^+\}, \{S_n^-\}$ converge, then $\lim\limits_{n\to\infty} S_n^* = \lim\limits_{n\to\infty} S_n^+ - \lim\limits_{n\to\infty} S_n^-$, i.e., $\sum |a_n|$ converges.
(b) If $\{S_n^*\}$ converges then the comparison test (Theorem 7.2.1) implies the convergence of $\{S_n^+\}$, $\{-S_n^-\}$ and the convergence of $\{-S_n^-\}$ guarantees the convergence of $\{S_n^-\}$.

Corollary 7.3.1. If one of the sequences $\{S_n^+\}, \{S_n^-\}$ converges while the other one diverges, then both $\sum a_n$ and $\sum |a_n|$ diverge. If both $\{S_n^+\}, \{S_n^-\}$ diverge, $\sum |a_n|$ diverges, while $\sum a_n$ may or may not converge.

Proof.

Let for example $\{S_n^+\}$ converge while $\{S_n^-\}$ diverges. By Theorem 7.3.3 the series $\sum |a_n|$ must diverge, since its convergence would imply the convergence of *both* $\{S_n^+\}$ and $\{S_n^-\}$. Also, $\sum a_n$ must diverge since its convergence, the convergence of $\{S_n^+\}$ and the relation $S_n^- = S_n - S_n^+$, implies the convergence of $\{S_n^-\}$ - which was assumed to diverge. The proof of the second part is left as an exercise for the student.

Definition 7.3.3. The series $\sum a_n$ is *conditionally convergent* if $\sum a_n$ converges while $\sum |a_n|$ diverges.

Example 7.3.7. The series $1-\frac{1}{2}+\frac{1}{3}-\frac{1}{4}+\ldots$ is conditionally convergent while the series $1-\frac{1}{2^2}+\frac{1}{3^2}-\frac{1}{4^2}+\ldots$ is not since it is absolutely convergent.

Example 7.3.8. Let $\sum a_n = 2-\frac{1}{2}+\frac{2}{3}-\frac{1}{4}+\frac{2}{5}-\frac{1}{6}+\ldots$. Here both $\{S_n^+\}, \{S_n^-\}$ diverge and $\{S_n\}$ diverges as well, i.e., $\sum a_n$ is not conditionally convergent. In the case of the series $1-\frac{1}{2}+\frac{1}{3}-\frac{1}{4}+\ldots$ the sequences $\{S_n^+\}, \{S_n^-\}$ diverge, but $\{S_n\}$ converges.

Theorem 7.3.4. If $\sum a_n$ is conditionally convergent, both sequences $\{S_n^+\}, \{S_n^-\}$ must diverge.

Proof.

Let $\{S_n^+\}$ converge. The relation $S_n^- = S_n - S_n^+$ implies the convergence of $\{S_n^-\}$ and by virtue of Theorem 7.3.3 $\sum |a_n|$ converges. Thus $\sum a_n$ is not conditionally convergent but rather absolutely convergent.

Definition 7.3.4. Let $\sum a_n$ be an infinite series and let the relation $l = l(k)$ denote a one-to-one correspondence from the set of all positive integers onto itself. The series $\sum b_k$ where $b_k = a_{l(k)}$ for all k, is called a *rearrangement* of $\sum a_n$.

Example 7.3.9. The series $1+\frac{1}{3}-\frac{1}{2}-\frac{1}{4}+\frac{1}{5}+\frac{1}{7}-\frac{1}{6}-\frac{1}{8}+\ldots$ is a rearrangement of the series $1-\frac{1}{2}+\frac{1}{3}-\frac{1}{4}+\frac{1}{5}-\frac{1}{6}+\frac{1}{7}-\frac{1}{8}\ldots$. The relation $l = l(k)$ is

$$l(4j+1) = 4j+1\,,\ l(4j+2) = 4j+3\,,\ l(4j+3) = 4j+2\,,\ l(4j+4) = 4j+4$$

for an arbitrary integer $j \ge 0$.

Two fundamental results related to rearranged series are given next.

Theorem 7.3.5. If $\sum a_n$ is absolutely convergent and $\sum b_k$ is a rearrangement of $\sum a_n$, then $\sum b_k$ converges to $\sum a_n$.

Proof.

Clearly $\sum a_n$ converges since the series is absolutely convergent. Let us distinguish between two cases.

(a) $a_n \ge 0$ for all $n \ge 1$. Consider an arbitrary partial sum of $\sum b_k$, say, $\sum_{k=1}^{m} b_k$. Since all the terms of $\sum a_n$ are nonnegative, we clearly have $\sum_{k=1}^{m} b_k = \sum_{k=1}^{m} a_{l(k)} \le \sum_{n=1}^{N} a_n$ for all $N \ge \max\{l(1), l(2), \ldots, l(m)\}$. Consequently the rearrangement converges and $\sum b_k \le \sum a_n$. Since $\sum a_n$ is also a rearrangement of $\sum b_k$ we get $\sum a_n \le \sum b_k$ as well which yields $\sum b_k = \sum a_n$.

(b) The terms a_n are not necessarily nonnegative. In this case we generate the sequences $\{a_n^+\}, \{a_n^-\}$ and $\{b_k^+\}, \{b_k^-\}$. Consider a partial sum $\sum_{k=1}^{m} b_k$ of the rearranged series. Then

$$\sum_{k=1}^{m} b_k = \sum_{k=1}^{m} b_k^+ + \sum_{k=1}^{m} b_k^- \tag{7.3.3}$$

The series $\sum a_n^+$ is clearly absolutely convergent and its terms are all nonnegative. Therefore, by the proof of part (a) $\sum b_k^+$ is absolutely convergent and converges to $\sum a_n^+$. Similarly, $-\sum a_n^-$ is absolutely convergent and its terms all nonnegative, which yields the absolute convergence of $-\sum b_k^-$ to $-\sum a_n^-$. Consequently, $\sum b_k^-$ converges to $\sum a_n^-$ and by virtue of Eq. (7.3.3), $\sum b_k$ converges to $\sum a_n$.

Theorem 7.3.6. If each arbitrary rearrangement of a series $\sum a_n$ is convergent, then $\sum a_n$ is absolutely convergent.

Proof.

First, since $\sum a_n$ converges in its original arrangement, we must have $\lim_{n\to\infty} a_n = \lim_{n\to\infty} a_n^+ = \lim_{n\to\infty} a_n^- = 0$. If $\sum a_n$ is not absolutely convergent then, by virtue of Theorem 7.3.3 one of series $\sum a_n^+$, $\sum a_n^-$ does not converge. For example, let $\sum a_n^+$ diverge and we will generate a rearrangement that diverges, in contradiction to the given information. Since zeros in the original series appear in both $\sum a_n^+$ and $\sum a_n^-$ we will rearrange *all* the elements of $\sum a_n^+$ and only the *negative* ones in $\sum a_n^-$. Define an increasing sequence of positive integers $\{n_1, n_2, \ldots, n_k, \ldots\}$ such that

$$
\begin{aligned}
&a_1^+ + a_2^+ + \ldots + a_{n_1}^+ + a_{j_1}^- > 1 \\
&a_{n_1+1}^+ + a_{n_1+2}^+ + \ldots + a_{n_2}^+ + a_{j_2}^- > 1 \\
&\vdots \\
&a_{n_{k-1}+1}^+ + a_{n_{k-1}+2}^+ + \ldots + a_{n_k}^+ + a_{j_k}^- > 1 \\
&\vdots
\end{aligned}
$$

where $a_{j_k}^-$ is the k-th negative term in $\{a_n^-\}$. This is possible since $\sum a_n^+$, composed of nonnegative elements, diverges, and since $\lim_{n\to\infty} a_n^- = 0$. Now, rearrange the series $\sum a_n$ as

$$
\begin{aligned}
a_1^+ + a_2^+ + \ldots + a_{n_1}^+ + a_{j_1}^- + a_{n_1+1}^+ + a_{n_1+2}^+ + \ldots + a_{n_2}^+ + a_{j_2}^- + \ldots \\
+ a_{n_{k-1}+1}^+ + a_{n_{k-1}+2}^+ + \ldots + a_{n_k}^+ + a_{j_k}^- + \ldots
\end{aligned}
$$

and denote its partial sums by T_n. Then $T_{n_1+1} > 1, T_{n_2+2} > 2, \ldots, T_{n_k+k} > k, \ldots$ which implies that the sequence $\{T_n\}$ diverges. Thus, both series $\sum a_n^+$ and $\sum a_n^-$ converge, i.e., $\sum a_n$ is absolutely convergent as stated.

Lemma 7.3.1. Let $\sum a_n$, $\sum b_n$ denote two arbitrary series and let $S_n = \sum_{i=1}^{n} a_n$.

Then

$$\sum_{k=1}^{n} a_k b_k = S_1(b_1 - b_2) + S_2(b_2 - b_3) + \ldots + S_{n-1}(b_{n-1} - b_n) + S_n b_n \qquad (7.3.4)$$

Proof.

The proof is by induction. For $n = 1$ both sides of Eq. (7.3.4) equal to $a_1 b_1$ ($S_1 = a_1$). Assume the lemma to hold for some $n > 0$. Then

$$\sum_{k=1}^{n+1} a_k b_k = \sum_{k=1}^{n} a_k b_k + a_{n+1} b_{n+1} = \sum_{k=1}^{n-1} S_k (b_k - b_{k+1}) + S_n b_n + a_{n+1} b_{n+1}$$

$$= \sum_{k=1}^{n-1} S_k (b_k - b_{k+1}) + S_n b_n - S_n b_{n+1} + S_n b_{n+1} + a_{n+1} b_{n+1}$$

$$= \sum_{k=1}^{n} S_k (b_k - b_{k+1}) + S_{n+1} b_{n+1}$$

i.e., Eq. (7.3.4) holds for $n+1$ as well.

Lemma 7.3.1 provides another useful convergence test.

Theorem 7.3.7 (Dirichlet's test). If a series $\sum a_n$ has bounded partial sums and if $\{b_n\}$ is a decreasing sequence which converges to 0, then $\sum a_n b_n$ converges.

Proof.

Let $|S_n| \le M$, $n \ge 1$ where $S_n = \sum_{i=1}^{n} a_k$. By Cauchy criterion we must show that given $\varepsilon > 0$ an integer $n_0 > 0$ can be found such that $|a_{n+1} b_{n+1} + a_{n+2} b_{n+2} + \ldots + a_m b_m| < \varepsilon$ for all $m > n > n_0$. By Lemma 7.3.1, since the terms of $\{b_n\}$ are all nonnegative and $b_k \ge b_{k+1}$, we get

$$\begin{aligned}\left|a_{n+1}b_{n+1}+a_{n+2}b_{n+2}+\ldots+a_m b_m\right| &= \left|\sum_{k=n}^{m-1} S_k(b_k-b_{k+1})+S_m b_m - S_n b_n\right| \\ &\le \sum_{k=n}^{m-1}\left|S_k\right|(b_k-b_{k+1})+\left|S_m\right|b_m+\left|S_n\right|b_n \\ &\le M(b_n-b_m)+Mb_m+Mb_n=2Mb_n\end{aligned}$$

Since $\lim_{n\to\infty} b_n = 0$ an $n_0 > 0$ exists such that $b_n < \frac{\varepsilon}{2M}$, $n > n_0$ and this yields

$$\left|a_{n+1}b_{n+1}+a_{n+2}b_{n+2}+\ldots+a_m b_m\right| < \varepsilon \, , m > n > n_0$$

which completes the proof.

Example 7.3.10. A well known result from trigonometry is the identity

$$\sin(x)+\sin(2x)+\ldots+\sin(nx)=\frac{\sin\left(\frac{nx}{2}\right)\sin\left(\frac{(n+1)x}{2}\right)}{\sin\left(\frac{x}{2}\right)} \tag{7.3.5}$$

which holds for all $x \ne 0$. The right-hand side of Eq. (7.3.5) is bounded in absolute value by $\left(\left|\sin(x/2)\right|\right)^{-1}$ for all n. Since the harmonic series satisfies the requirements imposed on $\{b_n\}$ in Theorem 7.3.7, the series $\sum_{n=1}^{\infty}\frac{\sin(nx)}{n}$ converges for all $x \ne 0$ while it converges to 0 for $x = 0$.

The requirement that $\{b_n\}$ is a decreasing sequence which converges to 0, is quite fundamental in showing the convergence of $\sum a_n b_n$. In the next example both $\sum a_n$ and $\sum b_n$ converge, yet $\sum a_n b_n$ diverges.

Example 7.3.11. Let $\sum a_n = \sum b_n = 1 - \frac{1}{\sqrt{2}} + \frac{1}{\sqrt{3}} - \frac{1}{\sqrt{4}} + \ldots$. Both series converge by Leibniz's test, yet $\sum a_n b_n = 1 + \frac{1}{2} + \frac{1}{3} + \frac{1}{4} + \ldots$ diverges.

Theorem 7.3.8 (Abel's test). Let $\sum a_n$ converge and let $\{b_n\}$ be a bounded monotone sequence. Then $\sum a_n b_n$ converges.

Proof.

Clearly the sequence $\{b_n\}$ converges to some B and without loss of generality we may assume that it is increasing. Consider the sequence $\{B-b_n\}$ which is decreasing and converges to 0. Since $\sum a_n$ converges, its partial sums must be bounded and by Theorem 7.3.7 the series $\sum a_n(B-b_n)$ converges. Clearly, the series $\sum a_n B$ converges as well and the relation

$$\sum a_n b_n = \sum a_n B - \sum a_n (B-b_n)$$

guarantees the convergence of $\sum a_n b_n$ as a difference between two converging series.

PROBLEMS

1. Show that the series $\frac{1}{2\sqrt{2}} - \frac{1}{3\sqrt{3}} + \frac{1}{4\sqrt{4}} - \dots$ converges.
2. Show that the series $\frac{1}{2\ln(2)} - \frac{1}{3\ln(3)} + \frac{1}{4\ln(4)} - \dots$ converges.
3. Which of the following is absolutely convergent series:

(a) $1 - \frac{1}{2\ln(2)} + \frac{1}{3\ln(3)} - \frac{1}{4\ln(4)} + \dots$ (b) $\sin(1) - \sin\left(\frac{1}{4}\right) + \sin\left(\frac{1}{8}\right) - \sin\left(\frac{1}{16}\right) + \dots$

4. Consider the series $1 - \frac{1}{2} + \frac{1}{4} - \frac{1}{8} + \frac{1}{16} - \frac{1}{32} + \frac{1}{64} - \frac{1}{128} + \dots$ and its rearrangement $1 - \frac{1}{8} - \frac{1}{2} + \frac{1}{4} + \frac{1}{16} - \frac{1}{128} - \frac{1}{32} + \frac{1}{64} + \dots$ (a) Are they convergent? (b) Do they converge to the same value?
5. By theorem 7.3.6 there must be a rearrangement of the series $1 - \frac{1}{2} + \frac{1}{3} - \frac{1}{4} + \dots$ that diverges (why?). Find one.

7.4 Multiplication of Series and Infinite Products

In this section we will denote an arbitrary series by $\sum_{n=0}^{\infty} a_n$ instead of $\sum_{n=1}^{\infty} a_n$. The reason is the association made here between $\sum_{n=0}^{\infty} a_n$ and the power series (treated in detail in Section 7.5) $\sum_{n=0}^{\infty} a_n x^n = a_0 + a_1 x + a_2 x^2 + \ldots + a_n x^n + \ldots$. Consider two formal power series $\sum_{n=0}^{\infty} a_n x^n$ and $\sum_{n=0}^{\infty} b_n x^n$. If we *formally* multiply these expressions and for each power x^n denote the sum of all the coefficients which multiply this power by c_n, we get

$$\begin{aligned} c_0 &= a_0 b_0 \\ c_1 &= a_0 b_1 + a_1 b_0 \\ &\vdots \\ c_n &= \sum_{k=0}^{n} a_k b_{n-k} \\ &\vdots \end{aligned} \tag{7.4.1}$$

Thus, we may formally write

$$\left(\sum_{i=0}^{\infty} a_n x^n\right)\left(\sum_{i=0}^{\infty} b_n x^n\right) = \sum_{n=0}^{\infty} c_n x^n \tag{7.4.2}$$

where the coefficients c_n are determined by Eq. (7.4.1). If we now substitute $x = 1$ we obtain the formal relation

$$\left(\sum_{i=0}^{\infty} a_n\right)\left(\sum_{i=0}^{\infty} b_n\right) = \sum_{n=0}^{\infty} c_n \tag{7.4.3}$$

which provides the motivation for the following definition.

Definition 7.4.1. The *Cauchy product* of $\sum_{n=0}^{\infty} a_n$ and $\sum_{n=0}^{\infty} b_n$, is the series $\sum_{n=0}^{\infty} c_n$, where

$$c_n = \sum_{k=0}^{n} a_k b_{n-k} \,,\; n \ge 0 \tag{7.4.4}$$

Notice that nothing related to the convergence of the two given series, was assumed.

Example 7.4.1. Let $\sum a_n = 1 + \frac{1}{2} + \frac{1}{4} + \frac{1}{8} + \ldots$ and $\sum b_n = 1 + \frac{1}{3} + \frac{1}{9} + \frac{1}{27} + \ldots$.

Then

$$c_0 = 1 \cdot 1 = 1\,,\; c_1 = 1 \cdot \frac{1}{3} + \frac{1}{2} \cdot 1 = \frac{2}{3}\,,\; c_2 = 1 \cdot \frac{1}{9} + \frac{1}{2} \cdot \frac{1}{3} + \frac{1}{4} \cdot 1 = \frac{19}{36}\,, \ldots$$

Theorem 7.4.1. If $\sum a_n$ and $\sum b_n$ are absolutely convergent, their Cauchy product $\sum c_n$ is also absolutely convergent and Eq. (7.4.3) holds.

Proof.

Let A_n, B_n, C_n denote the partial sums of $\sum a_n$, $\sum b_n$, $\sum c_n$ respectively. We distinguish between two cases:

(a) $a_n \geq 0$, $b_n \geq 0$ for all $n \geq 0$. Consider the product

$$A_n \cdot B_n = (a_0 + a_1 + \ldots + a_n)(b_0 + b_1 + \ldots + b_n)$$

It certainly contains *all* the terms $a_i b_j$, $i + j \leq n$, which implies $C_n \leq A_n B_n$. On the other hand *each* term of the product is of the form $a_i b_j$, $i + j \leq 2n$ which yields $A_n B_n \leq C_{2n}$. The sequence $\{C_n\}$ is thus increasing and bounded by AB where

$$A = \lim_{n \to \infty} A_n\,,\; B = \lim_{n \to \infty} B_n$$

Consequently, $\sum c_n$ converges to some $C = \lim_{n \to \infty} C_n$ where $C \leq AB$. Since all c_n are nonnegative, $\sum c_n$ is also absolutely convergent. Finally, by virtue of the relations

$$\lim_{n \to \infty} C_n = \lim_{n \to \infty} C_{2n} = C \;,\; A_n B_n \leq C_{2n}$$

we get $AB \leq C$, i.e., $C = AB$.

The general case. Since the two series are absolutely convergent, $\sum |a_n|$ and $\sum |b_n|$ converge. Denote the Cauchy product of these series by $\sum \overline{c}_n$. Then

$$\bar{c}_0 = |a_0||b_0|, \bar{c}_1 = |a_0||b_1| + |a_1||b_0|, \ldots, \bar{c}_n = \sum_{k=0}^{n} |a_k||b_{n-k}|, \ldots$$

and by the proof of part (a) the series

$$\sum \bar{c}_n = |a_0||b_0| + |a_0||b_1| + |a_1||b_0| + |a_0||b_2| + |a_1||b_1| + |a_2||b_0| + \ldots \quad (7.4.5)$$

converges. Notice that since *all* the terms of each $\bar{c}_n$ are nonnegative we do not need any pair of parenthesis at the right-hand side of Eq. (7.4.5). Hence, the series defined as

$$\sum c'_n = a_0 b_0 + a_0 b_1 + a_1 b_0 + a_0 b_2 + a_1 b_1 + a_2 b_0 + \ldots$$

converges as well ($c'_0 = a_0 b_0$, $c'_1 = a_0 b_1$, $c'_2 = a_1 b_0$, ...). Notice that this series *is not* $\sum c_n$ which was defined as

$$\sum c_n = a_0 b_0 + (a_0 b_1 + a_1 b_0) + \ldots + \left(\sum_{k=0}^{n} a_k b_{n-k} \right) + \ldots \quad (7.4.6)$$

but rather obtained by removing all pairs of parenthesis from the right-hand side of Eq. (7.4.6). Since $\sum c'_n$ converges absolutely, any rearrangement of its terms yields a series with identical sum. Also, the partial sums of $\sum c_n$ are also partial sums of $\sum c'_n$ and therefore $\sum c_n$ converges and equals to $\sum c'_n$. We now perform a rearrangement of $\sum c'_n$ such that for each n the first $(n+1)^2$ terms will be those obtained by the expansion of $A_n B_n = (a_0 + a_1 + \ldots + a_n)(b_0 + b_1 + \ldots + b_n)$. This is feasible since all the terms in $A_n B_n$ appear also in $A_{n+1} B_{n+1}$. The sum of the rearranged series equals to $\sum c'_n$ but since $A_n B_n$ are a subset of its partial sums we also get

$$\sum c_n = \sum c'_n = \lim_{n \to \infty} A_n B_n = \lim_{n \to \infty} A_n \cdot \lim_{n \to \infty} B_n = \left(\sum a_n \right) \left(\sum b_n \right) \quad (7.4.7)$$

Thus, the Cauchy product of the two series converges to the product of their sums. The absolute convergence of $\sum c_n$ follows from the inequality

$$|c_n| \le d_n = |a_0 b_n| + |a_1 b_{n-1}| + \ldots + |a_n b_0|$$

combined with the comparison test (Theorem 7.2.1) applied to the series $\sum |c_n|$ and $\sum d_n$. This completes the proof.

Example 7.4.2. Let $\sum a_n = \sum b_n = 1 + \frac{1}{2} + \frac{1}{4} + \frac{1}{8} + \ldots = 2$. Both series are absolutely convergent. Hence their Cauchy product converges and equals to 4. In particular we obtain the identity

$$\sum_{n=0}^{\infty} \frac{n+1}{2^n} = 4$$

as may be shown by the student.

The next results present additional relations between $\sum a_n$, $\sum b_n$ and their Cauchy product $\sum c_n$.

Theorem 7.4.2 (Merten's theorem). If $\sum_{n=0}^{\infty} a_n$, $\sum_{n=0}^{\infty} b_n$ converge to A, B respectively and one of the series, say $\sum_{n=0}^{\infty} a_n$, is absolutely convergent, then Cauchy product of the two series converge to AB.

Proof.

Let $A_n = \sum_{k=0}^{n} a_k$, $B_n = \sum_{k=0}^{n} b_k$ denote arbitrary partial sums of two series. The n-th partial sum of the Cauchy product is

$$\begin{aligned} C_n = c_0 + c_1 + \ldots + c_n &= a_0 b_0 + (a_0 b_1 + a_1 b_0) + \ldots + (a_0 b_n + a_1 b_{n-1} + \ldots + a_n b_0) \\ &= a_0 B_n + a_1 B_{n-1} + \ldots + a_n B_0 \end{aligned} \tag{7.4.8}$$

and by denoting $\beta_n = B_n - B$ we obtain

$$\begin{aligned} C_n &= a_0 (B + \beta_n) + a_1 (B + \beta_1) + \ldots + a_n (B + \beta_0) \\ &= A_n B + a_0 \beta_n + a_1 \beta_{n-1} + \ldots + a_n \beta_0 = A_n B + r_n \end{aligned} \tag{7.4.9}$$

Since $\lim_{n \to \infty} A_n B = AB$, it is sufficient to show

$$\lim_{n\to\infty} r_n = \lim_{n\to\infty}(a_0\beta_n + a_1\beta_{n-1} + \ldots + a_n\beta_0) = 0$$

The convergence of $\sum_{n=0}^{\infty} b_n$ yields $\lim_{n\to\infty}\beta_n = 0$. Consequently, $\{\beta_n\}$ is bounded, i.e., $|\beta_n| < K$, $n \ge 0$ for some finite K . Given an arbitrary $\varepsilon > 0$, choose $n_1 > 0$ such that

$$|\beta_n| < \frac{\varepsilon}{2\overline{A}}\,,\ n \ge n_1 \qquad (7.4.10)$$

where $\overline{A} = \sum_{n=0}^{\infty}|a_n|$. Next, let $n_2 > 0$ be such that

$$|a_n| + |a_{n+1}| + \ldots + |a_m| < \frac{\varepsilon}{2K}\,,\ m > n > n_2 \qquad (7.4.11)$$

The feasibility of both choices follows from the convergence and absolute convergence of $\sum_{n=0}^{\infty} b_n$ and $\sum_{n=0}^{\infty} a_n$ respectively. Now, let $n > n_1 + n_2 - 1$. Then

$$|r_n| = |a_0\beta_n + a_1\beta_{n-1} + \ldots + a_n\beta_0| \le |a_0||\beta_n| + |a_1||\beta_{n-1}| + \ldots + |a_{n-n_1}||\beta_{n_1}|$$

$$+|a_{n-n_1+1}||\beta_{n_1-1}| + |a_{n-n_1+2}||\beta_{n_1-2}| + \ldots + |a_n||\beta_0| \quad (7.4.12)$$

The first term of the right-hand side of Eq. (7.4.12) satisfies

$$|a_0||\beta_n| + |a_1||\beta_{n-1}| + \ldots + |a_{n-n_1}||\beta_{n_1}| < \left(|a_0| + |a_1| + \ldots + |a_{n-n_1}|\right)\frac{\varepsilon}{2\overline{A}} \le \frac{\varepsilon}{2} \qquad (7.4.13)$$

while the second term, since $n - n_1 + 1 > n_2$ (why?) provides

$$|a_{n-n_1+1}||\beta_{n_1-1}| + |a_{n-n_1+2}||\beta_{n_1-2}| + \ldots + |a_n||\beta_0| < \left(|a_{n-n_1+1}| + |a_{n-n_1+2}| + \ldots + |a_n|\right)\cdot K$$

$$< \frac{\varepsilon}{2K}\cdot K = \frac{\varepsilon}{2} \qquad (7.4.14)$$

Therefore, $|r_n| < \varepsilon$ for $n > n_1 + n_2 - 1$ and the proof is completed.

The next result is given without proof.

Theorem 7.4.3 (Abel's theorem). If $\sum_{n=0}^{\infty} a_n$, $\sum_{n=0}^{\infty} b_n$ converge to A, B and Cauchy product of these series converge to C, then $C = AB$.

The last topic in this section is *infinite products* defined as follows.

Definition 7.4.1. For an arbitrary sequence $\{a_n\}$ denote

$$P_1 = \prod_{k=1}^{1} a_1 = a_1 \quad ; \quad P_n = \prod_{k=1}^{n} a_k = \left(\prod_{k=1}^{n-1} a_k \right) \cdot a_n \text{ , } n \geq 2 \tag{7.4.15}$$

The sequence $\{P_n\}$ is a sequence of *partial products* which converges to $\prod_{k=1}^{\infty} a_k$ if and only if $\lim_{n \to \infty} P_n = P$ for some finite P. In this case we write $\prod_{k=1}^{\infty} a_k = P$. If $\{P_n\}$ converges then the *infinite product* $\prod_{k=1}^{\infty} a_k$ *converges*, otherwise it *diverges*.

Example 7.4.3. Let $a_n = \frac{n}{n+1}$, $n \geq 1$. Here $P_n = \prod_{k=1}^{n} a_k = \frac{1}{n+1}$, i.e., $\lim_{n \to \infty} P_n = 0$ which yields $\prod_{k=1}^{\infty} a_k = 0$.

Theorem 7.4.4. If $\prod_{k=1}^{\infty} a_k$ converges to $P \neq 0$ then $\lim_{n \to \infty} a_n = 1$.

Proof.

Since $a_n = \frac{P_n}{P_{n-1}}$ we get $\lim_{n \to \infty} a_n = \frac{\lim_{n \to \infty} P_n}{\lim_{n \to \infty} P_{n-1}} = \frac{P}{P} = 1$.

Theorem 7.4.5. If $a_n > 0$ for all n then $\prod_{k=1}^{\infty} a_k$ converges if and only if $\sum_{k=1}^{\infty} \ln(a_k)$.

The proof is left as an exercise for the student.

PROBLEMS

1. Find the Cauchy product of $1+\frac{1}{2}+\frac{1}{4}+\frac{1}{8}+\ldots$ and $1+\frac{1}{3}+\frac{1}{9}+\frac{1}{27}+\ldots$
2. Find the Cauchy product of $1+\frac{1}{2}+\frac{1}{4}+\frac{1}{8}+\ldots$ and $1+\frac{1}{2^2}+\frac{1}{3^2}+\frac{1}{4^2}+\ldots$
3. Let $\sum a_n = \sum b_n = 1-\frac{1}{\sqrt{2}}+\frac{1}{\sqrt{3}}-\frac{1}{\sqrt{4}}+\ldots$ (a) Find the Cauchy product of the given series. (b) Is the Cauchy product convergent?

7.5 Power Series and Taylor Series

Definition 7.5.1. An infinite series

$$\sum_{n=0}^{\infty} a_n (x-a)^n = a_0 + a_1(x-a) + \ldots + a_n(x-a)^n + \ldots \tag{7.5.1}$$

where the coefficients $a_n, n \ge 0$ and a are real numbers and $-\infty < x < \infty$ is called a *power series* about a.

Example 7.5.1. The series $x + x^3 + x^5 + \ldots + x^{2n+1} + \ldots$ is a power series about 0. All the odd coefficients are zero, while the even ones equal to 1.

The next result relates to the convergence of a power series. For sake of simplicity we assume $a = 0$.

Theorem 7.5.1. Assume that a power series converges for some fixed $x = x_0 \ne 0$. Then, it converges absolutely for arbitrary $x < x_0$ and converges uniformly for $x \le R$ for any fixed $R < x_0$.

Proof.

Since $\sum_{n=0}^{\infty} a_n x_0^n$ converges, its general term converges to 0. Consequently, an integer N such that $\left|a_n x_0^n\right| \le 1$, $n \ge N$ exists. Hence

$$\sum_{n=0}^{\infty}\left|a_n x^n\right|=\sum_{n=0}^{N-1}\left|a_n x^n\right|+\sum_{n=N}^{\infty}\left|a_n x^n\right|=\sum_{n=0}^{N-1}\left|a_n x^n\right|+\sum_{n=N}^{\infty}\left|a_n x_0^n\right|\left|\frac{x}{x_0}\right|^n \tag{7.5.2}$$

Since $|x/x_0|<1$ and by virtue of $\left|a_n x_0^n\right|\le 1$, $n \ge N$, the right-hand side of Eq. (7.5.2) converges. The second part of Theorem 7.4.1 is now easily concluded and is left as an exercise for the reader.

Example 7.5.2. Consider the power series

$$x+\frac{x^2}{2!}+\ldots+\frac{x^n}{n!}+\ldots$$

It is easily seen to converge for $x=1$. Consequently it converges absolutely for all $x<1$ and converges uniformly for $x \le R$ where R is an arbitrary number less than 1.

Example 7.5.3. The power series

$$\sum_{n=1}^{\infty} nx^n = x+2x^2+\ldots+nx^n+\ldots \tag{7.5.3}$$

converges for all arbitrary x, $|x|<1$. Indeed, let a_n denote the general term of the series and $|x|<1$. Then

$$\left|\frac{a_{n+1}}{a_n}\right|=\frac{(n+1)|x|}{n}<\sqrt{|x|} \tag{7.5.4}$$

provided that (why?)

$$n>\frac{\sqrt{|x|}}{1-\sqrt{|x|}} \tag{7.5.5}$$

Consequently, since $\sqrt{|x|}<1$, the power series converges absolutely and given an arbitrary fixed $R<1$, converges uniformly for all $x \le R$.

If a power series diverges for some fixed $x=x_0 \ne 0$, then it must diverge for all x such that $|x|>|x_0|$. Indeed, if it converges for such x, then, by implementing Theorem 7.4.1 it must converge for x_0 as well! We thus get the following result.

Theorem 7.5.2. For an arbitrary power series $\sum_{n=0}^{\infty} a_n x^n$ there are two possible cases:

1. $\sum_{n=0}^{\infty} a_n x^n$ converges for all x, $-\infty < x < \infty$.

2. A finite $R \geq 0$ exists such that $\sum_{n=0}^{\infty} a_n x^n$ converges for $|x| < R$ and diverges for $|x| > R$. The number R is called the *radius of convergence.*

The next result provides a procedure for calculating the radius of convergence.

Theorem 7.5.3. Given an arbitrary power series $\sum_{n=0}^{\infty} a_n x^n$, its radius of convergence R is obtained by

$$\frac{1}{R} = \overline{\lim_{n\to\infty}} \sqrt[n]{|a_n|} \tag{7.5.6}$$

Proof.

Let $K = \overline{\lim_{n\to\infty}} \sqrt[n]{|a_n|}$. Then, for arbitrary fixed x we have

$$\overline{\lim_{n\to\infty}} \sqrt[n]{|a_n x^n|} = |x| K \tag{7.5.7}$$

and consequently, by Cauchy's root test (Theorem 7.2.9) it implies that $\sum_{n=0}^{\infty} a_n x^n$ converges for $|x|K < 1$ and diverges if $|x|K > 1$. Thus, if $K > 0$ the radius of convergence is

$$R = \frac{1}{K}$$

while in the case $K = 0$ the series converges for all x, $-\infty < x < \infty$.

Example 7.5.4. Consider the power series

$$1 + x + 2x^2 + \frac{x^3}{3} + 4x^4 + \ldots + \frac{x^{2n+1}}{(2n+1)} + 2nx^{2n} + \ldots \tag{7.5.8}$$

The reader can easily obtain

$$\overline{\lim_{n\to\infty}} \sqrt[n]{|a_n|} = 1$$

which implies that the series is convergent for $|x| < 1$ and divergent for $|x| > 1$. Clearly, the series is not convergent for $|x| = 1$ (why?).

The next theorem, given without proof, relates to the derivative of a power series.

Theorem 7.5.4. Consider a power series $\sum_{n=0}^{\infty} a_n x^n$ with radius of convergence R.

Then, the function $f(x) = \sum_{n=0}^{\infty} a_n x^n$, $|x| < R$ possesses a derivative at this domain which satisfies

$$f'(x) = \sum_{n=1}^{\infty} n a_n x^{n-1}, |x| < R \tag{7.5.9}$$

Moreover, $f(x)$ is infinitely differentiable at $|x| < R$ and as in Eq. (7.5.9) the various derivatives are obtained by repeatedly differentiating the power series term by term, i.e.,

$$f^{(k)}(x) = \sum_{n=k}^{\infty} n(n-1)\cdots(n-k+1)x^{n-k} , |x| < R \tag{7.5.10}$$

Example 7.5.5. Let $f(x) = x + 2x^2 + 3x^3 + \ldots + nx^n + \ldots$. Its radius of convergence is 1. Hence

$$f'(x) = 1 + 4x + 9x^2 + \ldots + n^2 x^{n-1} + \ldots , |x| < 1$$
$$f''(x) = 4 + 18x + 48x^2 + \ldots + n^2(n-1)x^{n-2} + \ldots , |x| < 1$$

etc.

We end this chapter by introducing the power series somewhat differently, using the concept of Taylor series.

Taylor series

Let a function $f(x)$ and its derivatives $f'(x), f''(x), \ldots, f^{(n)}(x)$ be defined and continuous in a closed interval $[a,b]$ with $f^{(n+1)}(x)$ continuous at the open interval (a,b). By Taylor's theorem (see Section 5.4)

$$f(x) = f(a) + f'(a)(x-a) + \frac{f''(a)}{2!}(x-a)^2 + \ldots + \frac{f^{(n)}(a)}{n!}(x-a)^n + R_n \quad (7.5.11)$$

where the remainder

$$R_n = \frac{f^{(n+1)}(\xi)}{(n+1)!}(x-a)^{n+1}, \; a < \xi < x \quad (7.5.12)$$

is called *remainder by Lagrange*. Another form for the remainder

$$R_n = \frac{f^{(n+1)}(\xi')}{n!}(x-\xi')^n (x-a), \; a < \xi' < x \quad (7.5.13)$$

is called *remainder by Cauchy*, and usually $\xi \neq \xi'$. Clearly, if n varies, ξ usually varies as well. If $\lim_{n \to \infty} R_n = 0$, then $f(x)$ can be expressed by a power series as

$$f(x) = f(a) + f'(a)(x-a) + \frac{f''(a)}{2!}(x-a)^2 + \ldots + \frac{f^{(n)}(a)}{n!}(x-a)^n + \ldots \quad (7.5.14)$$

This series is called *Taylor series* of $f(x)$ about a. The existence of a formal Taylor series (as a converging series!) of a function is *no guarantee* that Eq. (7.5.14) holds. In addition to its existence, the condition $\lim_{n \to \infty} R_n = 0$ must hold as well.

The advantage of representing a function by its Taylor series is obvious. An infinite power series as are polynomials, is easy to manipulate and study its basic properties and behavior.

Example 7.5.6. Let $f(x) = \sin(x)$, $-\infty < x < \infty$. The remainder by Lagrange is

$$|R_n| = \frac{|x^{n+1} A(\xi)|}{(n+1)!}$$

where $A(\xi)$ is either $\sin(\xi)$ or $\cos(\xi)$, i.e., $|A(\xi)| \leq 1$. Since, for all real x

$$\lim_{n\to\infty} \frac{x^n}{n!} = 0$$

we get $\lim_{n\to\infty} R_n = 0$ and consequently

$$\sin(x) = x - \frac{x^3}{3!} + \frac{x^5}{5!} - \ldots + (-1)^{n-1}\frac{x^{2n-1}}{(2n-1)!} + \ldots \quad , \quad -\infty < x < \infty \tag{7.5.15}$$

Similarly we obtain

$$\cos(x) = 1 - \frac{x^2}{2!} + \frac{x^4}{4!} - \ldots + (-1)^{n-1}\frac{x^{2n-2}}{(2n-2)!} + \ldots \quad , \quad -\infty < x < \infty \tag{7.5.16}$$

Example 7.5.7. Consider the exponential function $f(x) = e^x$, $-\infty < x < \infty$. Since $f'(x) = e^x$ (see Corollary 5.3.2) we easily get

$$f^{(n)}(x) = e^x \ , n \ge 1$$

and the function's Taylor series is

$$1 + x + \frac{x^2}{2!} + \ldots + \frac{x^n}{n!} + \ldots$$

The remainder by Lagrange

$$R_n = \frac{x^{n+1}}{(n+1)!} e^{\xi} \ , 0 < \xi < x$$

converges to 0 for arbitrary fixed x and thus

$$e^x = 1 + x + \frac{x^2}{2!} + \ldots + \frac{x^n}{n!} + \ldots \ -\infty < x < \infty \tag{7.5.17}$$

A case of a function $y = f(x)$ whose Taylor series exists and converges – but not to $f(x)$ is given next.

Example 7.5.8. Consider the function

$$y = \begin{cases} e^{-1/x^2} \ , x \ne 0 \\ 0 \ , x = 0 \end{cases} \tag{7.5.18}$$

It is infinitely differentiable for all $x \neq 0$ and by using induction we get

$$y^{(n)} = e^{-1/x^2} P_n\left(\frac{1}{x}\right), x \neq 0 \tag{7.5.19}$$

where $P_n(u)$ is a polynomial in u. By Eq. (7.5.17) we get $e^x > x^n/n!$ for all $x > 0$ and consequently can easily derive $\lim_{x\to 0} \frac{e^{-1/x^2}}{x^m} = 0$ for all positive integer m, which, combined with Eq. (7.5.19) leads to

$$\lim_{x\to 0} (1/x) y^{(n)}(x) = 0 , n \geq 1 \tag{7.5.20}$$

Finally, by using Eq. (7.5.20) and induction, we get $y^{(n)} = 0 , n \geq 0$. The Taylor series in this case is thus identically zero, while the function itself is not. This *implies* that Lagrange remainder of the function for arbitrary x *does not converge* to zero as $n \to \infty$.

PROBLEMS

1. Let $f(x) = 1 + x + \frac{x^2}{2} + \frac{x^3}{3} + \dots$. Find the radius of convergence, and the first two derivatives.
2. Find Taylor series of $\ln|1+x|$ about $x = 0$. Find the radius of convergence and show that the remainder approaches zero.
3. (a) Repeat problem 2 for $f(x) = \ln|1-x|$. (b) What can be said about $y = \ln\left|\frac{1+x}{1-x}\right|$?
4. Obtain the derivative of $\sin(x)$ directly and by using its Taylor series.
5. Obtain the Taylor series $\tan^{-1}(x)$ about $x = 0$ and find its radius of convergence.
6. Obtain the Taylor series of e^x about $x = 0$ and show that it converges uniformly for $|x| \leq 100$. Does it converge uniformly for $|x| \leq 200$? Explain.

8 Fourier Series

In the previous chapter we have studied how functions which satisfy certain requirements can be replaced by an infinite series of powers of x. The main advantage of this fact is that a power series or a polynomial, representing a function, is easily manipulated for various applications.

There are however other ways to decompose a function into simple elements. The most popular is given by the Fourier Series Theory, named after the French mathematician Joseph Fourier (1768 – 1830), who, in 1822, started to investigate this theory. The first main result is that under certain conditions, an arbitrary periodic function $f(x)$, can be replaced by a trigonometric series, which includes sine and cosine functions. This representation is very useful in mathematics, physics and engineering sciences.

We open this chapter with basics of Fourier Series Theory.

8.1 Trigonometric Series

Definition 8.1.1. Consider an integrable function $f(x)$ over the interval $[a,b]$ and let $w(x)$ denote a *weight* function that satisfies $w(x)>0$, $a \le x \le b$. The number

$$\|f\| = \left(\int_a^b w(x)[f(x)]^2 dx \right)^{1/2} \tag{8.1.1}$$

is called the *norm* of $f(x)$ over $[a,b]$ with respect to $w(x)$. It has the following properties:

(a) $\|f+g\| \le \|f\| + \|g\|$ for arbitrary $f(x)$ and $g(x)$.

(b) $\|cf\| = |c|\|f\|$ for arbitrary function $f(x)$ and constant c.

(c) If $\|f\|=0$ and $f(x)$ is continuous, then $f(x)=0$, $a \le x \le b$.

M. Friedman and A. Kandel: Calculus Light, ISRL 9, pp. 217–232.
springerlink.com

Example 8.1.1. Let $w(x)=1$, $a=0$, $b=1$. Then $\|x^2\|=\frac{1}{\sqrt{5}}$, $\|e^x\|=\sqrt{\frac{e^2-1}{2}}$ and $\|x-x^2\|=\frac{1}{2\sqrt{3}}$. The proof is left for the reader.

The norm defined by Eq. (8.1.1) can be used for determining how close two functions are to each other. We call $\|f-g\|$ the *distance* between $f(x)$ and $g(x)$ and say that a sequence of functions $\{f_n(x)\}$ converges to $f(x)$ *in norm*, if

$$\lim_{n\to\infty}\|f_n-f\|=0$$

We next introduce the formal expression

$$\frac{a_0}{2}+\sum_{n=1}^{\infty}\left[a_n\cos\left(\frac{n\pi x}{l}\right)+b_n\sin\left(\frac{n\pi x}{l}\right)\right] \tag{8.1.2}$$

which is called a *trigonometric series*. At this point, nothing is said about whether the series converges.

One of the most important features of trigonometric functions is their *periodicity*.

Definition 8.1.2. A function $f(x)$, $x\in D$ is called *periodic* if

$$f(x+T)=f(x)\,,\ x\in D \tag{8.1.3}$$

for some constant T. The number T is called a *period* of $f(x)$. Every integer multiple of T is also a period.

We will always assume $T>0$. Otherwise, reverse the direction of the $x-$ axis. The smallest T which is still a period of $f(x)$, is called the *fundamental period* of $f(x)$. If two functions $f(x)$ and $g(x)$ share the same period T, then T is also a period of $f(x)+g(x)$. The next result presents some of the properties of the trigonometric functions in Eq. (8.1.2).

Theorem 8.1.1. The functions $\cos\left(\frac{n\pi x}{l}\right)$, $\sin\left(\frac{n\pi x}{l}\right)$; $n\ge 1$ are periodic with fundamental period $\frac{2l}{n}$ and together with the constant function, say $\frac{1}{2}$

(the coefficient of a_0 in Eq. (8.1.2)), are mutually orthogonal over the interval $[-l,l]$, i.e.

$$\int_{-l}^{l}\cos\left(\frac{n\pi x}{l}\right)\cos\left(\frac{m\pi x}{l}\right)dx=\begin{cases}0\,, & m\neq n\\ l\,, & m=n\end{cases}$$

$$\int_{-l}^{l}\sin\left(\frac{n\pi x}{l}\right)\sin\left(\frac{m\pi x}{l}\right)dx=\begin{cases}0\,, & m\neq n\\ l\,, & m=n\end{cases}$$

$$\int_{-l}^{l}\sin\left(\frac{n\pi x}{l}\right)\cos\left(\frac{m\pi x}{l}\right)dx$$

$$\int_{-l}^{l}\cos\left(\frac{n\pi x}{l}\right)dx=\int_{-l}^{l}\sin\left(\frac{n\pi x}{l}\right)dx=0$$

$$\int_{-l}^{l}\frac{1}{2}dx=l$$

The proof follows from basic trigonometric identities and is left as an exercise for the reader.

An immediate result of this theorem is that *if* the series of Eq. (8.1.2) *converges uniformly* to a function $f(x)$, which therefore must have a period $2l$ (why?), then the coefficients are

$$a_n=\frac{1}{l}\int_{-l}^{l}f(x)\cos\left(\frac{n\pi x}{l}\right)dx\,,\ n\geq 0$$

$$b_n=\frac{1}{l}\int_{-l}^{l}f(x)\sin\left(\frac{n\pi x}{l}\right)dx\,,\ n\geq 1 \tag{8.1.4}$$

To obtain the Eq. (8.1.4) we multiply the equality

$$f(x)=\frac{a_0}{2}+\sum_{n=1}^{\infty}\left[a_n\cos\left(\frac{n\pi x}{l}\right)+b_n\sin\left(\frac{n\pi x}{l}\right)\right]$$

by $\cos\left(\frac{n\pi x}{l}\right)$, $n \ge 0$ and by $\sin\left(\frac{n\pi x}{l}\right)$, $n \ge 1$, integrate over $[-l,l]$ and apply the orthogonality of the trigonometric functions. The formal series of Eq. (8.2.1), where a_n, b_n are taken from Eq. (8.1.4), is called the *Fourier series* of $f(x)$, and its coefficients are the *Fourier coefficients* of $f(x)$. Whether this series converges and if so, does it converge to $f(x)$, is a question of utmost importance that will be later investigated.

Example 8.1.2. Consider the function

$$f(x) = \begin{cases} -x, & -l \le x \le 0 \\ x, & 0 < x < l \end{cases}$$

which we extend periodically by $f(x+2l) = f(x)$ for all x. The result is a periodic function with period $2l$, called a *triangular wave* (Fig. 8.1.1).

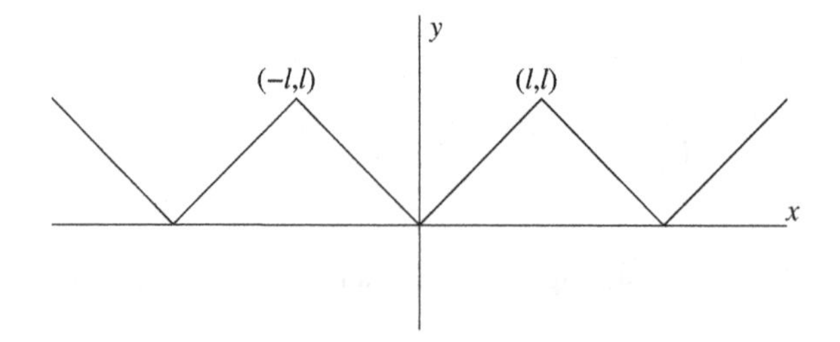

Fig. 8.1.1. A triangular wave.

Its Fourier coefficients are

$$a_0 = l\,;\; a_n = \frac{2l}{n^2\pi^2}[\cos(n\pi) - 1],\; n \ge 1\,;\; b_n = 0,\; n \ge 1$$

and the attached formal Fourier series is

$$f(x) \sim \frac{l}{2} - \frac{4l}{\pi^2}\sum_{n=1}^{\infty}\frac{1}{(2n-1)^2}\cos\left[\frac{(2n-1)\pi x}{l}\right]$$

As stated, whether the right-hand side converges to $f(x)$ remains to be seen. We can however, calculate partial sums S_k, $k \ge 1$ of the first k terms of the series and compare with $f(x)$. A comparison with S_2 and S_3 for $l = \pi$, illustrated in Fig. 8.1.2, demonstrates how well even S_2 approximates the function.

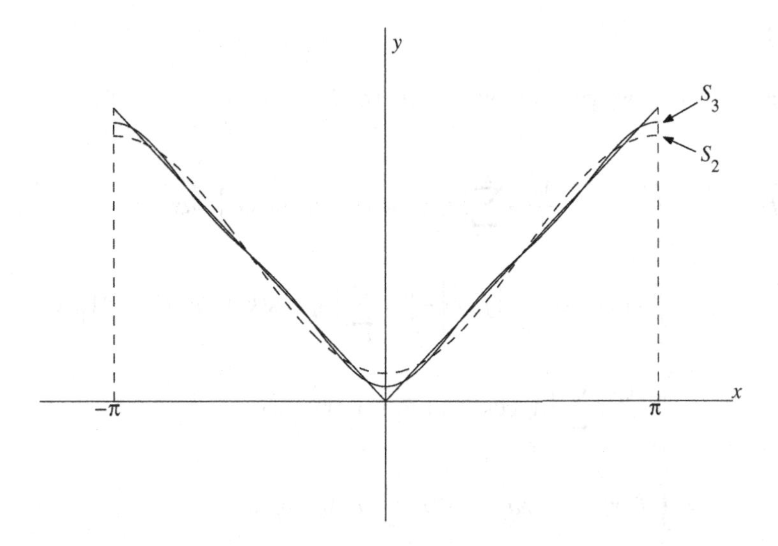

Fig. 8.1.2. The triangular wave for $l = \pi$; the approximates S_2 (dashed) and S_3 .

Before discussing the *convergence problem*, it should noted that if $f(x)$ is just integrable and periodic, valuable information about its Fourier coefficients may be already derived. For the sake of simplicity we will assume $l = \pi$ in Eq. (8.1.2), i.e. a function $f(x)$ with period 2π is considered. Let G denote the set of all the linear combinations

$$g(x) = \frac{A_0}{2} + \sum_{k=1}^{n} [A_k \cos(kx) + B_k \sin(kx)] \tag{8.1.5}$$

and we search for a function $g \in G$ which is closest to $f(x)$.

Theorem 8.1.2. If $f(x)$ is an integrable function over $[-\pi, \pi]$ and

$$S_n = \frac{a_0}{2} + \sum_{k=1}^{n} [a_n \cos(nx) + b_n \sin(nx)]$$

where $a_0, a_1, b_1, \ldots, a_n, b_n$ are given by Eq. (8.1.4). Then $\|f - S_n\| \leq \|f - g\|$ for all $g(x)$ of Eq. (8.1.5), i.e. among the linear combinations of the trigonometric functions $\{1, \cos(x), \sin(x), \ldots, \cos(nx), \sin(nx)\}$ the partial sum of Fourier series is *always* the best approximate to $f(x)$ in the norm defined by Eq. (8.1.1).

Proof.

Using Eq. (8.1.4), we get for arbitrary $g(x) \in G$

$$\|f-g\|^2 = \int_{-\pi}^{\pi} \left\{ f(x) - \frac{A_0}{2} - \sum_{k=1}^{n} [A_k \cos(kx) + B_k \sin(kx)] \right\}^2 dx$$

$$= \int_{-\pi}^{\pi} [f(x)]^2 dx - 2\int_{-\pi}^{\pi} f(x) \left(\frac{A_0}{2} + \sum_{k=1}^{n} [A_k \cos(kx) + B_k \sin(kx)] \right) dx$$

$$+ \int_{-\pi}^{\pi} \left(\frac{A_0}{2} + \sum_{k=1}^{n} [A_k \cos(kx) + B_k \sin(kx)] \right)^2 dx$$

$$= \int_{-\pi}^{\pi} [f(x)]^2 dx - \pi a_0 A_0 - 2\pi \sum_{k=1}^{n} (a_k A_k + b_k A_k)$$

$$+ \int_{-\pi}^{\pi} \left(\frac{A_0}{2} + \sum_{k=1}^{n} [A_k \cos(kx) + B_k \sin(kx)] \right)^2 dx$$

Furthermore, due to $\int_{-\pi}^{\pi} \cos^2(kx)dx = \int_{-\pi}^{\pi} \sin^2(kx)dx = \pi$ and the mutual orthogonality of the functions $\{1, \cos x, \sin x, \ldots, \cos(nx), \sin(nx)\}$, we conclude

$$\|f-g\|^2 = \int_{-\pi}^{\pi} [f(x)]^2 dx - \pi a_0 A_0 - 2\pi \sum_{k=1}^{n} (a_k A_k + b_k A_k) + \frac{\pi A_0^2}{2} + \pi \sum_{k=1}^{n} (A_k^2 + B_k^2)$$

By adding and subtracting an identical expression to this last equality, one gets

$$\|f-g\|^2 = \int_{-\pi}^{\pi} [f(x)]^2 dx - \frac{\pi a_0^2}{2} - \pi \sum_{k=1}^{n} (a_k^2 + b_k^2)$$

$$+ \frac{\pi a_0^2}{2} - \pi a_0 A_0 + \frac{\pi A_0^2}{2} + \pi \sum_{k=1}^{n} [(a_k - A_k)^2 + (b_k - B_k)^2]$$

$$= \int_{-\pi}^{\pi} [f(x)]^2 dx - \frac{\pi a_0^2}{2} - \pi \sum_{k=1}^{n} (a_k^2 + b_k^2) + \pi \underbrace{\left[\frac{(a_0 - A_0)^2}{2} + \sum_{k=1}^{n} [(a_k - A_k)^2 + (b_k - B_k)^2] \right]}_{\geq 0}$$

Consequently, the particular choice $A_k = a_k$, $0 \le k \le n$ and $B_k = b_k$, $1 \le k \le n$ yields

$$\|f - S_n\|^2 = \int_{-\pi}^{\pi} [f(x)]^2 dx - \frac{\pi a_0^2}{2} - \pi \sum_{k=1}^{n} (a_k^2 + b_k^2) \tag{8.1.6}$$

and the proof is concluded since $\|f - g\|^2 = \|f - S_n\|^2 + \sigma^2 \ge \|f - S_n\|^2$, where

$$\sigma^2 = \pi \left[\frac{(a_0 - A_0)^2}{2} + \sum_{k=1}^{n} [(a_k - A_k)^2 + (b_k - B_k)^2 \right]$$

Since $\|f - S_n\|^2 \ge 0$, Eq. (8.1.6) implies the *Bessel's inequality*

$$\frac{1}{\pi} \int_{-\pi}^{\pi} [f(x)]^2 dx \ge \frac{a_0^2}{2} + \sum_{k=1}^{n} (a_k^2 + b_k^2) \tag{8.1.7}$$

Example 8.1.3. Let $f(x) = x^2, -\pi \le x \le \pi$. The Fourier coefficients are

$$a_0 = \frac{1}{\pi} \int_{-\pi}^{\pi} x^2 dx = \frac{2\pi^2}{3} \quad ; \quad a_n = \frac{4\cos(n\pi)}{n^2}, \ b_n = 0, \ n \ge 1$$

and Bessel's inequality yields

$$\frac{1}{\pi} \int_{-\pi}^{\pi} x^4 dx = \frac{2\pi^4}{5} \ge \frac{2\pi^4}{9} + 16 \sum_{k=1}^{n} \frac{1}{k^4} \quad , \quad n \ge 1$$

or

$$\pi^4 \ge 90 \sum_{k=1}^{n} \frac{1}{k^4} \quad , \quad n \ge 1$$

PROBLEMS

1. Find the norms of $\cos(nx), \sin(nx), xe^x$ over $[-\pi, \pi]$.
2. Let $f(x)$, $-\infty < x < \infty$ denote an integrable function with period T. Show that for arbitrary a,b:

(i) $$\int_0^a f(x)dx = \int_T^{a+T} f(x)dx$$

(ii) $$\int_0^T f(x)dx = \int_a^{a+T} f(x)dx$$

(iii) $$\int_a^{a+T} f(x)dx = \int_b^{b+T} f(x)dx$$

3. Let $f(x) = x$, $-\pi \le x < \pi$. Extend the function periodically to all x and obtain its formal Fourier series
4. Let $f(x) = \sin^2 x$, $0 \le x < \pi$. Extend it periodically, obtain its Fourier series and compare between $f(x)$ and the partial sums S_2, S_3, S_4 over the interval $0 \le x < \pi$.
5. Find the Fourier series of $f(x) = |\sin x|$, $-\pi < x < \pi$, extended as a periodic function with a period 2π.
6. Obtain the Fourier series generated by the following functions:

(a) $f(x) = x^3$, $-1 \le x < 1$; $f(x+2) = f(x)$

(b) $f(x) = \cos^2 x$, $-\pi \le x < \pi$; $f(x+2\pi) = f(x)$

(c) $f(x) = \begin{cases} \cos^2 x, & 0 \le x \le \pi \\ 0, & -\pi < x \le 0 \end{cases}$; $f(x+2\pi) = f(x)$

(d) $f(x) = \begin{cases} \cos x, & 0 \le x < \pi \\ \sin x, & -\pi < x \le 0 \end{cases}$; $f(x+2\pi) = f(x)$

7. Obtain the Bessel's inequality for $f(x) = x^3$, $-\pi \le x < \pi$.

8. Show that if the Fourier series of a continuous function $f(x)$ over the interval $-l \le x \le l$, converges uniformly to $f(x)$, than the Parseval's identity

$$\frac{1}{l}\int_{-l}^{l} [f(x)]^2 dx = \frac{a_0^2}{2} + \sum_{n=1}^{\infty} (a_n^2 + b_n^2) \quad (8.1.8)$$

holds.

8.2 Convergence

The formal Fourier series generated by a periodic function $f(x)$ may or may not converge to the function. In this section we will present a set of sufficient conditions, that if satisfied by $f(x)$, guarantee the convergence of the series. First we introduce the concept of piecewise continuity.

Definition 8.2.1. A function $f(x)$, defined over an interval $[a,b]$ except maybe at a finite number of points, is called *piecewise continuous* if for some given partition

$$a = x_0 < x_1 < \cdots < x_{n-1} < x_n = b \tag{8.2.1}$$

of the interval $[a,b]$, the following requirements are fulfilled for all $i : 0 \le i \le n-1$:

1. $f(x)$ is continuous over the open interval (x_i, x_{i+1}).
2. $f(x)$ converges from inside to finite limits at both endpoints, i.e.

$$\lim_{x \downarrow x_i} f(x) = f_i^+ \, , \quad \lim_{x \uparrow x_{i+1}} f(x) = f_{i+1}^- \tag{8.2.2}$$

Thus, a piecewise continuous function may not be defined at some partition point x_i, but it must possess there left and right limits if $x_i \in (a,b)$, right limit if $x_i = a$ and left limit if $x_i = b$. We may also define a piecewise continuous function as a function defined and continuous over an interval, except for a finite number of points where the function has finite jumps.

Example 8.2.1. The function

$$f(x) = \begin{cases} x \, , 0 < x < 1 \\ 2 \, , 1 \le x < 3 \\ x \, , 3 < x < 4 \end{cases}$$

which is illustrated in Fig. 8.2.1 is piecewise continuous over the interval $[0,4]$. It is not defined at $x = 0,3,4$ but has finite limits everywhere at the partition points.

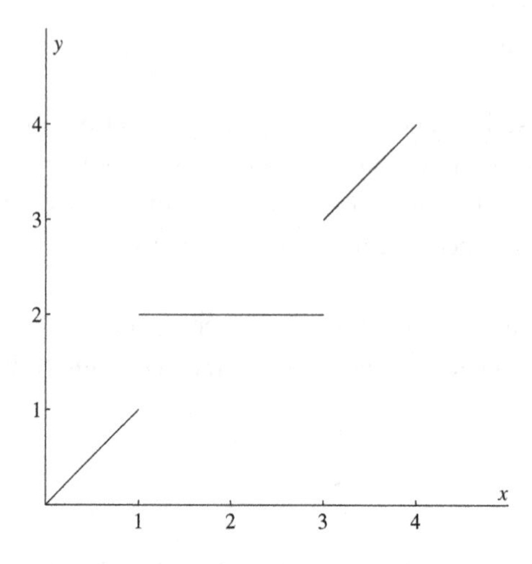

Fig. 8.2.1. The piecewise continuous function of Example 8.2.1.

Example 8.2.2. The function

$$f(x) = \begin{cases} \dfrac{1}{2-x}, & 0 < x < 2 \\ \dfrac{4}{x}, & 2 < x < 8 \end{cases}$$

illustrated in Fig. 8.2.2 is not piecewise continuous over $[0,8]$ since it has an infinite jump at $x = 2$.

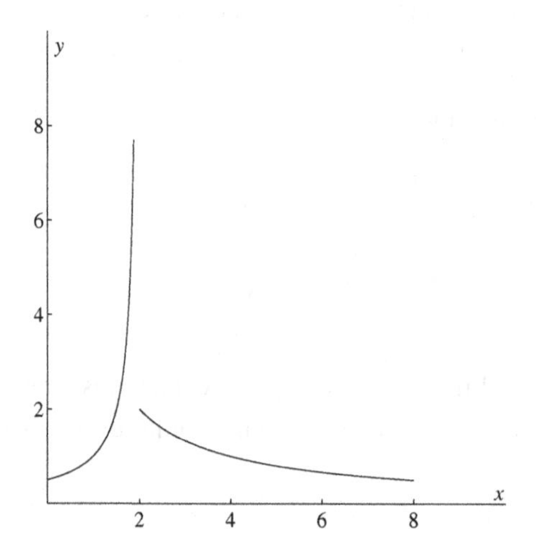

Fig. 8.2.2. A non piecewise continuous function.

The next result which is presented without proof, provides quite general sufficient conditions for the convergence of Fourier series.

Theorem 8.2.1 (The convergence theorem). Let $f(x)$ and $f'(x)$ be piecewise continuous over the interval $[-l,l]$ and let $f(x)$ be extended periodically for $-\infty < x < \infty$. Then, the generated Fourier series

$$\frac{a_0}{2}+\sum_{n=1}^{\infty}\left[a_n\cos\left(\frac{n\pi x}{l}\right)+b_n\sin\left(\frac{n\pi x}{l}\right)\right]$$

where the coefficients are given by Eq. (8.1.4), converges to

$$\frac{1}{2}\left[\lim_{t\downarrow x}f(t)+\lim_{t\uparrow x}f(t)\right],-l\le x\le l \tag{8.2.3}$$

uniformly in x. In particular, if f is continuous at x, then

$$\frac{a_0}{2}+\sum_{n=1}^{\infty}\left[a_n\cos\left(\frac{n\pi x}{l}\right)+b_n\sin\left(\frac{n\pi x}{l}\right)\right]=f(x)$$

Corollary 8.2.1. Let $f(x)$ and $f'(x)$ be continuous functions over a closed interval and let $f(x)$ be extended periodically for all x. Then, the generated Fourier series converges uniformly to $f(x)$.

Example 8.2.3. Consider the piecewise continuous *sawtooth* function (Fig. 8.2.3) defined over a single period as

$$f(x)=\begin{cases}x+1\,,-1<x<0\\ x\,,0<x<1\end{cases}$$

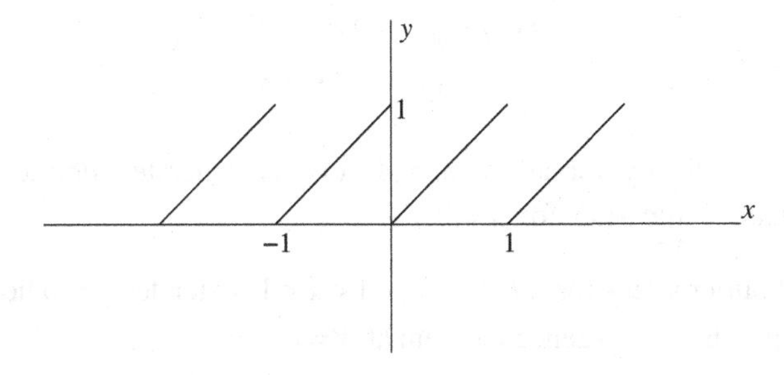

Fig. 8.2.3. The sawtooth function of Example 8.2.3.

whose derivative is $f'(x)=1$, $x \neq 0$. The coefficients of the generated Fourier series are $a_0=1; a_n=0$, $n \geq 0$ and $b_n=-\frac{1+\cos(n\pi)}{n\pi}$, $n \geq 1$. Therefore,

$$f(x) \sim \frac{1}{2}-2\left(\frac{\sin(2\pi x)}{2\pi}+\frac{\sin(4\pi x)}{4\pi}+\frac{\sin(6\pi x)}{6\pi}+\ldots\right)$$

The average of the right and left limits of $f(x)$ at $x=0$ is $\frac{1}{2}$ and as we see it is also the value of the Fourier series. A familiar result is derived by substituting $x=\frac{1}{4}$:

$$f\left(\frac{1}{4}\right)=\frac{1}{4}=\frac{1}{2}-\frac{1}{\pi}\left(1-\frac{1}{3}+\frac{1}{5}-\ldots\right)$$

which implies

$$\frac{\pi}{4}=1-\frac{1}{3}+\frac{1}{5}-\ldots$$

Similar identities all of which are consequences of Theorem 8.2.1 may be obtained by taking other values of x.

PROBLEMS

1. Let $f(x)$ denote a continuous periodic function over $(-\infty,\infty)$ with period $2l$. What can you tell about the generated Fourier series of $f(x)$?
2. Consider the function

$$3. \quad f(x)=\begin{cases} x\,, -1<x<0 \\ 1\,, 0<x<1 \\ 2-x\,, 1<x<2 \end{cases}$$

extended periodically for all x and denote its generated Fourier series by $S(x)$. Calculate $\lim_{x\to a} S(x)$ for $a=0,1,2$.

3. Generate Fourier series for $f(x)=x^2$, $-1<x<1$, extended periodically for all x, and apply the convergence theorem at $x=1$.

4. Are the requirements of Theorem 8.2.1 satisfied by $f(x)=\sqrt{x}$, $0<x<2$? By $f(x)=\sqrt{x}$, $0.001<x<2$?

8.3 Even and Odd Functions

Quite often a Fourier series of a function consists either of only cosine or sine terms. This occurs when the function is even or odd respectively. Let us recall the basic properties of such functions:

(a) The sum, difference, product and quotient of two even functions are even functions.
(b) The sum and difference of two odd functions are odd functions.
(c) The product and quotient of two odd functions are even functions.
(d) Let $f(x)$, $g(x)$ denote even and odd integrable functions respectively over the interval $[-l,l]$. Then

$$\int_{-l}^{l} f(x)dx = 2\int_{0}^{l} f(x)dx \quad , \quad \int_{-l}^{l} g(x)dx = 0$$

8.3.1 Even Functions

Consider an even periodic function $f(x)$ with period $2l$. Since the trigonometric functions $\cos\left(\frac{n\pi x}{l}\right)$, $\sin\left(\frac{n\pi x}{l}\right)$ are even and odd respectively, so are the functions $f(x)\cos\left(\frac{n\pi x}{l}\right)$, $f(x)\sin\left(\frac{n\pi x}{l}\right)$. Consequently, the Fourier coefficients of $f(x)$ are

$$a_n = \frac{2}{l}\int_{0}^{l} f(x)\cos\left(\frac{n\pi x}{l}\right)dx \quad , \quad n \ge 0$$

$$b_n = 0 \quad , \quad n \ge 1 \tag{8.3.1}$$

and the expansion

$$f(x) \sim \frac{a_0}{2} + \sum_{n=1}^{\infty} a_n \cos\left(\frac{n\pi x}{l}\right)$$

is called a *Fourier cosine series.*

Example 8.3.1. Let $f(x)$ denote the triangular wave of Example 8.1.2. Here

$$a_0 = \frac{2}{l}\int_0^l x dx = l \; ; \; a_n = \frac{2}{l}\int_0^l x \cos\left(\frac{n\pi x}{l}\right) dx = \frac{2l}{n^2\pi^2}[\cos(n\pi) - 1] \, , \; n \ge 1$$

The particular choice $l = 1$ yields

$$f(x) \sim \frac{1}{2} - \frac{4}{\pi^2}\left[\cos(\pi x) + \frac{1}{3^2}\cos(3\pi x) + \frac{1}{5^2}\cos(5\pi x) + \ldots\right]$$

Since $f(x)$ is continuous, the convergence theorem guarantees the convergence of the Fourier series to $f(x)$ for arbitrary x. In particular

$$f(0) = 0 = \frac{1}{2} - \frac{4}{\pi^2}\left[1 + \frac{1}{3^2} + \frac{1}{5^2} + \ldots\right]$$

and consequently

$$\frac{\pi^2}{8} = \sum_{n=1}^{\infty} \frac{1}{(2n-1)^2} \tag{8.3.2}$$

8.3.2 *Odd Functions*

Consider now an odd periodic function $f(x)$ with period $2l$. In this case the functions $f(x)\cos\left(\frac{n\pi x}{l}\right)$, $f(x)\sin\left(\frac{n\pi x}{l}\right)$ are odd and even respectively and the Fourier coefficients of $f(x)$ are

$$\begin{aligned} b_n &= \frac{2}{l}\int_0^l f(x)\sin\left(\frac{n\pi x}{l}\right) dx \quad , \quad n \ge 1 \\ a_n &= 0 \quad , \quad n \ge 0 \end{aligned} \tag{8.3.3}$$

and the expansion

$$f(x) \sim \sum_{n=1}^{\infty} b_n \sin\left(\frac{n\pi x}{l}\right)$$

is called a *Fourier sine series*.

Example 8.3.2. Let $f(x)=x\,,-1<x<1$ extended periodically for all x. Integration by parts yields

$$b_n = 2\int_0^1 x\sin(n\pi x)dx = -\frac{2\cos(n\pi)}{n\pi}\,,\ n\ge 1$$

and the generated Fourier sine series is

$$f(x) \sim \frac{2}{\pi}\left[\sin(\pi x) - \frac{1}{2}\sin(2\pi x) + \frac{1}{3}\sin(3\pi x) - \ldots\right]$$

Since $f(x)$ is continuous at $x=\frac{1}{2}$, the convergence theorem yields

$$\frac{1}{2} = \frac{2}{\pi}\left[1 - \frac{1}{3} + \frac{1}{5} - \ldots\right] \tag{8.3.4}$$

a relation already obtained in Example 8.2.3.

Examples 8.3.1 and 8.3.2 illustrate that the function $f(x)=x\,,0<x<1$ can be expanded as two different Fourier series, namely its Fourier cosine and sine series. This does not contradict anything since the cosine series represents the function *extended as an even function* over the complete interval $[-1,1]$, while the sine series represents a different function, namely the *odd extension* of $f(x)$. In other words the two Fourier series represent the same function over one half of its domain, but represent a different function over the other half.

The Fourier cosine and sine series were obtained by choosing specific modes for extending a function $f(x)\,,0<x<l$ to an additional interval $-l<x<0$ and then carrying the extension periodically for all x. However, sometimes we may want to extend a function differently, for example $f(x)=0\,,-l<x<0$. In such a case we obtain a mixed Fourier series, consisting of both cosine and sine terms.

Example 8.3.3. Consider the function

$$f(x) = \begin{cases} 0\,, -1<x<0 \\ x\,, 0<x<1 \end{cases}$$

Integration by parts yield

$$a_0 = \frac{1}{2}\,;\, a_n = \frac{\cos(n\pi)-1}{n^2\pi^2}\,,\ n\ge 1\,;\, b_n = -\frac{\cos(n\pi)}{n\pi}\,,\ n\ge 1$$

i.e.,

$$f(x) \sim \frac{1}{4} + \sum_{n=1}^{\infty} \left[\frac{\cos(n\pi) - 1}{n^2 \pi^2} \cos(n\pi x) - \frac{\cos(n\pi)}{n\pi} \sin(n\pi x) \right]$$

Substituting $x = 0$ (where the function is continuous) yields the familiar Eq. (8.3.2). What do we have at $x = \frac{1}{2}$?

PROBLEMS

1. Which of the following functions are even, odd or neither.

   (a) $|x|^3$
   (b) $|x+1| + |x-1|$
   (c) $e^{|x|} \sin x$
   (d) $e^{-x^2}(1+x)$
   (e) $x^2 + x^3$

2. Let $f(x)$, $g(x)$ denote even and odd smooth functions respectively. Show $f'(0) = 0$ and $g(0) = 0$.
3. Show that the derivatives of even and odd functions are odd and even functions respectively.
4. Find the Fourier cosine series for the following functions:

   (a)

$$f(x) = \begin{cases} x \, , 0 < x < 1 \\ 2 \, , 1 < x < 2 \end{cases} , \quad \text{period 4}$$

   (b)

$$f(x) = \begin{cases} 0 \, , 0 < x < 1 \\ x \, , 1 < x < 2 \\ 1 \, , 2 < x < 3 \end{cases} , \quad \text{period 6}$$

5. Find the Fourier sine series for the functions in problem 4.

9 Elementary Numerical Methods

Calculus is certainly one of the pillars of modern science and the cornerstone to most applications of mathematics in other disciplines. In order to apply its basic ideas to real world problems, additional study of subjects such as differential equations and numerical analysis is eventually needed. However, the authors feel that short introductions to these topics, rather than scare the students, may show them the benefits of mastering the mathematical computational tools presented in this book.

9.1 Introduction

A *numerical method* is a model for solving a problem for which a solution consists of calculating one or several numbers. For example, a numerical method to find an approximate to the derivative $f'(x)$ is to calculate $f(x+h)$, $f(x-h)$ for a "small" h and take

$$f'(x) \approx \frac{f(x+h)-f(x-h)}{2h}$$

The quality of this approximation depends on the size of h and we can show that

$$\lim_{h\to 0}\frac{f(x+h)-f(x-h)}{2h} = f'(x)$$

Usually, we want to approximate a solution to a problem within some desired degree of *precision* or *tolerance*. The first step is finding a numerical method that can perform that task. Then, we design an *algorithm*, based on this method, which is a *finite* sequence of steps that need to be executed in order to obtain the approximate solution. We emphasize the word *finite* since whether we use a pencil, a calculator or a computer, we must confine ourselves to a finite number of arithmetic operations. The number of operations may increase if a higher accuracy is requested, but it will always stay finite.

Example 9.1.1. In order to solve the linear equation $ax+b=c$ with arbitrary coefficients a,b,c we can use the following algorithm:

M. Friedman and A. Kandel: Calculus Light, ISRL 9, pp. 233–261.
springerlink.com

Step 1. Calculate $d = c - b$.

Step 2. If $a = 0$ write 'there is no solution' and stop; If $a \neq 0$ calculate $x = \frac{d}{a}$ and stop.

Example 9.1.2. Consider the problem of finding the square root of a given $a > 1$. An efficient numerical method for calculating an approximate to $\sqrt{a}$, is to create the sequence

$$x_0 = a\,;\, x_n = \frac{1}{2}\left(x_{n-1} + \frac{a}{x_{n-1}}\right),\, n \geq 1 \tag{9.1.1}$$

The process of calculating each x_n is called *iteration* and we often use this term for the number x_n as well. Since $\{x_n\}$ is a monotone decreasing sequence (why?) and $\lim_{n\to\infty} x_n = \sqrt{a}$ (for details see Section 9.3 below), this method enables us to approximate $\sqrt{a}$ to any given degree of precision. The algorithm designed for this purpose will be formulated next, after we set an agreeable definition for convergence for our problem.

Given an approximate s to $\sqrt{a}$, we measure its accuracy by the distance $| s - \sqrt{a} |$. Let $\varepsilon > 0$ denote a prefixed desired tolerance. Unfortunately, we do not know $\sqrt{a}$ and are unable to compute $| s - \sqrt{a} |$. However, since

$$| s^2 - a | = | s - \sqrt{a} || s + \sqrt{a} | \geq 2\sqrt{a}\, | s - \sqrt{a} | > 2 | s - \sqrt{a} |$$

we may replace the *convergence test* defined here by the inequality $| s - \sqrt{a} | < \varepsilon$, by the stronger request $| s^2 - a | < 2\varepsilon$ which implies $| s - \sqrt{a} | < \varepsilon$ (why?). The algorithm can be represented as follows:

Step 1. Given an arbitrary positive number $a > 1$ and a desired tolerance $\varepsilon > 0$, choose an *initial approximate* to $\sqrt{a}$ (from above), say $x_0 = a$ and set $n = 0$.

Step 2. Use Eq. (9.1.1) to get the iteration x_{n+1}.

Step 3. If $\left|x_{n+1}^2 - a\right| < 2\varepsilon$ the computation is over and x_{n+1} is taken as the final approximate to $\sqrt{a}$. The total number of iterations needed for *convergence* is $n+1$; If $\left|x_{n+1}^2 - a\right| \geq 2\varepsilon$ set $n \leftarrow n+1$ and go to Step 2.

Since in fact $\lim_{n\to\infty} x_n = \sqrt{a}$, this is indeed an algorithm, i.e. the process must end and produce the desired accuracy after a finite number of steps.

If for example $a = 10$, $\varepsilon = 10^{-3}$ we get $x_1^2 - 10 = 20.2500$, $x_2^2 - 10 = 3.3889$, $x_3^2 - 10 = 0.2144$ and $x_4^2 - 10 = 0.0011 < 2 \cdot 10^{-3}$. Therefore, the approximate $x_4 = 3.1625$ satisfies $\left|x_4 - \sqrt{10}\right| < 10^{-3}$. The number of iterations needed for convergence is 4.

The process used in Example 9.1.2 is a typical *iterative process*. During such process, in order to calculate an unknown quantity a, we usually generate a sequence $\{x_0, x_1, x_2, \ldots, x_n, \ldots\}$ of iterations that converges to a. Generally, a complete convergence requires an infinite number of iterations (as in Example 9.1.2) which we are unable to perform. Therefore we set some tolerance $\varepsilon > 0$ and stop the process when we first obtain $|x_n - a| < \varepsilon$. However, as already stated, since a is unknown, this convergence test is impractical. In most cases we therefore apply a different test and stop the iteration when the inequality $|x_{n+1} - x_n| < \varepsilon$ first occurs. If a relation $|x_n - a| < C|x_{n+1} - x_n|$ can be found (see Section 9.2), we simply iterate until $|x_{n+1} - x_n| < \frac{\varepsilon}{C}$ and thus guarantee $|x_n - a| < \varepsilon$.

PROBLEMS

1. Present an algorithm to obtain the roots of a general quadratic equation $ax^2 + bx + c = 0$.
2. Find an algorithm to obtain $\sqrt{a}$ for arbitrary $a > 0$. Assume that the sequence of Eq. (9.1.1) converges to $\sqrt{a}$.
3. Show that for arbitrary $a > 0$, the sequence created by Eq. (9.1.1) is monotone decreasing for all $a \ge 1$. What can be derived when $a < 1$?
4. Archimedes believed at some point that $\pi = 3.14159\cdots$ is exactly $\sqrt{10}$. Use Eq. (9.1.1) and problem 3 to show that he was wrong.
5. Find the number of iterations needed to obtain $\sqrt{9}$ with accuracy 10^{-8}, using the algorithm given by Eq. (9.1.1).

9.2 Iteration

In this section a detailed case study of an iteration problem and algorithms for solving it are introduced.

We are concerned with the equation

$$x = f(x) \tag{9.2.1}$$

where $f(x)$ is a real-valued function defined over the interval $[a,b]$. Let us first present a formal algorithm for solving this equation. By 'solving' we mean, finding an approximate value to the exact solution s of Eq. (9.2.1). The algorithm presented here is called the *Standard Iteration Method* (SIM).

Algorithm 1 (SIM)

Step 1. Choose a first approximation x_0 to the exact unknown solution s, a tolerance $\varepsilon > 0$ and a maximum number of iterations N that we allow for the process. Set $n = 0$, where n denotes the current number of iterations already performed.

Step 2. Evaluate the next iteration $x_{n+1} = f(x_n)$.

Step 3. If $|x_{n+1} - x_n| > \varepsilon$ and $n+1 < N$ set $n \leftarrow n+1$ and go to Step 2; If $|x_{n+1} - x_n| > \varepsilon$ and $n+1 \geq N$ write 'maximum number of iterations exceeded' and stop; If $|x_{n+1} - x_n| \leq \varepsilon$ set $n \leftarrow n+1$, write 'convergence ended successfully after n iterations with an approximate x_n' and stop.

Notice that we have not yet discussed whether Eq. (9.2.1) *possesses* a solution and whether this solution is *unique*. We also did not state the requirements that need to be satisfied to assure that Algorithm 1 provides a real approximate to s. The next result provides sufficient conditions for the existence of a solution to Eq. (9.2.1).

Theorem 9.2.1.

Let $f(x)$, a real-valued function over the interval $[a,b]$, satisfy:

1. $a \leq f(x) \leq b$ for all $x : a \leq x \leq b$.
2. $f(x)$ is continuous over $[a,b]$.

Then, Eq. (9.2.1) has at least one solution s.

Imposing the first requirement is essential if Algorithm 1 is to be applied. Indeed, if $f(x_n) \notin [a,b]$ for some n, then x_{n+1} is simply not defined.

Proof.

Let $g(x) = x - f(x)$. Since $a \leq f(a), f(b) \leq b$ by requirement 1, we get

$$g(a) = a - f(a) \leq 0$$
$$g(b) = b - f(b) \geq 0$$

If $g(a)=0$ or $g(b)=0$ then $s=a$ or $s=b$ respectively is a solution. Otherwise, $g(a)<0$, $g(b)>0$ and the continuity of $g(x)$ guarantees the existence of $s: a<s<b$ such that $g(s)=0$, i.e. $s=f(s)$. This completes the proof.

A typical function that satisfies the requirements of Theorem 9.2.1 is illustrated in Fig. 9.2.1. Note that there may be more than one solution to Eq. (9.2.1).

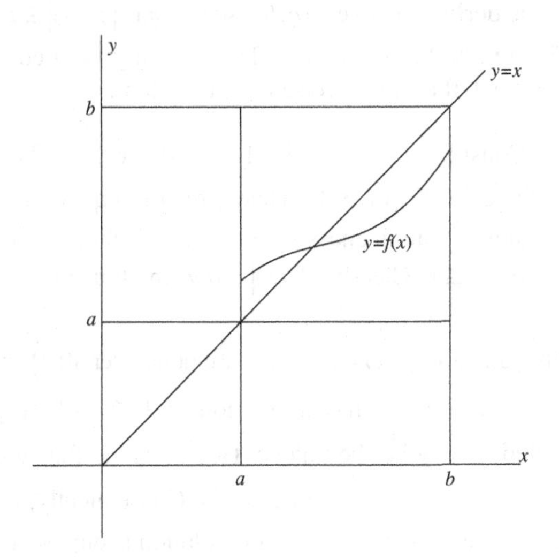

Fig. 9.2.1. Illustrating $x=f(x)$.

The next result presents sufficient conditions for the existence of a *unique* solution to Eq. (9.2.1).

Theorem 9.2.2.

Let $f(x)$, a real-valued function over $[a,b]$, satisfy the requirements of Theorem 9.2.1 and assume the existence of a constant $L: 0<L<1$ such that

$$|f(x_1)-f(x_2)| \le L|x_1-x_2| \;,\quad x_1,x_2 \in [a,b] \tag{9.2.2}$$

Then, Eq. (9.2.1) has a unique solution s in $[a,b]$.

If $|f(x_1)-f(x_2)| \le L|x_1-x_2|$ for all $x_1,x_2 \in [a,b]$, we say that a Lipschitz condition holds for $f(x)$. For uniqueness we request a Lipschitz condition with $0<L<1$.

Proof.

The existence of at least one solution is guaranteed by Theorem 9.2.1. Let s_1, s_2 denote two solutions to Eq. (9.2.1). Then $s_1-s_2=f(s_1)-f(s_2)$. Consequently

$$|s_1 - s_2| = |f(s_1) - f(s_2)| \le L|s_1 - s_2|$$

and if $s_1 \ne s_2$ we divide the inequality by $|s_1 - s_2|$ and get $L \ge 1$ which contradicts the third requirement of Theorem 9.2.2. Hence, $s_1 = s_2$ as stated.

Let $f(x)$ possess a derivative over $[a,b]$ such that $|f'(x)| \le L < 1$. Then, the requirement of Theorem 9.2.2 is satisfied. This is easily verified using the mean-value theorem, and it is left as an exercise for the student.

Example 9.2.1. Consider $f(x) = 0.9x(1-x)$, $0 \le x \le 1$. This function is continuous and satisfies $0 \le f(x) \le 1$. Therefore, the equation $x = f(x)$ has a solution. To show uniqueness let us check the derivative $f'(x) = 0.9 - 1.8x$, $0 \le x \le 1$. Clearly $|f'(x)| \le 0.9$ and thus the solution is unique.

Example 9.2.2. The function $f(x) = e^{-x}$ is continuous over $[0,1]$ and $0 \le e^{-x} \le 1$. Therefore, the equation $x = e^{-x}$ has a solution $s: 0 \le s \le 1$ (Fig. 9.2.1). Since $|f'(0)| = 1$ the solution may not be unique. However, the functions x and $f(x)$ are strictly increasing and decreasing, respectively. Consequently, it is impossible to have more than one solution to $x = e^{-x}$, i.e. the solution is unique (Fig. 9.2.1).

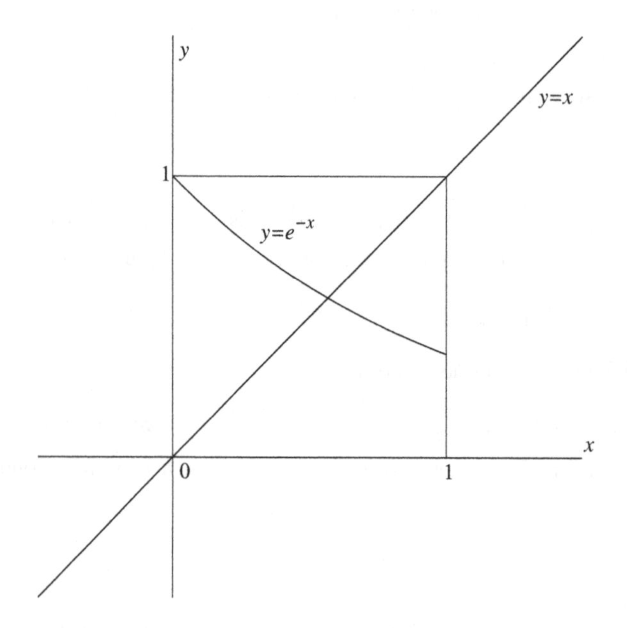

Fig. 9.2.2. A unique solution to $x = e^{-x}$, $0 \le x \le 1$.

The *error* at the n-th iteration is defined as the difference between the approximate x_n and the exact solution s, i.e.

$$\varepsilon_n = x_n - s \tag{9.2.3}$$

Clearly, the algorithm SIM is meaningful only if x_n converges to s, i.e. if $\lim_{n\to\infty} \varepsilon_n = 0$. We will now show that this is the case if $f(x)$ satisfies the requirements of Theorem 9.2.1 and a Lipschitz condition with $L < 1$. Indeed,

$$|x_n - s| = |f(x_{n-1}) - f(s)| \le L|x_{n-1} - s| \le \cdots \le L^n |x_0 - s|$$

Since $L < 1$ we have $\lim_{n\to\infty} L^n = 0$ and consequently $\lim_{n\to\infty} |x_n - s| = 0$ as well.

Example 9.2.3. Consider the equation $x = f(x)$ where $f(x) = 1 - x^3/6$ over the interval $[0,1]$. The requirements of Theorem 9.2.1 are satisfied and a Lipschitz condition with $L = 1/2$ holds. Since $s \le 1$, the choice $x_0 = 0$ provides $|x_n - s| \le 1/2^n$. Thus, to guarantee an error of less then 0.001, ten iterations are needed.

The most important property of an algorithm is whether or not this algorithm leads to convergence or simply *converges*. However, from practical point of view one should also consider the *speed* of convergence. It makes a difference whether 9 or 9,000 iterations are needed to obtain an error below a given tolerance. For example, a popular expression for π is

$$\pi = 4\left(1 - \frac{1}{3} + \frac{1}{5} - \frac{1}{7} + \ldots\right)$$

but the student, correctly, will find it unattractive: to reach accuracy of 10^{-3} no less than a thousand terms are needed. On the other hand, another well known relation

$$\frac{\pi^4}{90} = \sum_{n=1}^{\infty} \frac{1}{n^4}$$

that yields

$$\pi = \left(90 \sum_{n=1}^{\infty} \frac{1}{n^4}\right)^{0.25}$$

is more appealing: the same accuracy is obtained by adding only six terms!

Given an iteration method that approximates the solution s via a sequence $\{x_n\}_{n=0}^{\infty}$, we now define the convergence rate of the sequence to s, as follows:

Definition 9.2.1. Let $\lim_{n\to\infty} x_n = s$ and let the error sequence $\{\varepsilon_n\}_{n=0}^{\infty}$ defined by Eq. (9.2.3), satisfy

$$\lim_{n\to\infty}\left|\frac{\varepsilon_{n+1}}{|\varepsilon_n|^{\alpha}}\right| = K\ ,\ K > 0 \tag{9.2.4}$$

for some constants α and K. Then α is called the *order of convergence* of the given method. If $\alpha = 1$ and $K < 1$ the method is called *linear*. In the case $\alpha = 2$ the convergence is called *quadratic*. The case $\alpha < 1$ is impossible if the procedure converges (why?).

If the error sequence is such that $|\varepsilon_{n+1}| \le L|\varepsilon_n|^{\alpha}$ for sufficiently large n, the order of convergence is at least α (i.e. no worse than α). For example, the standard iteration method for solving $x = f(x)$ is at least linear, provided that a Lipschitz condition with $L < 1$ exists for $f(x)$. However, a careful analysis reveals that if $f(x)$ has a continuous derivative, SIM is exactly linear with $K = f'(s)$.

Theorem 9.2.3. Let $f(x)$ satisfy the requirements of Theorem 9.2.1 and in addition:

(1) $f(x) \in C^1[a,b]$.
(2) $0 < |f'(x)| < 1$ for all $x : a \le x \le b$.

Then, for arbitrary x_0, the sequence $\{x_n\}, x_n = f(x_{n-1})$ converges linearly to the unique solution of $x = f(x)$. In particular

$$\lim_{n\to\infty}\frac{\varepsilon_{n+1}}{\varepsilon_n} = f'(s) \tag{9.2.5}$$

and since $0 < f'(s) < 1$, the convergence is linear.

We will now show how by adding very little in terms of computation, we may accelerate the convergence of SIM.

Let $f(x)$ satisfy the requirements of Theorem 9.2.3 and construct the sequence $\{x_n\}, x_n = f(x_{n-1})$. Due to Eq. (9.2.5) we have

$$\varepsilon_{n+1} = (M + \theta_n)\varepsilon_n \quad , \quad \lim_{n \to \infty} \theta_n = 0 \tag{9.2.6}$$

where $M = f'(s)$. If we ignore θ_n for large n, we have approximately

$$x_{n+1} - s = M(x_n - s)$$
$$x_{n+2} - s = M(x_{n+1} - s)$$

which implies $M = (x_{n+2} - x_{n+1})/(x_{n+1} - x_n)$ and consequently

$$s = x_n - \frac{(x_{n+1} - x_2)^2}{x_{n+2} - 2x_{n+1} + x_n} \tag{9.2.7}$$

This equation holds only approximately, but the right-hand side can be used as a far better approximation to s than x_n. In fact, if we denote

$$x_n' = x_n - \frac{(x_{n+1} - x_2)^2}{x_{n+2} - 2x_{n+1} + x_n} \, , \, \varepsilon_n' = x_n' - s$$

it can be shown that

$$\lim_{n \to \infty} \frac{\varepsilon_{n+1}'}{\varepsilon_n'} = M^2 \tag{9.2.8}$$

and since $0 < M^2 < M < 1$, the sequence $\{x_n'\}$ converges faster than $\{x_n\}$ to s. Notice though, that this improved procedure, called the *Aitken's method* is also linear. The improvement is by decreasing the constant K of Eq. (9.2.4) from M to M^2. This by itself already implies that $\lim_{n \to \infty} \varepsilon_n'/\varepsilon_n = 0$ as can be easily shown by the reader.

Example 9.2.4. Consider the equation $x = \cos^2(x)/2$ over the interval $[0,1]$. All the requirements of Theorem 9.2.3 are satisfied and let us take an initial approximation $x_0 = 1$. Table 9.2.1 provides a comparison between SIM and Aitken's method and demonstrates the superiority of the latter. While SIM requests 15 iterations to obtain six accurate digits, only 6 iterations are needed by Aitken's method for the same precision.

Table 9.2.1. SIM vs. Aitken's method for $x = \cos^2(x)/2$, $x_0 = 1$.

n	x_n	x'_n
0	1.000000	0.390914
1	0.145963	0.412016
2	0.489423	0.417241
3	0.389495	0.417643
4	0.427906	0.417705
5	0.413901	0.417713
6	0.419124	0.417715
7	0.417192	0.417715
⋮	⋮	
15	0.417715	
16	0.417715	

PROBLEMS

1. Show that neither requirement of Theorem 9.2.1 is *necessary* for having a solution to $x = f(x)$.
2. The following functions are continuous over the interval $[0,1]$, satisfy $0 \le f(x) \le 1$ and a Lipschitz condition exists for each of them. Find the solution(s) to $x = f(x)$ and compare the results with the conclusions obtained by observing the Lipschitz constants: (a) $f(x) = 1 - x^2/3$ (b) $f(x) = 0.5 + 0.5\sin(10x)$ (c) $f(x) = x$ (d) $f(x) = 0.5 + 0.2\tan^{-1}[20(x - 0.5)]$
3. Repeat and solve Example 9.2.4 with initial approximations $x_0 = 0.75, 0.5$.
4. Explain the advantage of using Aitken's method to solve $x = f(x)$ in the case of $f(x) = e^{-5x}$, $0 \le x \le 1$.
5. Let $f(x) = 1 - x^3$, $0 \le x \le 1$. Generate the sequence $x_n = f(x_{n-1})$ where $x_0 = 0.5$ and analyze the results.

9.3 The Newton - Raphson Method

We now present one of the most popular algorithms: a simple, easily manipulated and accurate scheme for solving $f(x) = 0$.

Consider the general problem of solving the equation

$$f(x) = 0 \tag{9.3.1}$$

where $f(x)$ is a real-valued function defined over the closed bounded interval $I = [a,b]$ and twice continuously differentiable. Let s denote a solution to

Eq. (9.3.1) within the interval and assume that $x_0 \in I$ is a 'close' point to s, i.e., $s = x_0 + h$ where h is 'small'. By applying Taylor's theorem we get

$$0 = f(s) = f(x_0 + h) = f(x_0) + hf'(x_0) + \frac{h^2}{2} f''(c) \tag{9.3.2}$$

where c is an interim point between x_0 and s. Thus, if h is sufficiently small, we are justified in ignoring the term $(h^2/2)f''(c)$ and representing an approximation of h as

$$h \approx -\frac{f(x_0)}{f'(x_0)} \tag{9.3.3}$$

Thus, the value $x_1 = x_0 - f(x_0)/f'(x_0)$ can be expected to provide a better approximation to s than x_0. Now we can use x_1 as an initial approximation and continue the process until we are sufficiently close to the exact solution s. The iterative procedure

$$x_{n+1} = x_n - \frac{f(x_n)}{f'(x_n)} \tag{9.3.4}$$

which is based on Eq. (9.3.3) is called *Newton – Raphson Method* (NRM) and is probably the most popular and powerful tool for solving Eq. (9.3.1).

A geometrical illustration of NRM is given in Fig. 9.3.1: The tangent to the curve $y = f(x)$ at $(x_n, f(x_n))$ intersects the x- axis at x_{n+1}, which is taken as the next approximation to s.

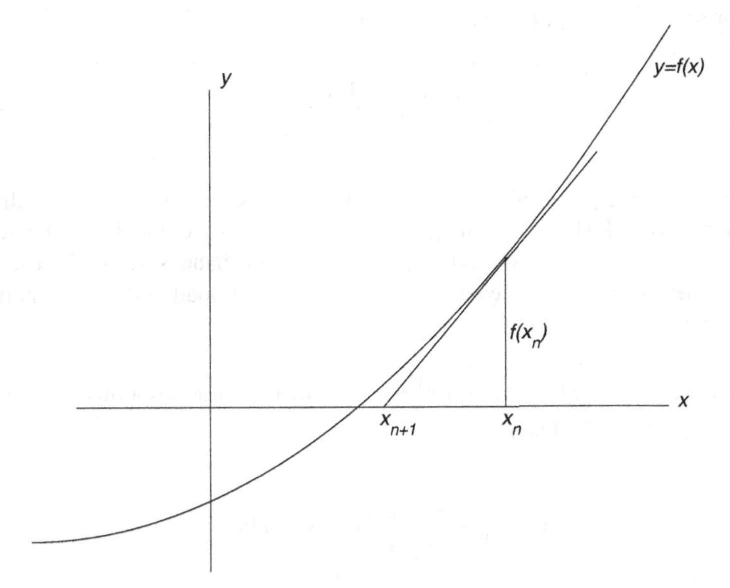

Fig. 9.3.1. The Newton – Raphson method for solving $f(x) = 0$.

Example 9.3.1. Let $f(x) = 2x^4 - x^3 - 1$. One of the solutions of $f(x) = 0$ is $s = 1$. Consider an initial approximation $x_0 = 1.2$. By applying Eq. (9.3.4) we obtain

$$x_1 = x_0 - \frac{2x_0^4 - x_0^3 - 1}{8x_0^3 - 3x_0^2} = 1.050673$$

and by repeating the process $x_2 = 1.004185$, $x_3 = 1.000031$, $x4 = 1.000000$.

The speed of convergence in Example 9.3.1, as compared to that of SIM or even the Aitken's method, is quite impressive. This is no coincidence but follows from the following result.

Theorem 9.3.1. Let $f(x)$ be defined over the closed interval $I = [a,b]$, satisfying the following requirements:

1. $f(x) \in C^2[a,b]$, i.e. $f(x)$ possesses at least two continuous derivatives within I .
2. $f(s) = 0$ for some $s : a < s < b$.
3. $f'(s) \neq 0$

Then, there exists a neighborhood I_η of s: $[s - \eta, s + \eta]$, within I, such that NRM converges to s for arbitrary initial approximation x_0 in I_η. Furthermore, the error sequence $\{\varepsilon_n\}$, $\varepsilon_n = x_n - s$ satisfies

$$\lim_{n \to \infty} \frac{\varepsilon_{n+1}}{\varepsilon_n^2} = \frac{1}{2} \frac{f''(s)}{f'(s)} \tag{9.3.5}$$

Theorem 9.3.1. the proof of which is beyond the scope of this book, states that the convergence of NRM is *at least* quadratic. Indeed, if by coincidence the second derivative vanishes at the solution point s, the right-hand side of Eq. (9.3.5) is zero and the convergence is even faster than the usual quadratic rate as shown in the next example.

Example 9.3.2. Let $f(x) = x^4 - 2x^3 + 1$. One of the solution points of $f(x) = 0$ is $s = 1$. Let $x_0 = 1.2$. Then

$$x_1 = x_0 - \frac{x_0^4 - 2x_0^3 + 1}{4x_0^3 - 6x_0^2} = 0.978704$$

and $x_2 = 1.000019$, $x_3 = 1.000000$. The second derivative is $f''(x) = 12x^2 - 12x$ and the convergence rate which is higher than quadratic, is a consequence of $f''(1) = 0$.

The quadratic convergence of NRM is certainly advantageous. However, there is no guarantee that NRM converges for arbitrary x_0, unless $x_0 \in I_\eta$. Unfortunately, since we do not *a priori* know either s or η, there is no safe choice (without further investigation) for x_0, i.e., a random choice of the initial approximation *may not* lead to convergence to s. The next result (presented without a proof, which is beyond the scope of this text) removes the necessity of getting a good first approximation x_0.

Theorem 9.3.2. Let $f(x)$, a real-valued function, defined over the closed interval $I = [a,b]$, satisfy the following requirements:

1. $f(x) \in C^2[a,b]$.
2. $f(a)f(b) < 0$.
3. $f'(x) \neq 0$, $x \in I$.
4. $f''(x)$ does not change sign within I.
5. If c denotes the endpoint (either a or b) with the smaller value of $|f'(x)|$, then

$$\left|\frac{f(c)}{f'(c)}\right| \le b - a \tag{9.3.6}$$

Then, NRM converges at least quadratically, to the unique solution s of $f(x) = 0$ within $[a,b]$, for *any* first approximation x_0.

Thus, if the requirements of Theorem 9.3.2 are satisfied, no information about the location of s is necessary. Nevertheless, validating requirements 3 and 4 could be tiresome, and one of them or both may not hold at all. Therefore, NRM does not always provide a full-proof convergence procedure for iteratively solving $f(x) = 0$. Yet, in many cases, a simple observation may provide an appropriate choice of x_0.

Example 9.3.3. Let $f(x) = x^4 - x - 1$, $1 \le x \le 2$. The first four requirements of Theorem 9.3.2 are easily validated. The fifth requirement holds as well: $f'(1) = 3$, $f'(2) = 31$ i.e. $c = 1$ and $|f(1)/f'(1)| = 1/3 \le b - a = 1$. Consequently, the NRM converges to the unique solution of $f(x)=0$ at the interval $[1,2]$ for any initial approximation x_0 within this interval.

The requirements stated in Theorem 9.3.2 are sufficient conditions for the convergence of the NRM. If one or more requirements are not satisfied, the process still may or may not lead to convergence. Usually, the iterations will converge to the solution s, provided that $|x_0 - s|$ is sufficiently small. Otherwise, they may converge to a different solution (i.e. if $f(x)=0$ has more than one solution), or diverge, as shown in the next example.

Example 9.3.4. Consider the equation $f(x)=\sin(x)=0$ over the interval $[-\pi/2,\pi/2]$. It has a unique solution $s=0$ which is easily obtained by the NRM, if $x_0 \approx 0$. However, if, for example, we choose as initial approximation the positive solution of the equation $2x=\tan(x)$, i.e. $x_0 = 1.16556\ldots$, the NRM yields $x_1 = -x_0$ since

$$x_1 = x_0 - \frac{f(x_0)}{f'(x_0)} = x_0 - \frac{\sin(x_0)}{\cos(x_0)} = x_0 - 2x_0 = -x_0$$

Similarly, $x_2 = x_0$ and the process oscillates. For smaller values of x_0 the NRM converges to zero and for larger values it converges to a different solution of $\sin(x)=0$, i.e., a solution outside the given interval. Thus, the NRM does not converge for arbitrary $x_0 \in [-\pi/2,\pi/2]$ and therefore some of the requirements of Theorem 5.3.2. (4 and 5) are not satisfied.

PROBLEMS

1. Solve $x^5 - 15 = 0$ using Newton's method. Start with $x_0 = 2$ and stop the iteration when $|x_{n+1} - x_n| < 10^{-8}$.
2. Determine the functions which fulfill the five requirements of Theorem 9.3.2: (a) $x^3 - 6, -1 \le x \le 2$ (b) $|x|, -1 \le x \le 1$ (c) $x^2 \ln(x), 2/3 \le x \le 2$
3. Solve $x^4 - 3x^3 = 0$ using Newton's method with $x_0 = 0.7$. Show that the convergence is slower than expected and provide an explanation.
4. Use the NRM to solve $x - e^{-2x} = 0$ with $x_0 = 1$. Find the number of iterations needed to get $|x_n - x_{n-1}| < 10^{-8}$.
5. Repeat and solve Problem 4 with $x_0 = 1000$ and discuss the two sets of results. Is there a significant difference between them?

6. Replace $f'(x_n)$ by $\dfrac{f(x_n)-f(x_{n-1})}{x_n-x_{n-1}}$ in Eq. (9.3.4) to get

$$x_{n+1} = x_n - \frac{f(x_n)[x_n - x_{n-1}]}{f(x_n)-f(x_{n-1})}$$

Use this algorithm (called the secant method) to solve Problem 4 and compare the results with those obtained by the NRM.

9.4 Interpolation Methods

Quite often it is desirable to replace a given complex function $f(x)$ by a simpler, easily manipulated function $g(x)$, which approximates $f(x)$ over the domain of interest. If the approximate $g(x)$ is chosen such that $f(x)=g(x)$ over a given set of *representative* points, it is called *interpolator*.

Definition 9.4.1. Consider a function $f(x), a \le x \le b$ and a set of points $x_0, x_1, \ldots, x_n \in [a,b]$. The function $g(x), a \le x \le b$ is called an *interpolator* of $f(x)$ at $x_0, x_1, \ldots, x_n$ if

$$g(x_i) = f(x_i), 0 \le i \le n \tag{9.4.1}$$

If the number of points is *sufficiently large* and if they are *spread* over the whole interval, we may expect $g(x)$ to *approximate* $f(x)$ everywhere, i.e.,

$$g(x) \approx f(x), a \le x \le b \tag{9.4.2}$$

A particular useful case is when the interpolator is a polynomial, since polynomials are easily manipulated. We next discuss an example of a polynomial interpolator, show its existence and derive the accuracy guaranteed by replacing an arbitrary function with this interpolator.

9.4.1 Lagrange Polynomial

The first question related to polynomial interpolators is whether they exist. The most trivial interpolator is a straight line, which intersects a given function at two points. This is a *linear interpolator* shown in Fig. 9.4.1. It does not usually provide a useful approximate for the function, unless this particular function is almost linear.

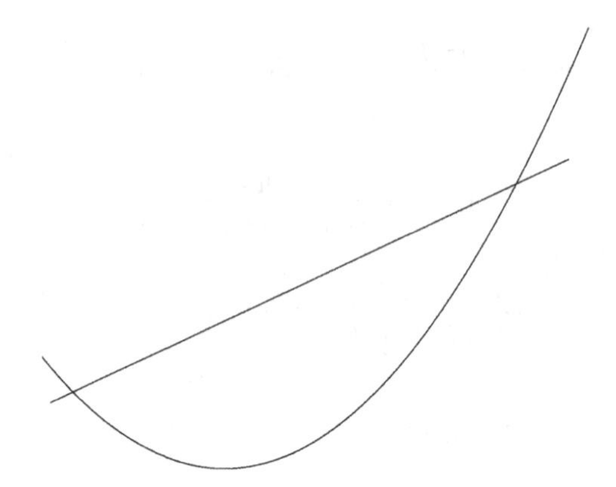

Fig. 9.4.1. A linear interpolator for arbitrary function.

The next result confirms the existence of a polynomial interpolator in the general case.

Theorem 9.4.1. Let $y = f(x)$, $a \le x \le b$. For every data set $\{(x_i, y_i), 0 \le i \le n\}$ where $x_0, x_1, \ldots, x_n \in [a,b]$ are distinct points, there exists a unique n-th order polynomial $p_n(x)$, which satisfies

$$p_n(x_i) = y_i = f(x_i)\,,\ 0 \le i \le n \tag{9.4.3}$$

This is a *Lagrange polynomial* of order n which interpolates $f(x)$ at $x_0, x_1, \ldots, x_n$.

Proof.

Consider the n-th order polynomials

$$l_i(x) = \frac{(x-x_0)(x-x_1)\cdots(x-x_{i-1})(x-x_i)(x-x_{i+1})\cdots(x-x_n)}{(x_i-x_0)(x_i-x_1)\cdots(x_i-x_{i-1})(x_i-x_{i+1})\cdots(x_i-x_n)} \tag{9.4.4}$$

They are well defined since all the points are distinct. Each $l_i(x)$ vanishes at all the points except at x_i, where $l_i(x_i) = 1$. Consequently, the polynomial

$$p_n(x) = \sum_{i=0}^{n} l_i(x) y_i \tag{9.4.5}$$

satisfies Eq. (9.4.3). The uniqueness of Lagrange polynomial is easily seen. Let $q_n(x)$, another n-th order polynomial, satisfy $q_n(x_i) = y_i = f(x_i)$, $0 \le i \le n$. Then, the polynomial $r_n(x) = p_n(x) - q_n(x)$ is of order n and vanishes at $n+1$ distinct points. Therefore $r_n(x)$ is identically zero, i.e. $p_n(x) = q_n(x)$ which concludes the proof.

Naturally, one would expect the quality of a Lagrange polynomial approximation to increase with n. However, quite often, this is not the case. It is somewhat surprising but easily follows from the next result which is presented here without a proof and provides the error obtained by replacing a given function by its interpolator.

Theorem 9.4.2. Consider the function and the data set of Theorem 9.4.1. Then, its Lagrange polynomial satisfies

$$f(x) - p_n(x) = \frac{1}{(n+1)!} l(x) f^{(n+1)}(\xi_x) \, , \; a \le x \le b \tag{9.4.6}$$

where $l(x) = (x - x_0)(x - x_1) \cdots (x - x_n)$ and ξ_x depends on x and is an interim point between a and b.

Thus, the error obtained by replacing a function by its Lagrange interpolator, depends on three quantities. Two of them, $(n+1)!$ and on the $(n+1)-th$ derivative of $f(x)$, are determined by the given function and by the order of the polynomial. The third one, $l(x)$, depends on the structure of the interpolating points $x_0, x_1, \ldots, x_n$.

Example 9.4.1. Consider the data set

$$\{(-2.4, 8.3776), (-2.1, 1.3981), (-1, 0), (0.8, 1.2096), (2.3, 5.5341)\}$$

The 4th order Lagrange interpolator, $y = x^4 - 5x^2 + 4$, which is plotted in Fig. 9.4.2, does not seem to be ideal for representing the data set, since it oscillates while the actual data does not. In fact, the data seems to represent a fast monotonic decreasing function, which after reaching a minimum, increases moderately.

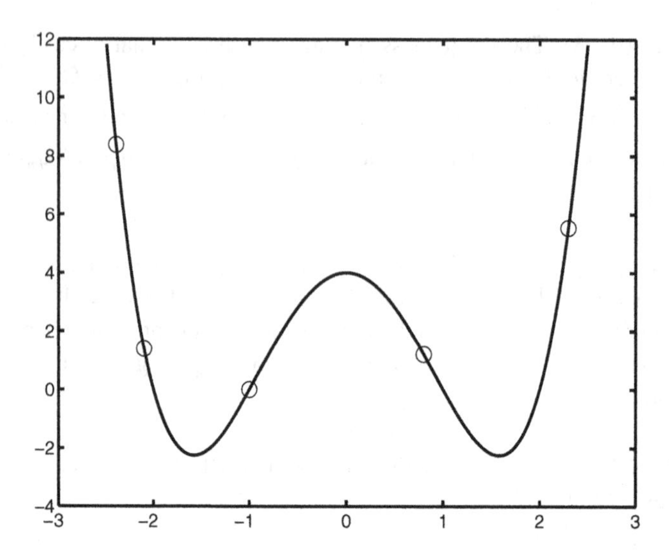

Fig. 9.4.2. An unsuccessful Lagrange interpolation for 5-point data set.

The possible oscillation of high order polynomials is unlikely to occur when low order interpolators are applied. The most popular low order polynomial interpolators are *cubic splines*.

9.4.2 Cubic Splines

A cubic spline is a sequence of combined cubic polynomials, connected in a way that guarantees continuity and two continuous derivatives of the spline at the connecting points:

Definition 9.4.2. A piecewise cubic polynomial with two continuous derivatives is called a *cubic spline*.

Example 9.4.2. Consider the function

$$f(x) = \begin{cases} x^3 + 1 \,,\, 0 \le x < 1 \\ 2x^3 - 3x^2 + 3x \,,\, 1 \le x \le 2 \end{cases}$$

It is a piecewise cubic polynomial everywhere and the first two left and right derivatives at $x = 1$ are the same. Hence, $f(x)$ is a cubic spline over the whole interval $[0,2]$.

Given a data set $\{(x_i, f(x_i)), 0 \le i \le n\}$ we seek a function $s(x)$ which satisfies the following requirements:

(a) $s(x_i) = f(x_i)$, $0 \le i \le n$, i.e., $s(x)$ is an interpolator of $f(x)$.

(b) $s(x)$ is a cubic polynomial at each interval $[x_{i-1}, x_i]$, $1 \le i \le n$.

(c) $s(x) \in C^2[x_0, x_n]$.

(d) $s''(x_0) = s''(x_n) = 0$.

If such $s(x)$ exists, it is by definition a cubic spline and approximates $f(x)$ via interpolation. Requirement (d) suggests that this interpolator is not likely to oscillate.

Theorem 9.4.3. Given a data set $\{(x_i, y_i), 0 \le i \le n\}$ where $y_i = f(x_i)$, $0 \le i \le n$, there exists a *unique* cubic spline (CS) which satisfies requirements (a) – (d).

Proof.

Let $s(x)$ be an arbitrary piecewise cubic polynomial over $[x_0, x_n]$ such that $s''(x)$ is continuous everywhere. Denote $M_i = s''(x_i)$, $0 \le i \le n$. Since $s(x)$ is a cubic polynomial over each subinterval, $s''(x)$ must be linear there. In fact we easily get

$$s''(x) = \frac{(x_i - x)M_{i-1} + (x - x_{i-1})M_i}{x_i - x_{i-1}}, \quad x_{i-1} \le x \le x_i \tag{9.4.7}$$

By integrating twice both sides of Eq. (9.4.7) and substituting the two boundary conditions $s(x_{i-1}) = y_{i-1}$, $s(x_i) = y_i$ we obtain

$$\begin{aligned} s(x) = {} & \frac{(x_i - x)^3 M_{i-1} + (x - x_{i-1})^3 M_i}{6(x_i - x_{i-1})} + \frac{(x_i - x)y_{i-1} + (x - x_{i-1})y_i}{x_i - x_{i-1}} \\ & - \frac{1}{6}(x_i - x_{i-1})[(x_i - x)M_{i-1} + (x - x_{i-1})M_i], \quad x_{i-1} \le x \le x_i \end{aligned} \tag{9.4.8}$$

Clearly, the functions $s(x)$ and $s''(x)$ are continuous everywhere over $[x_0, x_n]$. However, it is *a priori* assumed that $s'(x)$ is continuous as well we must also have (validation of this claim is left for the reader)

$$\begin{aligned} & (x_i - x_{i-1})M_{i-1} + 2(x_{i+1} - x_{i-1})M_i + (x_{i+1} - x_i)M_{i+1} \\ & + 6\left(\frac{y_i - y_{i-1}}{x_i - x_{i-1}} - \frac{y_{i+1} - y_i}{x_{i+1} - x_i}\right) = 0, \quad 1 \le i \le n-1 \end{aligned} \tag{9.4.9}$$

We thus get a system of $(n-1)$ linear equations with $(n+1)$ variables $M_i, 0 \le i \le n$. By imposing two additional conditions $M_0 = M_n = 0$ (requirement (d)) the system can be shown to possess a unique solution given by Eq. (9.4.8).

Example 9.4.3. Consider the data set $\{(0,0),(1,1),(2,3)\}$. The cubic spline that interpolates the data and approximates its function is generated by the unique solution of the linear system

$$\begin{aligned} &M_0 + 4M_1 + M_2 = 6 \\ &M_0 = 0 \\ &M_2 = 0 \end{aligned}$$

which is $M_0 = 0$, $M_1 = 1.5$, $M_2 = 0$. The CS is composed of two cubic polynomials obtained by Eq. (9.4.8) as

$$s(x) = \begin{cases} (x^3 + 3x)/4 \\ (-x^3 + 6x^2 - 3x + 2)/4 \end{cases}$$

PROBLEMS

1. Generate the Lagrange polynomial of the data set

$$\{(-2,19),(-1,3),(0,1),(1,1),(2,15)\}$$

2. Generate the Lagrange polynomial of the data set

$$\{(0,0),(\pi/4,\sin(\pi/4)),(\pi/2,1),(3\pi/4,\sin(3\pi/4)),(\pi,0)\}$$

   What is the maximum difference between the polynomial and the function $\sin(x)$?
3. Calculate the cubic spline of the data set $\{(0,0),(1,1),(2,1),(3,0)\}$.
4. Calculate the cubic spline of the data set $\{(0,0),(1,1),(2,1),(3,1)\}$.
5. Repeat and solve Problem 4 where the requirement $s''(x_0) = s''(x_n) = 0$ is replaced with $s''(x_0) = s''(x_n) = 1$ ($s(x)$ represents the cubic spline generated for the data set). Compare between the solutions.

9.5 Least – Squares Approximations

Consider some experimental data related to unknown function $y = f(x)$. If the data points resemble linear or quadratic curves, then rather than trying to approximate the data by high order Lagrange interpolator or even by the low order piecewise cubic spline, we could find the straight line or parabola that best approaches the data.

9.5.1 Linear Least – Squares Method

Given a data set $\{(x_i, f(x_i)), 1 \le i \le n\}$ and an arbitrary straight line $y = ax + b$, we present the linear discrete least – squares approach as finding the pair $(a,b) = (a_0, b_0)$ which minimizes the *error sum*

$$E = \sum_{i=1}^{n} [f(x_i) - ax_i - b]^2 \tag{9.5.1}$$

The quantity E represents the deviation of the *measurements* from the straight line. In order to get the best linear fit to the data we simply solve

$$\frac{\partial E}{\partial a} = 0 \, , \, \frac{\partial E}{\partial b} = 0 \tag{9.5.2}$$

i.e., a system of two linear equations given by

$$a\sum_{i=1}^{n} x_i^2 + b\sum_{i=1}^{n} x_i = \sum_{i=1}^{n} x_i y_i$$

$$a\sum_{i=1}^{n} x_i + bn = \sum_{i=1}^{n} y_i$$

where $y_i = f(x_i)\,, 1 \le i \le n$.

Example 9.5.1. Consider the following five measurements:

$$\{(-2,-4.3),(-1,-1.8),(0,0),(1,2),(2,3.8)\}$$

which may represent for example a small deviation from the straight line $y = 2x$. The linear least – squares (LLS) procedure provides the equations $10a = 20$ and $5b = -0.3$, i.e., $a = 2$, $b = -0.06$. The approximation's quality is illustrated in Fig. 9.5.1.

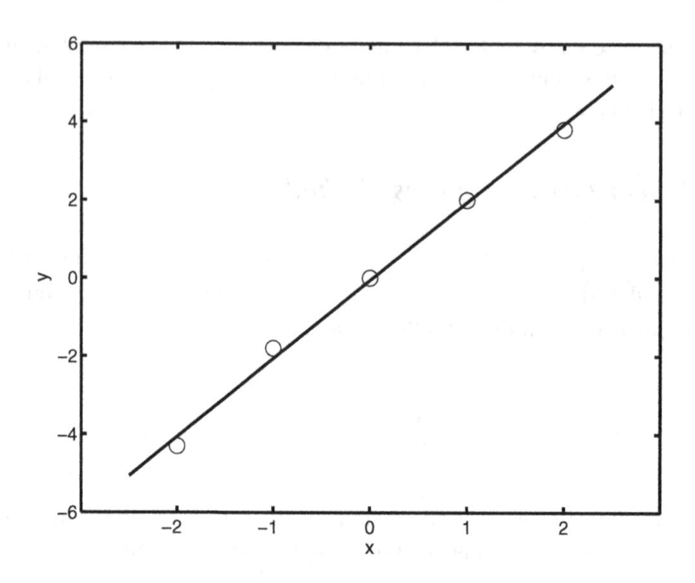

Fig. 9.5.1. Least – squares approximation for Example 9.5.1.

9.5.2 Quadratic Least – Squares Method

Sometimes a quadratic approximation is more likely to represent the data, the measurements resemble a curve such as $y = ax^2 + bx + c$. In this case we minimize the error expression

$$E = \sum_{i=1}^{n} [f(x_i) - ax_i^2 - bx_i - c]^2 \tag{9.5.3}$$

by solving $\dfrac{\partial E}{\partial a} = \dfrac{\partial E}{\partial b} = \dfrac{\partial E}{\partial c} = 0$,i.e.,

$$a\sum_{i=1}^{n} x_i^4 + b\sum_{i=1}^{n} x_i^3 + c\sum_{i=1}^{n} x_i^2 = \sum_{i=1}^{n} y_i x_i^2$$

$$a\sum_{i=1}^{n} x_i^3 + b\sum_{i=1}^{n} x_i^2 + c\sum_{i=1}^{n} x_i = \sum_{i=1}^{n} y_i x_i$$

$$a\sum_{i=1}^{n} x_i^2 + b\sum_{i=1}^{n} x_i + nc = \sum_{i=1}^{n} y_i$$

Example 9.5.2. Consider the 9 – point data set

$$\{(0,1.1),(0.5,2.2),(1,3.8),(1.5,7),(2,10.7),(2.5,16.3),(3,22.1),(3.5,28.5),(4,38)\}$$

The equations for determining the quadratic least – squares approximation are

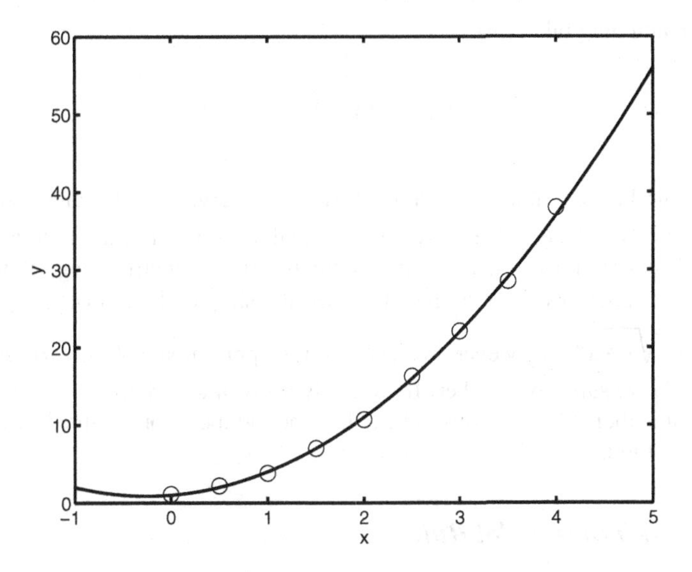

Fig. 9.5.2. Quadratic least – squares approximation.

PROBLEMS

1. Find the linear least-squares approximation for the data set

$$\{(0,1),(2,1),(3,3),(4,5),(6,6)\}$$

2. Use the quadratic least-squares method to approximate the data set of Problem 1.
3. Find the quadratic least-squares approximation for the data set

$$\{(0,0),(1,1),(3,3),(6,6)\}$$

without solving three linear equations.

4. (a) Solve Problem 3 for the data set $\{(0,0),(1,1),(3,3.2),(6,6)\}$.

   (b) Could you anticipate the result of (a) without solving the linear system?
5. Find the linear least-squares approximation for the function $y = e^x$ between $x = 1$ and $x = 2$ using the function values at $x = 1, 1.5, 2$.

6. Find the quadratic least-squares approximation for $y = e^x$ between $x = 1$ and $x = 3$ using the function values at $x = 1, 1.5, 2, 2.5, 2.75, 3$.

9.6 Numerical Integration

The Riemann integral

$$I(f;a,b) = \int_a^b f(x)dx \tag{9.6.1}$$

of an integrable function $f(x)$ defined over an interval $[a,b]$, cannot always be calculated *analytically*, i.e. it may not be possible to obtain it as finite number of *standard* expressions such as polynomials, trigonometric, logarithmic end exponential functions. This is the case even for simple choices of $f(x)$ such as e^x/x or $\sqrt{x+x^6}$. However, we can always approximate $I(f;a,b)$ using one *numerical procedure* or another. If we apply a given algorithm for evaluating the integral, it is then advantageous to be able to bound the error obtained by using the specific method.

9.6.1 *The Trapezoidal Rule*

Let an integrable function $y = f(x)$ be defined over the interval $[a,b]$ and consider the subintervals generated by

$$a = x_0 < x_1 < \ldots < x_{n-1} < x_n = b \tag{9.6.2}$$

Over each $[x_{i-1}, x_i]$ the function is approximated by the secant connecting between (x_{i-1}, y_{i-1}) and (x_i, y_i), and the *area*

$$\int_{x_{i-1}}^{x_i} f(x)dx$$

is approximated by area of the trapezoid defined by $(x_{i-1},0),(x_{i-1},y_{i-1}),(x_i,y_i),(x_i,0)$, i.e., by

$$T_{(i-1,i)} = (x_i - x_{i-1})[f(x_{i-1}) + f(x_i)]/2$$

The whole integral is thus approximated by

$$T_n = \sum_{i=1}^{n} (x_i - x_{i-1})[f(x_{i-1} + f(x_i)] \tag{9.6.3}$$

The procedure given by Eq. (9.6.3) is called the *trapezoidal rule*. Its error estimate is our next result, given here without a proof.

Theorem 9.6.1. If $f(x) \in C^2[a,b]$ and if $x_0, x_1, \ldots, x_{n-1}, x_n$ are evenly spaced, then

$$I(f;a,b) - T_n = -\frac{h^2(b-a)}{12} f''(\xi) \tag{9.6.4}$$

where h is the length of each subinterval and ξ an interim point between a and b.

Hence, if $| f''(x) | \le M$ for all $x \in [a,b]$ then

$$\left| I(f;a,b) - T_n \right| \le \frac{M(b-a)}{12} h^2$$

Example 9.6.1. Let $f(x) = e^x$, $0 \le x \le 2$ and we apply the trapezoidal rule to evaluate $\int_0^2 e^x dx$ using 20 equal subintervals. Here $b - a = 2$, $h = 0.1$ and since $f''(x) = e^x$ we may take $M = e^2 \approx 7.389$. The error bound obtained is 0.0123. In fact, the exact integral is $e^2 - 1 \approx 6.389$ while $T_{20} = 6.394$, providing a real error of only 0.005.

9.6.2 Simpson Rule

The trapezoidal rule basically replaces $f(x)$ by its linear interpolator between each two consecutive points, before performing an exact evaluation of the total area of the generated trapezoids. The next method goes one step further and replaces the function between x_{i-1} and x_{i+1} by a second order Lagrange polynomial which interpolates $f(x)$ at the three consecutive points x_{i-1}, x_i, x_{i+1}. The method is called Simpson Rule and in order to apply it we divide the interval $[a,b]$ into an even number of subintervals.

The second order Lagrange interpolator is (Eqs. (9.4.4) and (9.4.5))

$$p(x;x_{i-1},x_i,x_{i+1})=\frac{(x-x_i)(x-x_{i+1})}{(x_{i-1}-x_i)(x_{i-1}-x_{i+1})}f(x_{i-1})+\frac{(x-x_{i-1})(x-x_{i+1})}{(x_i-x_{i-1})(x_i-x_{i+1})}f(x_i)$$

$$+\frac{(x-x_{i-1})(x-x_i)}{(x_{i+1}-x_{i-1})(x_{i+1}-x_i)}f(x_{i+1}) \qquad (9.6.5)$$

If $x_{i+1}-x_i=x_i-x_{i-1}=h$ we obtain (left as an exercise for the reader)

$$\int_{x_{i-1}}^{x_{i+1}} f(x)dx \approx \int_{x_{i-1}}^{x_{i+1}} p(x;x_{i-1},x_i,x_{i+1})dx=\frac{h}{3}[f(x_{i-1})+4f(x_i)+f(x_{i+1})] \qquad (9.6.6)$$

If the partition of $[a,b]$ is composed of equally spaced points

$$a=x_0<x_1<\ldots<x_{n-1}<x_n=b$$

where $n=2m$, i.e., an even number, the integral, using Simpson rule, is approximated by

$$I(f;a,b)\approx S_n=\frac{h}{3}[f_0+4f_1+2f_2+4f_3+2f_4+\ldots+2f_{n-2}+4f_{n-1}+f_n] \qquad (9.6.7)$$

with an error

$$I(f;a,b)-S_n=-\frac{h^4(b-a)}{180}f^{(4)}(\xi) \qquad (9.6.8)$$

where ξ is some interim point between a and b. Obviously, Eq. (9.6.8) is applicable only if $f(x)$ is sufficiently smooth, which here means possessing four continuous derivatives.

Example 9.6.2. Let $f(x)=\sin(x)$, $0\le x\le\pi$ and consider Simpson rule for calculating $I(\sin(x);0,\pi)$ using 10 equal subintervals. Thus, $b-a=\pi$, $h=\pi/10$ and since $\left|f^{(4)}(x)\right|=|\sin(x)|\le 1$, the error (in absolute value) cannot exceed

$$\frac{(\pi/10)^4\pi}{180}=0.00017$$

However, the exact integral is 2 while the Simpson approximation is found as 2.00011 with an actual error of 0.00011.

Since the error bound of Simpson rule contains h^4 at the numerator and 180 at the denominator an enormously small error is expected even if h is not too small (see previous example where $h \approx 0.3$). However, if $f^{(4)}(x)$ happens to be unbounded, the method may not perform well and other algorithms should be considered.

Example 9.6.3. Consider $f(x) = 1/\sqrt{x}$, $0 < x \le 1$ which we define as 0 at $x = 0$. The function is integrable over the interval $[0,1]$ and

$$\int_0^1 \left(1/\sqrt{x}\right)dx = 2$$

Applying Simpson rule with 10 equal subintervals provides $S_{10} \approx 1.60$ which differs by 20% from the exact value. By decreasing h to 0.05 we obtain $S_{20} = 1.72$ which still differs significantly from the correct answer.

9.6.3 Gaussian Integration

Since in many applications we replace the integrated function by some approximating polynomial, it might be beneficial to apply integration method which provides exact value whenever the integrand is a polynomial. This motivated the Gaussian integration method which is given next.

Theorem 9.6.2. Let $y = f(x)$, $-1 \le x \le 1$. Then, for each integer n there exist a set of distinct points $\{x_i, -1 < x_i < 1; 1 \le i \le n\}$ and a set of *weights* $\{w_i > 0, 1 \le i \le n\}$ such that

$$\int_{-1}^1 f(x)dx = \sum_{i=1}^n w_i f(x_i) \tag{9.6.9}$$

provided that $f(x)$ is a polynomial of order $\le 2n-1$. Furthermore, the two sets are unique.

The proof of this theorem is beyond the scope of this book. Tables of $\{x_i, w_i\}$ for various values of n can be found in every textbook of mathematical tables. The integration scheme presented by Eq. (9.6.9) is called the Gaussian integration method of order n.

Since every polynomial is a linear combination of powers of x, it follows that in order for Eq. (9.6.9) to be valid it must hold for $f(x) = 1, x, x^2, \ldots, x^{2n-1}$. Consequently we obtain a set of $2n$ equations that x_i, w_i need to satisfy:

$$
\begin{aligned}
&f(x)=1:2=w_1+w_2+\ldots+w_n\\
&f(x)=x:0=w_1x_1+w_2x_2+\ldots w_nx_n\\
&f(x)=x^2:2/3=w_1x_1^2+w_2x_2^2+\ldots+w_nx_n^2\\
&\vdots\\
&f(x)=x^{2n-2}:2/(2n-1)=w_1x_1^{2n-2}+w_2x_2^{2n-2}+\ldots+w_nx_n^{2n-2}\\
&f(x)=x^{2n-1}:0=w_1x_1^{2n-1}+w_2x_2^{2n-1}+\ldots+w_nx_n^{2n-1}
\end{aligned}
\tag{9.6.10}
$$

As stated by Theorem 9.6.2, this system of $2n$ nonlinear equations with $2n$ variables possesses a unique solution.

Example 9.6.4. In order to obtain the second order Gaussian method we need to solve four equations, namely

$$
\begin{aligned}
w_1+w_2&=2\\
w_1x_1+w_2x_2&=0\\
w_1x_1^2+w_2x_2^2&=2/3\\
w_1x_1^3+w_2x_2^3&=0
\end{aligned}
$$

It is easily seen that if any of four variables is zero, so are the others. This contradicts the first equation and consequently $x_i\neq 0, w_i\neq 0\,;\, i=1,2$. Since $w_1x_1=-w_2x_2$ and $w_1x_1^3=-w_2x_2^3$ we get $x_1^2=x_2^2$. The choice $x_1=x_2$ leads to contradiction (why?) and the other choice $x_1=-x_2$ when substituted in the second equation, leads to $w_1-w_2=0$. Thus $w_1=w_2=1$ and finally (third equation) $x_1=-1/\sqrt{3}$, $x_2=1/\sqrt{3}$.

Example 9.6.5. Let $f(x)=\cos(x)\,,\,-1\le x\le 1$. Gaussian integration with two points provide

$$G_2(\cos(x);-1,1)=\cos(-1/\sqrt{3})+\cos(1/\sqrt{3})\approx 1.676$$

while the exact integral is

$$\int_{-1}^{1}\cos(x)dx=2\sin(1)\approx 1.683$$

In this case second order Gaussian scheme is sufficient to obtain an error of 0.4%.

PROBLEMS

1. Use the trapezoidal rule with 10 equal subintervals to approximate $\int_0^{\pi} x\sin(x)\,dx$.
2. Calculate the integral of Problem 1 analytically and compare with the approximated value. Verify Eq. (9.6.4).
3. (a) Use the trapezoidal rule with 10 equal subintervals to approximate $\int_0^1 f(x)dx$, where

$$4.\quad f(x)=\begin{cases} 1/\sqrt{x}\,, & 0<x\le 1 \\ 0\,, & x=0 \end{cases}$$

   (b) Is the error estimate given by Eq. (9.6.4) valid in this case?

4. Obtain a sufficient condition for Simpson's rule to provide an exact answer.
5. Use Simpson's rule with 10 equal subintervals to approximate $\int_0^2 xe^x dx$.
6. How many equal subintervals are needed in the previous problem to guarantee accuracy of 10^{-8}?
7. Use the trapezoidal rule and Simpson's rule – each with 5 equal subintervals, and the Gaussian integration scheme with two points to approximate $\int_{-1}^1 xe^{-x}\,dx$. Compare the results with the exact value.

10 Special Topics

Certain topics in this chapter are included to demonstrate the beauty, which some people, even scientists, fail to find in mathematics. How many people really care whether or not the number e is irrational? After all, in performing everyday calculations, this number of royalty is cruelly stripped of most of its digits and is treated like a common rational number. Yet, no true researcher can stay indifferent in front of this amazingly simple yet brilliant proof of Theorem 10.1.1.

Other topics, such as Lagrange multipliers and numerical methods for differential equations, are meant to remind the reader, that at the end of the day, one of the most important requirements from the scientist is to *calculate*. Some topics from this category of *practicality*, are included in Chapter 10, and are characterized by being powerful computational tools and *not* beyond the scope of this book.

10.1 The Irrationality of e

The number $e = 2.718\ldots$ which is the basis to the natural logarithm and one of the most special numbers in mathematics, is a transcendental number, i.e., it is not a root of any polynomial with integer coefficients. The proof, presented by Hermite in 1873, is far beyond the scope of this book. In this section we present a simple elegant proof of a less pretentious result.

Theorem 10.1.1 The number e is irrational.

Proof.

Assume the opposite. Then, $e = \frac{p}{q}$ where p,q are integers, and since e is between 2 and 3, we must have $q > 1$. Since

$$e = \lim_{n\to\infty}\left(1+\frac{1}{n}\right)^n = 1+\frac{1}{1!}+\frac{1}{2!}+\cdots+\frac{1}{k!}+\cdots$$

M. Friedman and A. Kandel: Calculus Light, ISRL 9, pp. 263–281.
springerlink.com

we obtain

$$\frac{p}{q}=\sum_{k=0}^{\infty}\frac{1}{k!}=\sum_{k=0}^{q}\frac{1}{k!}+\sum_{k=q+1}^{\infty}\frac{1}{k!} \tag{10.1.1}$$

Multiplying Eq. (10.1.1) by $q!$ provides

$$p(q-1)!=q!\sum_{k=0}^{q}\frac{1}{k!}+q!\sum_{k=q+1}^{\infty}\frac{1}{k!}$$

and consequently

$$p(q-1)!-q!\sum_{k=0}^{q}\frac{1}{k!}=\frac{1}{q+1}+\frac{1}{(q+1)(q+2)}+\frac{1}{(q+1)(q+2)(q+3)}+\dots \tag{10.1.2}$$

The left hand side of Eq. (10.1.2) is clearly an integer. On the other hand, the infinite series at the right hand side is bounded by $\frac{1}{(q+1)}\left(1+\frac{1}{q}+\frac{1}{q^2}+\dots\right)$ which (since $q>1$) converges to $r=\frac{q}{(q+1)(q-1)}<1$. Thus, the assumption that e is rational leads to contradiction and the proof is completed.

PROBLEMS

1. Show that the numbers $\sqrt{e}$ and $\sqrt{\frac{e+1}{e-1}}$ is not rational.
2. Show that the number $e_1=1+\frac{1}{3!}+\frac{1}{5!}+\dots$ is irrational.
3. Show that the number $e_2=1+\frac{1}{2!}+\frac{1}{4!}+\dots$ is irrational.

10.2 Euler's Summation Formula

Consider a continuously differentiable function $f(x)$ over the interval $0\le x<\infty$ and denote $f_k=f(k)$ for arbitrary nonnegative integer k. The sum $\sum_{k=0}^{n}f_k$

approximates the integral $\int_0^n f(x)dx$ and we are interested in evaluating the difference

$$\sum_{k=0}^{n} f_k - \int_0^n f(x)dx \tag{10.2.1}$$

Theorem 10.2.1 If $f(x) \in C[0,\infty)$ then

$$\sum_{k=0}^{n} f_k - \int_0^n f(x)dx = \frac{1}{2}(f_0 + f_n) + \int_0^n \left(x - [x] - \frac{1}{2}\right) f'(x)dx \tag{10.2.2}$$

where $[x]$ denotes the largest integer less than or equal to x. This is *Euler's summation formula.*

Proof.

The relation

$$\sum_{k=1}^{n} \int_{k-1}^{k} kf'(x)dx = \sum_{k=1}^{n} k(f_k - f_{k-1}) = -\sum_{k=0}^{n} f_k + (n+1)f_n \tag{10.2.3}$$

can be easily verified and since $k = 1 + [x]$ for every $x \in [k-1, k)$, we may rewrite

$$-\sum_{k=0}^{n} f_k + (n+1)f_n = \int_0^n (1 + [x]) f'(x)dx \tag{10.2.4}$$

Also, integration by parts provides

$$\int_0^n xf'(x)dx = xf(x)\Big|_0^n - \int_0^n f(x)dx = nf_n - \int_0^n f(x)dx \tag{10.2.5}$$

By subtracting Eq. (10.2.4) from Eq. (10.2.5) we get

$$\sum_{k=0}^{n} f_k - \int_0^n f(x)dx = f_n + \int_0^n (x-[x]-1)f'(x)dx \tag{10.2.6}$$

and Eq. (10.2.2) follows by an appropriate rearrangement.

If in addition to being in $C[0,\infty)$ the function $f(x)$ is also positive and monotonic increasing, than the left-hand side of Eq. (10.2.2) is positive. Furthermore, the series

$$a_n = \sum_{k=0}^{n} f_k - \int_0^n f(x)dx\,,\ n \ge 0$$

is monotonic decreasing (why?) and its elements are bounded below by 0. Consequently it converges when $n \to \infty$. But so does the series $\{f_n\}$. Therefore, the integral $\int_0^n \left(x-[x]-\frac{1}{2}\right)f'(x)dx$ converges as $n \to \infty$.

Example 10.2.1. Let $f(x) = \frac{1}{x+1}\,, 0 \le x < \infty$ and replace n by $n-1$ in Eq. (10.2.2). Then

$$1+\frac{1}{2}+\cdots+\frac{1}{n} - \ln(n) = \frac{1}{2}+\frac{1}{2n} - \int_0^{n-1} \frac{g(x)}{(1+x)^2}dx = \frac{1}{2}+\frac{1}{2n} - \int_1^n \frac{g(x)}{x^2}dx$$

where $g(x) = x-[x]-\frac{1}{2}$ satisfies $g(x) = g(x-1)$ and $|g(x)| \le \frac{1}{2}$ for all x. As stated, the left-hand side converges and we obtain

$$\lim_{n\to\infty}\left(1+\frac{1}{2}+\cdots+\frac{1}{n} - \ln(n)\right) = \frac{1}{2} - \int_1^\infty \frac{g(x)}{x^2}dx \tag{10.2.7}$$

The constant at the right-hand side of Eq. (10.2.7) denoted by γ, is called *Euler's constant*. Since $-\frac{1}{2} \le g(x) \le \frac{1}{2}$ we easily get $0 \le \gamma \le 1$ but a more detailed computation yields $\gamma = 0.57721\ldots$.

Application

An important application of Euler's summation formula is deriving the *Stirling's formula*

$$\lim_{n\to\infty}\left(\frac{n}{e}\right)^n \frac{\sqrt{2\pi n}}{n!}=1 \tag{10.2.8}$$

This nontrivial relation is enormously used in many sciences, especially in analysis and statistics. Quite often we need to calculate or estimate an expression composed of one or several factorials, for example $\frac{(n!)(5n)!+n^n}{[(3n)!]^2}$. In such cases our only option is usually Stirling's formula. To obtain it, we first introduce *Wallis' formula*

$$\frac{\pi}{2}=\lim_{n\to\infty}\left[\frac{2\cdot 4\cdot 6\cdots(2n)}{1\cdot 3\cdot 5\cdots(2n-1)}\right]^2\frac{1}{2n+1} \tag{10.2.9}$$

which will be derived as a consequence of the following lemma.

Lemma 10.2.1. For every integer $n\geq 2$ the inequality

$$\frac{2\cdot 4\cdot 6\cdots(2n)}{1\cdot 3\cdot 5\cdots(2n+1)}<\frac{1\cdot 3\cdot 5\cdots(2n-1)}{2\cdot 4\cdot 6\cdots(2n)}\frac{\pi}{2}<\frac{2\cdot 4\cdot 6\cdots(2n-2)}{1\cdot 3\cdot 5\cdots(2n-1)} \tag{10.2.10}$$

holds.

Proof.

Recall that

$$\int_0^{\pi/2}\sin^n(x)dx=\begin{cases}\dfrac{2\cdot 4\cdot 6\cdots(n-1)}{1\cdot 3\cdot 5\cdots n}, & odd\ n\\[2ex] \dfrac{1\cdot 3\cdot 5\cdots(n-1)}{2\cdot 4\cdot 6\cdots n}\cdot\dfrac{\pi}{2}, & even\ n\end{cases} \tag{10.2.11}$$

Since $0<\sin(x)<1$ for all $x:0<x<\frac{\pi}{2}$ we get $\sin^{2n+1}(x)<\sin^{2n}x<\sin^{2n-1}(x)$ for all $n\in N$, $x\in(0,\pi/2)$. Integrating this inequality over the interval $[0,\pi/2]$ and applying Eq. (10.2.11), provides Eq. (10.2.10) which completes the proof.

Next we verify Eq. (10.2.9). The left inequality in Eq. (10.2.10) implies that (how?)

$$\left[\frac{2\cdot 4\cdot 6\cdots(2n)}{1\cdot 3\cdot 5\cdots(2n-1)}\right]^2\frac{1}{2n+1}<\frac{\pi}{2}$$

while the right one implies

$$\frac{2n}{2n+1}\cdot\frac{\pi}{2}<\left[\frac{2\cdot 4\cdot 6\cdots(2n)}{1\cdot 3\cdot 5\cdots(2n-1)}\right]^2\frac{1}{2n+1}$$

By combining these two results we get

$$\frac{2n}{2n+1}\cdot\frac{\pi}{2}<\left[\frac{2\cdot 4\cdot 6\cdots(2n)}{1\cdot 3\cdot 5\cdots(2n-1)}\right]^2\frac{1}{2n+1}<\frac{\pi}{2}$$

and Wallis' formula easily follows when $n\to\infty$.

To get Stirling's formula we apply Euler's summation formula for $f(x)=\ln(1+x)$. By virtue of Eq. (10.2.2)

$$\ln 1+\ln 2+\ldots+\ln n=\int_1^n \ln x dx+\frac{1}{2}\ln n+\int_1^n\frac{g(x)}{x}dx$$

$$=\left(n+\frac{1}{2}\right)\ln n-(n-1)+\int_1^n\frac{g(x)}{x}dx \tag{10.2.12}$$

where as before $g(x)=x-[x]-\frac{1}{2}$. Dirichlet's test for improper integrals guarantees that $\int_1^\infty\frac{g(x)}{x}dx$ converges (why?). Therefore, the sequence

$$a_n=\ln(n!)-\left(n+\frac{1}{2}\right)\ln n+n \tag{10.2.13}$$

converges to $A=\lim_{n\to\infty}a_n=\lim_{n\to\infty}\left(1+\int_1^n\frac{g(x)}{x}dx\right)$. Finding A is simple. Indeed, by applying Eq. (10.2.13) twice, for n and for $2n+1$, we get

$$2\ln[2\cdot 4\cdot 6\cdots(2n)] = 2\ln[2^n(1\cdot 2\cdot 3\cdots n)] = (2n+1)\ln(2n) - \ln 2 - 2n + 2a_n$$

$$\ln[(2n+1)!] = \left(2n+\frac{3}{2}\right)\ln(2n+1) - (2n+1) + a_{2n+1}$$

and subtracting the second equality from the first one, yields

$$\ln\frac{[2\cdot 4\cdot 6\cdots(2n)]}{[1\cdot 3\cdot 5\cdots(2n-1)\sqrt{2n+1}]} = (2n+1)\ln\left(1-\frac{1}{2n+1}\right) + 1 - \ln 2 + 2a_n - a_{2n+1}$$

Since $\lim_{n\to\infty}\left(1-\frac{1}{2n+1}\right)^{2n+1} = \frac{1}{e}$, the right-hand side of the last equation approaches $A-\ln 2$ when $n\to\infty$, while the left-hand side, by virtue of Wallis' formula, approaches $\ln(\sqrt{\pi/2})$. Consequently $A=\sqrt{2\pi}$ and Stirling's formula finally follows from Eq. (10.2.13) (how?).

Example 10.2.2. Let $A_n = \frac{(n!)^3}{(3n)!}$. Then

$$A_n \approx \frac{\left(\frac{n}{e}\right)^{3n}\left(\sqrt{2\pi n}\right)^3}{\left(\frac{3n}{e}\right)^{3n}\sqrt{6\pi n}} = \frac{2\pi n}{\sqrt{3}\cdot 3^{3n}}$$

which yields $\lim_{n\to\infty} A_n = 0$.

PROBLEMS

1. Use Stirling's formula to estimate the binomial coefficient $\binom{2n}{n}$ for large n.
2. Find $\lim_{n\to\infty}\frac{\sqrt[n]{n!}}{n}$.
3. Obtain the final part of Stirling's formula from Eq. (10.2.13).
4. Show that $\lim_{n\to\infty}\left(1+\frac{1}{2^\alpha}+\ldots+\frac{1}{n^\alpha}-\frac{n^{1-\alpha}}{1-\alpha}\right)$ exists for $0<\alpha<1$.
5. Use Wallis formula to approximate $\frac{40\cdot 42\cdots 78\cdot 80}{39\cdot 41\cdots 77\cdot 79}$.

10.3 Lagrange Multipliers

In order to present the next special topic – Lagrange multipliers, we shall first briefly introduce the concept of multi-variable functions, continuity of these functions and partial derivatives.

10.3.1 Introduction: Multi-variable Functions

The set

$$\mathfrak{R}^n = \{(x_1, x_2, \ldots, x_n), -\infty < x_i < \infty, 1 \le i \le n\} \tag{10.3.1}$$

is called the *real* n - *dimensional space*. It consists of all the elements $(x_1, x_2, \ldots, x_n)$ - called points, where each x_i is a real number. Two points $x = (x_1, x_2, \ldots, x_n)$ and $y = (y_1, y_2, \ldots, y_n)$ are equal if and only if they are identical, i.e., if $x_i = y_i, 1 \le i \le n$. We define addition of two points and multiplication of a point by a real number λ as

$$x + y = (x_1 + y_1, \ldots, x_n + y_n) \quad ; \quad \lambda x = (\lambda x_1, \ldots, \lambda x_n) \quad , \quad -\infty < \lambda < \infty \tag{10.3.2}$$

Subtraction is defined by $x - y = x + (-1) \cdot y$ and the distance between two arbitrary points x, y as

$$\|x - y\| = \sqrt{\sum_{i=1}^{n} (x_i - y_i)^2} \tag{10.3.3}$$

which equals 0 if and only if x and y are identical. Let D denote a subset of $\mathfrak{R}^n$. A rule f which assigns a unique real number z to each point of D is called a function. This *n-variable* (or *n-dimensional*) function is a relation from D into $\mathfrak{R}$. For the sake of simplicity we shall assume that D is an *open set*, i.e., for each $x_0 \in D$ there exists $\delta(x_0) > 0$ such that $|x - x_0| < \delta \Rightarrow x \in D$. In other words, Each point in D has some sphere around it which is completely in D.

Example 10.3.1. The relation $z = x^3 - \dfrac{1}{y}$ is a two dimensional function defined for all real x, y such that $y \ne 0$.

Example 10.3.2. The relation $u = \sqrt{x^2 + y^2 + z^4}$ defined over $\mathfrak{R}^3$ is not a function since each point (x, y, z) provides two values for u.

Definition 10.3.1 (continuity). A function $u = f(x_1, x_2, \ldots, x_n)$ defined over $D \subseteq \Re^n$ is continuous at a given point $x_0 = (x_{10}, \ldots, x_{n0}) \in D$ if for an arbitrary $\varepsilon > 0$, we can find $\delta > 0$ such that

$$\|x - x_0\| < \delta \Rightarrow \|f(x) - f(x_0)\| < \varepsilon \tag{10.3.4}$$

In this case we also write $\lim_{x \to x_0} f(x) = f(x_0)$.

Example 10.3.3. The function $f(x, y) = x^3 + \sin(xy)$ is defined and continuous over the whole $x - y$ plane.

Example 10.3.4. The function

$$f(x, y) = \begin{cases} \dfrac{1}{x^2 + y^2}, & (x, y) \neq (0,0) \\ 1, & x = y = 0 \end{cases}$$

is defined everywhere in $\Re^2$ and continuous everywhere except at $(0,0)$.

We now introduce the concept of partial derivative.

Definition 10.3.2. Let $f(x_1, x_2, \ldots, x_n)$ be an n-dimensional function defined over $D \subseteq \Re^n$ and let $x_0 = (x_{10}, \ldots, x_{n0}) \in D$. If

$$\lim_{h \to 0} \frac{f(x_{10}, \ldots, x_{i-1,0}, x_i + h, x_{i+1,0}, \ldots, x_{n0}) - f(x_{10}, \ldots, x_{i-1,0}, x_i, x_{i+1,0}, \ldots, x_{n0})}{h} \tag{10.3.5}$$

exists, we say that $f(x_1, x_2, \ldots, x_n)$ has a *first partial derivative* (or simply a *partial derivative*) with respect to x_i at the point $x_0 = (x_{10}, \ldots, x_{n0})$. The limit is denoted by $\dfrac{\partial f}{\partial x_i}(x_{10}, \ldots, x_{n0})$ or by $f_{x_i}(x_{10}, \ldots, x_{n0})$.

Thus, a partial derivative with respect to a specific variable x_i is obtained by treating the other variables as constants. In other words, during the process of calculating the partial derivative, f is considered function of the single variable x_i.

Example 10.3.5. Let $f(x,y,z)=x^2y-\cos(xz)$. The partial derivative $\frac{\partial f}{\partial x}$ is obtained by taking constant values for y,z, i.e.,

$$\frac{\partial f}{\partial x}=2xy+z\sin(xz)$$

If x,z are kept constants, we obtain the partial derivative with respect to y as

$$\frac{\partial f}{\partial y}=x^2$$

The third partial derivative with respect to z is

$$\frac{\partial f}{\partial z}=\sin(xz)+zx\cos(xz)$$

Second order partial derivatives are obtained using the same principle of treating all the variables but one as constants. Given an arbitrary $f(x_1,x_2,\ldots,x_n)$, we define its second order partial derivatives as

$$\frac{\partial^2 f}{\partial x_i \partial x_j}=\frac{\partial}{\partial x_i}\left(\frac{\partial f}{\partial x_j}\right) \tag{10.3.6}$$

We often write the left-hand side of Eq. (10.3.6) as $f_{x_i x_j}$. The relation between $f_{x_i x_j}$ and $f_{x_j x_i}$ is given next without proof.

Theorem 10.3.1. If $f_{x_i x_j}$ and $f_{x_j x_i}$ are continuous then

$$f_{x_i x_j}=f_{x_j x_i} \tag{10.3.7}$$

Example 10.3.6. Let $f(x,y)=xe^y-\frac{x}{y}$. Then

$$f_x=e^y-\frac{1}{y}\quad,\quad f_y=xe^y+\frac{x}{y^2}$$

and clearly

$$f_{xy} = f_{yx} = e^y + \frac{1}{y^2}$$

Definition 10.3.3. A function $f(x_1, x_2, \ldots, x_n)$ is said to possess a local maximum at $x_0 = (x_{10}, \ldots, x_{n0})$ if there exists an $\varepsilon > 0$ such that

$$\|x - x_0\| < \varepsilon \Rightarrow f(x) \le f(x_0) \tag{10.3.8}$$

and a local minimum if an $\varepsilon > 0$ exists such that

$$\|x - x_0\| < \varepsilon \Rightarrow f(x) \ge f(x_0) \tag{10.3.9}$$

A basic property of local maximum (minimum) points is given next.

Theorem 10.3.2. A necessary condition for $f(x_1, x_2, \ldots, x_n)$ to possess a local extremum (maximum or minimum) at $x_0 = (x_{10}, \ldots, x_{n0})$ is

$$\frac{\partial f}{\partial x_i}(x_0) = 0 \quad , \quad 1 \le i \le n \tag{10.3.10}$$

PROBLEMS

1. Find the domain of definition and the points of discontinuity for:

   (a) $f(x, y) = \dfrac{1}{x^2 - y}$

   (b) $f(x, y) = \begin{cases} \dfrac{1}{xy} & , \quad xy \ne 0 \\ 1 & , \quad xy = 0 \end{cases}$

2. Calculate the first partial derivatives and determine their domains of definition for:

   (a) $f(x, y) = \dfrac{1}{x} + \sqrt{x^2 + y^2}$

   (b) $f(x, y) = \dfrac{y + z}{x} + \sqrt{x^2 + y^2 - z^2}$

3. Calculate all the second partial derivatives for:

   (a) $f(x, y) = \sin(xy) + \cos(x + y)$

   (b) $f(x, y) = \tan(xyz)$

10.3.2 Lagrange Multipliers

The problem of maximum or minimum search for a given function is carried quite often under constraints. For example, if $T(x,y,z)$ denotes the temperature at an arbitrary point (x,y,z), we may be interested at the minimum (or maximum) of $T(x,y,z)$ over the surface of some three-dimensional body given by $f(x,y,z)=0$. The minimization process is thus *restricted* to a subset of the definition domain of $T(x,y,z)$. The Lagrange multipliers method is a simple elegant procedure for solving extremum problems under constraints.

Definition 10.3.4. Let $f(x), g_1(x), g_2(x), \ldots, g_m(x)$ denote continuous functions defined over a domain D in $\Re^n$. If for some $x^0 \in D$ that satisfies

$$g_i(x^0)=0 \ , \ 1 \le i \le m$$

there exists a neighborhood $D_0 \subset D$ such that $f(x) \ge f(x^0)$ for all $x \in D_0$ for which

$$g_i(x)=0 \ , \ 1 \le i \le m \tag{10.3.11}$$

the point x^0 is called a *relative minimum* of $f(x)$ under the *constraints* given by Eq. (10.3.11). We similarly define a *relative maximum* and in either case refer to a *relative extremum.*

Example 10.3.7. Let $f(x,y)=x^2+y^2$. We search for a relative minimum of $f(x,y)$ under the constraint $g(x,y)=x-y-1=0$. This can be done by finding the minimum of $F(x)=f(x,x-1)=x^2+(x-1)^2=2x^2-2x+1$. We differentiate $F(x)$ twice and obtain the minimum at $x=1/2$. Consequently $f(x,y)$ attains a relative minimum at $(1/2,-1/2)$.

Notice that we first eliminated y from $g(x,y)=0$, substituted it in $f(x,y)$ and then performed a standard procedure to find the extremum points. Usually, this approach can be tiresome and impractical. Instead, we present the Lagrange multipliers method implemented on the following example.

Example 10.3.8. Consider all the boxes centered at $(0,0,0)$ with edges each parallel to one of the axes, which are contained inside the ellipsoid

$$g(x,y,z)=\frac{x^2}{a^2}+\frac{y^2}{b^2}+\frac{z^2}{c^2}-1=0 \tag{10.3.12}$$

If (x,y,z) denotes the corner with three positive coordinates, the box volume is $V(x,y,z)=8xyz$. Since the corners of the box with the largest volume V_0 must be located on the ellipsoid itself (why?), it is evident that to obtain V_0 we have to maximize V under the constraint given by Eq. (10.3.12). We will now present a technique called *Lagrange multipliers* for treating this problem. Define

$$H(x,y,z,\lambda)=V(x,y,z)+\lambda g(x,y,z) \tag{10.3.13}$$

where λ is an additional variable, and perform an *unrestricted extremum search* for $H(x,y,z,\lambda)$, i.e. solve

$$\frac{\partial H}{\partial x}=0\,,\ \frac{\partial H}{\partial y}=0\,,\ \frac{\partial H}{\partial z}=0\,,\ \frac{\partial H}{\partial \lambda}=0 \tag{10.3.14}$$

or

$$\begin{aligned}
&\frac{\partial V}{\partial x}+\lambda\frac{\partial g}{\partial x}=4yz+\frac{2\lambda x}{a^2}=0\\
&\frac{\partial V}{\partial y}+\lambda\frac{\partial g}{\partial y}=4xz+\frac{2\lambda y}{b^2}=0\\
&\frac{\partial V}{\partial z}+\lambda\frac{\partial g}{\partial z}=4xy+\frac{2\lambda z}{c^2}=0\\
&g=\frac{x^2}{a^2}+\frac{y^2}{b^2}+\frac{z^2}{c^2}-1=0
\end{aligned} \tag{10.3.15}$$

Simple algebra yields the unique solution of $x=\frac{a}{\sqrt{3}}\,,\ y=\frac{b}{\sqrt{3}}\,,\ z=\frac{c}{\sqrt{3}}$. Obviously, this solution satisfies the constraint itself, which is the last part of Eq. (10.3.15). Is it also a relative maximum? The next result provides the answer.

Theorem 10.3.3. Let $f(x)$, $g(x)$ denote continuously differentiable functions over a domain D in $\Re^n$ and assume $\nabla g(x)\equiv\left(\frac{\partial g}{\partial x_1},\ldots,\frac{\partial g}{\partial x_n}\right)\neq 0$, $x\in D$, i.e., at each arbitrary point $x\in D$ at least one partial derivative of $g(x)$ is nonzero.. If $f(x)$ has a relative extremum under the constraint $g(x)=0$ at a point $x^0\in D$, then there exists a number λ such that

$$\frac{\partial f(x^0)}{\partial x_i}+\lambda\frac{\partial g(x^0)}{\partial x_i}=0 \quad , \quad 1\le i\le n \tag{10.3.16}$$

Since the equation

$$g(x^0)=0 \tag{10.3.17}$$

holds by definition, we end up with a system of $n+1$ equations and $n+1$ unknowns, namely $x_1,x_2,\ldots,x_n$ and λ.

Thus, if a problem possesses a relative extremum and if the system composed of Eqs. (10.3.16-10.3.17) has a unique solution, this solution must be a relative extremum. This is exactly the case in Example 10.3.8.

Proof. Since $\nabla g(x^0)\ne 0$ we have $\frac{\partial g(x^0)}{\partial x_k}\ne 0$ at least for one k between 1 and n. For the sake of simplicity assume $\frac{\partial g(x^0)}{\partial x_n}\ne 0$. Therefore, by implying the implicit function theorem, we can uniquely solve $g(x)=0$ at some neighborhood of x^0 and get

$$x_n=h(x_1,x_2,\ldots,x_{n-1}) \tag{10.3.18}$$

where h is continuously differentiable. Denote $x^0=(x_1^0,\ldots,x_{n-1}^0,x_n^0)=(x^*,x_n^0)$ where $x^*=(x_1^0,\ldots,x_{n-1}^0)$. Then the function $f[x_1,\ldots,x_{n-1},h(x_1,\ldots,x_{n-1})]$ is continuously differentiable at a neighborhood of x^* in $\Re^{n-1}$ and attains an *unrestricted* extremum at x^*. Consequently, by using the chain rule

$$\frac{\partial f(x^0)}{\partial x_i}+\frac{\partial f(x^0)}{\partial x_n}\frac{\partial h(x^*)}{\partial x_i}=0 \ , \ 1\le i\le n-1 \tag{10.3.19}$$

Now, the identity $g[x_1,\ldots,x_{n-1},h(x_1,\ldots,x_{n-1})]=0$ implies

$$\frac{\partial g(x^0)}{\partial x_i}+\frac{\partial g(x^0)}{\partial x_n}\frac{\partial h(x^*)}{\partial x_i}=0 \ , \ 1\le i\le n-1 \tag{10.3.20}$$

which, since $\frac{\partial g(x^0)}{\partial x_n}\ne 0$, yields

$$\frac{\partial h(x^*)}{\partial x_i} = -\frac{\partial g(x^0)}{\partial x_i} \Big/ \frac{\partial g(x^0)}{\partial x_n} \ , \ 1 \le i \le n-1 \tag{10.3.21}$$

By substituting in Eq. (10.3.19) we obtain that Eq. (10.3.16) is valid with the constant

$$\lambda = -\frac{\partial f(x^0)}{\partial x_n} \Big/ \frac{\partial g(x^0)}{\partial x_n} \tag{10.3.22}$$

This completes the proof of Theorem 10.3.3.

Application

The next example of maximizing the area of a triangle with constant given circumference, though simple, demonstrates the potential existing in the Lagrange multipliers method.

Example 10.3.9. Consider all the triangles with sides a,b,c such that $a+b+c=2p$ for some given p. The area of each triangle is given by Heron's formula as

$$S = \sqrt{p(p-a)(p-b)(p-c)} \tag{10.3.23}$$

and the problem of finding a triangle with maximum area is therefore maximizing S under the constraint $a+b+c-2p=0$. Clearly, we may maximize S^2 instead of S. Therefore, using the Lagrange multipliers method, we search for an unrestricted extremum point for $H = S^2 + \lambda(a+b+c-2p)$ and get the four-equation system

$$\begin{aligned} \frac{\partial H}{\partial a} &= -p(p-b)(p-c) + \lambda = 0 \\ \frac{\partial H}{\partial b} &= -p(p-c)(p-a) + \lambda = 0 \\ \frac{\partial H}{\partial c} &= -p(p-a)(p-b) + \lambda = 0 \\ \frac{\partial H}{\partial \lambda} &= a+b+c-2p = 0 \end{aligned} \tag{10.3.24}$$

This system possesses the unique solution, $a = b = c = \frac{2p}{3}$, i.e. an equilateral triangle and it is a relative maximum solution. The value of the single Lagrange multiplier, $\lambda = \frac{p^3}{9}$, is of no meaning or use to us.

We will now present the Lagrange multipliers procedure in the case of several constraints. In order to find a relative extremum of $f(x)$ under the constraints $g_i(x) = 0$, $1 \le i \le m$, we search for unrestricted extremum of the $(n+m)$-variable function

$$H(x;\lambda_1,\lambda_2,\ldots,\lambda_m) = f(x) + \sum_{i=1}^{m} \lambda_i g_i(x) \qquad (10.3.25)$$

where the Lagrange multipliers λ_i , $1 \le i \le m$ are additional variables to the original $x_1, x_2, \ldots, x_n$. The system that needs to be solved consists of the following $(n+m)$ equations:

$$\frac{\partial H}{\partial x_i} = \frac{\partial f}{\partial x_i} + \sum_{j=1}^{m} \lambda_j \frac{\partial g_j(x)}{\partial x_i} = 0 \quad , \quad 1 \le i \le n \qquad (10.3.26)$$

and

$$\frac{\partial H}{\partial \lambda_j} = g_j(x) = 0 \quad , \quad 1 \le j \le m \qquad (10.3.27)$$

Notice that due to Eq. (10.3.27), *any* solution of this system must satisfy all the constraints. The question of existence and nature of the solutions to Eqs. (10.3.26-10.3.27), is answered by the following result, which is an extension to Theorem 10.3.3, given here without a proof.

Theorem 10.3.4. Let $f(x)$, $g_1(x), g_2(x), \ldots, g_m(x)$ denote continuously differentiable functions over a domain D in $\Re^n$ and assume that for every $x \in D$ there exist distinct numbers $i_1, i_2, \ldots, i_m$ between 1 and n such that

$$\frac{\partial(g_1, g_2, \ldots, g_m)}{\partial(x_{i_1}, x_{i_2}, \ldots, x_{i_m})}(x) \ne 0 \qquad (10.3.28)$$

where the left-hand side of Eq. (10.3.28) is the determinant $|a_{jk}|, 1 \le j,k \le m$, $a_{jk} = \frac{\partial g_j}{\partial x_{i_k}}$. Then, if $f(x)$ attains a relative extremum at a point $x^0 \in D$, under the constraints $g_i(x)=0$, $1 \le i \le m$, there exist numbers λ_i , $1 \le i \le m$ such that

$$\frac{\partial f(x^0)}{\partial x_i} + \sum_{j=1}^{m} \lambda_j \frac{\partial g_j(x^0)}{\partial x_i} = 0 \quad , \quad 1 \le i \le n \tag{10.3.29}$$

This theorem illustrates that the general problem of finding a relative minimum or maximum of $f(x)$ can be usually replaced by a more familiar problem of finding unrestricted minimum or maximum respectively.

The next example is a typical problem in *Linear Programming*, where some of the constraints are formulated as inequalities rather than equalities. We first define equivalent constraints given as equalities, then use the Lagrange multipliers procedure.

Example 10.3.10. Consider a problem of minimizing the function

$$f(x_1,x_2,x_3) = x_1 + 2x_2 + x_3$$

under four constraints given by

$$x_1 + 3x_2 + 4x_3 - 10 = 0 \,,\, x_1 \ge 0 \,,\, x_2 \ge 0 \,,\, x_3 \ge 0$$

Three of the constraints are inequalities and can be replaced equivalently by

$$x_1 - y_1^2 = 0 \,,\, x_2 - y_2^2 = 0 \,,\, x_3 - y_3^2 = 0$$

The function for which we search an unrestricted minimum is

$$H(x_1,x_2,x_3,y_1,y_2,y_3,\lambda_1,\lambda_2,\lambda_3,\lambda_4) = x_1 + 2x_2 + x_3$$
$$+ \lambda_1(x_1 - y_1^2) + \lambda_2(x_2 - y_2^2) + \lambda_3(x_3 - y_3^2) + \lambda_4(x_1 + 3x_2 + 4x_3 - 10)$$

and the system to be solved is

$$\frac{\partial H}{\partial x_1} = 1 + \lambda_1 + \lambda_4 = 0 \,,\, \frac{\partial H}{\partial x_2} = 2 + \lambda_2 + 3\lambda_4 = 0 \,,\, \frac{\partial H}{\partial x_3} = 1 + \lambda_3 + 4\lambda_4 = 0$$
$$\frac{\partial H}{\partial y_1} = -2\lambda_1 y_1 = 0 \,,\, \frac{\partial H}{\partial y_2} = -2\lambda_2 y_2 = 0 \,,\, \frac{\partial H}{\partial y_3} = -2\lambda_3 y_3 = 0$$
$$\frac{\partial H}{\partial \lambda_1} = x_1 - y_1^2 = 0 \,,\, \frac{\partial H}{\partial \lambda_2} = x_2 - y_2^2 = 0 \,,\, \frac{\partial H}{\partial \lambda_3} = x_3 - y_3^2 = 0$$
$$\frac{\partial H}{\partial \lambda_4} = x_1 + 3x_2 + 4x_3 - 10 = 0$$

Since H can be rewritten as

$$H = H_1(x_1, x_2, x_3, \lambda_1, \lambda_2, \lambda_3, \lambda_4) - \lambda_1 y_1^2 - \lambda_2 y_2^2 - \lambda_3 y_3^2$$

then, in order to get a minimum we must have $\lambda_i \le 0$, $1 \le i \le 3$. The first three of the ten equations yield

$$\lambda_1 = -1 - \lambda_4$$
$$\lambda_2 = -2 - 3\lambda_4$$
$$\lambda_3 = -1 - 4\lambda_4$$

while the next three determine $\lambda_i y_i = 0$, $1 \le i \le 3$, i.e. $\lambda_i x_i = 0$, $1 \le i \le 3$. If all x_i vanish then the constraint $x_1 + 3x_2 + 4x_3 - 10 = 0$ is invalid. Therefore, at least one of them, say x_j, is nonzero. We therefore get $\lambda_j = 0$. Simple algebra yields that in order to guarantee $\lambda_i \le 0$, $1 \le i \le 3$, we must have $\lambda_3 = 0$. Consequently,

$$\lambda_4 = -\frac{1}{4}, \lambda_1 = -\frac{3}{4}, \lambda_2 = -\frac{5}{4}$$

and $x_1 = x_2 = 0$. Finally, the constraint $x_1 + 3x_2 + 4x_3 - 10 = 0$ yields $x_3 = 2.5$ and the relative minimum of $x_1 + 2x_2 + x_3$ is attained at $(0,0,2.5)$ and equals 2.5.

PROBLEMS

1. Find extremum values of $x_1 + x_2$, where (x_1, x_2) travels on the unit circle $x_1^2 + x_2^2 = 1$.
2. Find extremum values of $x_1 + x_2 + x_3$ where (x_1, x_2, x_3) travels on the unit sphere $x_1^2 + x_2^2 + x_3^2 = 1$.
3. Let a, b denote two positive numbers.

   (a) Find the extremum of $f(x, y) = \frac{x}{a} + \frac{y}{b}$ given $x^2 + y^2 = 1$.

   (b) Find the extremum of $g(x, y) = x^2 + y^2$ given $\frac{x}{a} + \frac{y}{b} = 1$.

4. Find the shortest distance from the point $(2,0)$ to the parabola $y^2 = x$.
5. Let a,b,c denote positive numbers. Find the maximum of $x^a y^b z^c$ under the constraint $x + y + z = 1$.

6. Find the extremum distances from the circle $x^2 + y^2 = 1$ to the straight line $x + 2y - 6 = 0$.

7. Find the minimum of $\sum_{i=1}^{n} x_i^2$ under the constraint $\sum_{i=1}^{n} a_i x_i = 1$.

8. Find the minimum of $x + y + z$ subject to the constraints $x + 2y + 3z = 6$ and $x \geq 0$, $y \geq 0$, $z \geq 0$.

Solutions to Selected Problem

Chapter 2

Section 2.1

1. The collections (a), (c) are equal setes. The collection (b) is not a set since the element -3 appears twice.
2. If $x \in A - B$ then $x \in A$, $x \notin B$, i.e., $x \notin B - A$. 4. (b) There are 8 elements in S.
7. Let $x \in A - (B \cup C)$. Then, $x \in A$, $x \notin B \cup C$. Consequently $x \notin B$, $x \notin C$, i.e., $x \in A - B$ and finally $x \in (A - B) - C$. Thus, $A - (B \cup C) \subseteq (A - B) - C$ and Similarly $A - (B \cup C) \supseteq (A - B) - C$, which completes the proof.

Section 2.2

8. (a) By definition $(n+1)^2 = (n+1)(n+1)$ which implies

$$(n+1)^2 = n \cdot (n+1) + 1 \cdot (n+1) = (n^2 + n \cdot 1) + (1 \cdot n + 1 \cdot 1)$$

and finally leads to $(n+1)^2 = n^2 + 2n + 1$

9. The inequality $n^2 > n + 3$ certainly holds for $n = 3$ since $3^2 = 9 > 3 + 3 = 6$. Assume that it holds for some arbitrary n.

Section 2.3

3. Define $2^1 = 2$ and $2^{n+1} = 2^n \cdot 2, n \ge 1$. The inequality $2^n > n$ holds for $n = 1$ since $2 > 1$. If it holds for arbitrary n, i.e., if $2^n > n$, then

$$2^{n+1} = 2^n \cdot 2 = 2^n + 2^n > n + n \ge n + 1$$

7. $t=\frac{r+s}{2}$

9. (a) $n=1$: $(1+r)^n=1+r=1+nr$ (b) Assume $(1+r)^n \geq 1+nr$ for arbitrary n, Then, since $1+r>0$, we get $(1+r)^{n+1}=(1+r)^n(1+r) \geq (1+nr)(1+r)$, i.e.,

$$(1+r)^{n+1} \geq 1+nr+r+r^2 = 1+(n+1)r+r^2 \geq 1+(n+1)r$$

Section 2.4

1. Let k be a prime number and consider a rational number $x=\frac{m}{n}$ such that m,n have no common factor and $m^l = kn^l$. Consequently, k must divide both m and n (why?) which contradicts the given input.

16. Substitute $a=b=1$ in Newton's binomial theorem (Eq. (2.4.3)).

Section 2.5

6. We consider four cases: (a) $x+5 \geq 0, x-7 \geq 0, x+5 < x+7 \Rightarrow$ no solution.
(b) $x+5 \geq 0, x-7<0, x+5<7-x \Rightarrow -5 \leq x < 1$. The other two cases are left for the reader.

7. The given inequality is equivalent, for all $a>0$, to $a^2+1 \geq 2a$ which holds since $a^2+1-2a=(a-1)^2 \geq 0$.

10. Clearly, $\sqrt[n]{1 \cdot 2 \cdot \ldots \cdot n} \leq \frac{1+2+\ldots+n}{n} = \frac{n(n+1)}{2n} = \frac{n+1}{2}$. Hence, $n! \leq \left(\frac{n+1}{2}\right)^n$.

Chapter 3

Section 3.1

4. The two straight lines $x=2$, $x=3$.

5. The particular part of the graph is the set $\{(2,2),(3,3),(3,7),(5,5),(5,7),(7,3),(7,5)\}$.

6. $S=\{(x,y) \mid -\infty < x < \infty, y>0\}$.

Section 3.2

1. (a) The relation is a function with domain $D=\{x \mid x \geq 1\}$.
(d) The relation has a domain $D=\{x \mid 0 \leq x \leq 1\}$ and it is not a function.

3. The domain is $D=\{x \mid x \neq 2,3\}$ with range $R=\{y \mid -\infty < y \leq -4 \,,\, y>0\}$

6. (b) $x=\dfrac{1+5y}{4-3y}$ (d) Show that the function does not have an inverse by obtaining a solution other than $x_1 = x_2$ to $x_1^2 + \dfrac{1}{x_1} = x_2^2 + \dfrac{1}{x_2}$.

10. (a) strictly increasing 12. (a) $f \circ g = \dfrac{|x|\sqrt{x^2+2}}{1+x^2}$.

Section 3.3

2. $y=\sqrt{x}+x$ implies $(y-x)^2 = x$ and $y^2-2xy+x^2-x=0$, i.e., y is an algebraic function
7. (a) $x=\sqrt{2}+\sqrt{3}$ implies $x^2 = 5+2\sqrt{6}$ and $(x^2-5)^2-24=0$, i.e. x is an Algebraic number.
12. The function is not defined for $x<0$, i.e., the domain is the single number $x=0$.
15. The function $f(x)g(x)$ is odd.

Section 3.4

2. For example choose $n_0 = \left[\dfrac{23}{7 \cdot 10^{-3}}\right]+1$. This is no guarantee that that the sequence converges to $\dfrac{1}{7}$ or to any limit.
4. The limit exists and equals $\dfrac{1}{1-q}$.
7. Clearly $\dfrac{1}{3}+\dfrac{1}{4} > \dfrac{1}{4}+\dfrac{1}{4}=\dfrac{1}{2}$, $\dfrac{1}{5}+\dfrac{1}{6}+\dfrac{1}{7}+\dfrac{1}{8} > \dfrac{1}{8}+\dfrac{1}{8}+\dfrac{1}{8}+\dfrac{1}{8}=\dfrac{1}{2}$ etc. Thus, a_n exceeds $\dfrac{k}{2}$ for all positive integer k as $n \to \infty$.
10. $a_n = 1+\left(1-\dfrac{1}{2}\right)+\left(\dfrac{1}{2}-\dfrac{1}{3}\right)+\ldots+\left(\dfrac{1}{n-1}-\dfrac{1}{n}\right)=2-\dfrac{1}{n}$ which implies $\lim_{n\to\infty} a_n = 2$.

Section 3.5

2. $\lim_{n\to\infty}(a_n b_n c_n) = \lim_{n\to\infty}((a_n b_n)c_n) = \lim_{n\to\infty}(a_n b_n) \lim_{n\to\infty} c_n = (AB)C = ABC$.

7. Without loss of generality we assume $a \ge 0$. Let m be an integer such that $m = [a] + 1$ which implies $m > a$. For an arbitrary $n > m$ we get

$$\frac{a^n}{n!} = \frac{a^{m-1}}{(m-1)!}\frac{a}{m}\frac{a}{m+1}\cdots\frac{a}{n} < \frac{a^{m-1}}{(m-1)!}\left(\frac{a}{m}\right)^{n-m+1} \to 0 \ as\ n \to \infty$$

11. $A = \frac{1+\sqrt{5}}{2}$ 12. Not necessarily: $\{a_n\} = \{0,1,0,1,\ldots\}$ and $\{b_n\} = \{1,0,1,0,\ldots\}$ are divergent while their sum clearly converges to 1.

13. For arbitrary r the sequence $a_n = r + \frac{\sqrt{2}}{n}$, $n \ge 1$ is a sequence of irrational numbers which converges to r.
14. No – use the example of problem 12.

Section 3.6

1. 0 3. (c) $\sup(S_3) = \frac{4}{3}$, $\inf(S_3) = -1$, $\overline{\lim}(S_3) = \underline{\lim}(S_3) = 1$.
4. (d) $\sup(S_4) = \frac{2}{3}$, $\inf(S_4) = -\frac{1}{2}$, $\overline{\lim}(S_4) = \underline{\lim}(S_4) = 0$.
9. The even and the odd terms separately, converge to 0. This is no guarantee for convergence. For example consider the sequence $\left\{1,0,1,\frac{1}{2},0,2,\frac{1}{3},0,3,\ldots\right\}$.
10. The sequence does not converge. $\sup(S) = 1.1$, $\inf(S) = 10^{-6}$, $\overline{\lim}(S) = \underline{\lim}(S) = 1$.

Section 3.7

1. Clearly $a_2 = 1 + \sqrt{1} = 2 > 1 = a_1$. Using Assume $a_{n+1} > a_n$. We then get $a_{n+2} = 1 + \sqrt{a_{n+1}} > 1 + \sqrt{a_n} = a_{n+1}$ (obviously $a_n > 0$, $n \ge 1$ - why?).
3. All the sequence elements are obviously positive and $a_{n+1} = a_n + \sqrt{a_n} > a_n$. If the sequence converges to some positive A then $A = A + \sqrt{A}$ which leads to $A = 0$ and to contradiction.
9. $a_n = \frac{n}{1}\cdot\frac{n}{2}\cdots\frac{n}{n-1}\cdot\frac{n}{n} \ge \frac{n}{1} = n \to \infty$

Chapter 4

Section 4.1

2. For an arbitrary $x_0 > 0$ write

$$\sqrt{x} - \sqrt{x_0} = \frac{x - x_0}{\sqrt{x} + \sqrt{x_0}}$$

from which we easily get that $f(x)$ has a limit at x_0. The case $x_0 = 0$ is straightforward. The function does not have a limit at $x = -1$ since it is not defined at the interval between -1 and 0.

4. The function has the limit 0 at $x = 0$.

Section 4.2

1. (b) The function is clearly continuous for $x > 0$ where $f(x) = x$ and for $x < 0$ where $f(x) = -x$. It is also continuous at $x = 0$: Indeed, given $\varepsilon > 0$ we look for $\delta > 0$ such that $|x - 0| < \delta$ implies $|f(x) - f(0)| < \varepsilon$ and the obvious choice is $\delta = \varepsilon$.

3. For any given x_0 and arbitrary $\varepsilon > 0$ choose $\delta = \frac{\varepsilon}{L}$.

Section 4.3

1. Clearly,
$f^n(x) - f^n(x_0) = [f^{n-1}(x) + f^{n-2}(x)f(x_0) + \ldots + f^{n-1}(x_0)][f(x) - f(x_0)]$.
Since $f(x)$ is continuous, it is bounded by some $M > 0$ which implies

$$|f^n(x) - f^n(x_0)| \le nM^{n-1}|f(x) - f(x_0)|$$

where nM^{n-1} is constant. Consequently, $x \to x_0$ implies $f^n(x) \to f^n(x_0)$ as well.

3. The function $g(x) = x^3 + \sin^2(x)$ is a sum of two functions which are continuous everywhere. Thus, it is continuous everywhere and so is $f(x)$ as a composite function.

4. (b) $x \ne 1$ (c) $-\infty < x < \infty$

Section 4.4

2. The inverse function is $f^{-1}(y)=\dfrac{1+\sqrt{8y-7}}{4}$ and it is continuous over $[2,7]$.

3. The inverse is $x=\sqrt{y^2-1}$ defined for $1<y<\infty$.

5. The function $y=\cos(x)$, $-\dfrac{\pi}{2}\le x\le\dfrac{\pi}{2}$ does not have an inverse since it is not one to one. For example $\cos(\pi/4)=\cos(-\pi/4)$.

Section 4.5

1. The function is both continuous and uniformly continuous.
2. (a) The function is continuous. (b) It is not uniformly continuous.
   (c) The function in problem 1 has a limit as $x\to\infty$.
3. A function defined over a *closed interval* and satisfies a Lipschitz condition, is continuous over this interval and is therefore uniformly continuous there.
7. (a) The function satisfies a Lipschitz condition:

$$|x^3-y^3|=|x-y||x^2+xy+y^2|\le 27|x-y|$$

Chapter 5

Section 5.1

1. The derivatives are $3x^2+2x$, 8.
5. The function is not continuous at any point and therefore derivative nowhere.
7. When two derivatives are identical the corresponding functions differ by a constant.

10. The function has derivative for $x>0$ and $f'(x)=\dfrac{1}{2\sqrt{x}}$.

12. At $x=0$ the function has a right derivative 0 and a left derivative 1. Thus it does not possess a derivative there.

Section 5.2

1. (c) $f'(x)=-3x^{-4}\sqrt{x}+\dfrac{1}{2\sqrt{x}}x^{-3}=-2.5x^{-4}\sqrt{x}$

2. (a) $f'(x)=3(x^2+x)^2(2x+1)$

4. The function is one – to – one and consequently the inverse exists and

$$\left[f^{-1}(x)\right]' = \frac{1}{1+3x^2}$$

7. Solve $\left(x^3+x^2\right)' = 3x^2+2x = 10$. The solutions are $\frac{-2\pm\sqrt{124}}{6}$ and the positive one is within the interval $[1,2]$.

9. $\xi = \frac{38}{15}$.

Section 5.3

1. $\left[\sin^2(x)+\cos^2(x)\right]' = 2\sin(x)\cos(x) + 2\cos(x)[-\sin(x)] = 0$.
3. The left-hand side of the relation is always positive! (why?).
8. (a) $\cosh^2(x) - \sinh^2(x) = \frac{e^{2x}+e^{-2x}+2}{4} - \frac{e^{2x}+e^{-2x}-2}{4} = 1$.
9. $\tanh'(x) = \frac{\cosh^2(x)-\sinh^2(x)}{\cosh^2(x)} = \frac{1}{\cosh^2(x)}$ and similarly

   $\coth'(x) = -\frac{1}{\sinh^2(x)}$.
11. (a) The domain is all x for which $\cosh(x) > 0$, i.e., $-\infty < x < \infty$.

Section 5.4

1. (a) $y'' = -\sin(e^x)e^{2x} + e^x\cos(e^x)$. 2. The stationary points are $x=0$, $x=n$.

4. The function is defined for all $x \neq 0$ and has a minimum at $x = \sqrt[3]{-\frac{1}{4}}$.

8. Clearly $e^x > 1$ for arbitrary $x > 0$. By the first mean-value theorem

$$e^x - 1 = e^x - e^0 = xe^{\xi}\ , 0 < \xi < x$$

and thus $e^x = 1 + xe^{\xi} > 1 + x\ , x > 0$.

Section 5.5

1. (a) $\frac{3}{4}$ 2. (b) −2 4. (b) 0.

Chapter 6

Section 6.1

1. (a) $\int_0^3 f(x)dx = 5$ (b) The integral still exists and its value unchanged.
4. No.

Section 6.2

1. (a) Differentiability (or continuity) (b) Continuity.

Section 6.3

2. Since $f(x)$ is continuous and $g(x)$ monotonic increasing and bounded by 0.001, all functions are integrable (Theorems 6.3.1, 6.3.5, 6.3.6).
5. Use Theorem 6.3.6.

Section 6.4

1. (a) 10 (c) $1+\ln(2)$ 2. (a) $x+\dfrac{x^3}{3}-x+\tan^{-1}(x)$.

5. (a) $\dfrac{7\pi}{12}$ (b) $\dfrac{(a+2)^8-3^8}{8}$ 6. (a) $\dfrac{1}{2}e^{(x^2)}$

Section 6.5

2. $\lambda=\dfrac{e^2-1}{2}$, $x_0=\ln\left(\dfrac{e^2-1}{2}\right)\approx 1.164$ and x_0 is unique.

3. $\lambda=\dfrac{2}{\pi}$, $x_0=\sin^{-1}\left(\dfrac{2}{\pi}\right)\approx 0.690, 2.451$

5. By Theorem 6.5.3: $\int_0^{\pi}\sin(x)\cos(x)dx=\cos(0)\int_0^{\xi}\sin(x)dx+\cos(\pi)\int_{\xi}^{\pi}\sin(x)dx$ for some $\xi: 0\le\xi\le\pi$. This implies $\cos(\xi)=0$, i.e., $\xi=\dfrac{\pi}{2}$.

Section 6.6

1. (b) $\int xe^x dx = xe^x - \int e^x dx = xe^x - e^x + C$ 4. Substitute $u = 1 - x^2$.

7. (a) $-\ln(\cos(x))$ 10. The integral is $\frac{1}{a-b}\ln\left(\frac{x-a}{x-b}\right)$ if $a \neq b$. Otherwise it is $\frac{1}{a-x}$.

Section 6.7

2. The integrals (b) and (c) are divergent. 3. The integrals (a) and (b) are divergent.

Chapter 7

Section 7.1

1. Yes 2. No 4. (a) Converges (b) Converges (d) Diverges

Section 7.2

2. $\int_1^\infty \frac{dx}{x\sqrt{x}} = \left[\frac{-2}{\sqrt{x}}\right]_1^\infty = 2$ and the series converges.

5. $\frac{a_{n+1}}{a_n} = \frac{\left(\frac{1}{2}\right)^{n+1} + \left(\frac{2}{3}\right)^{n+1}}{\left(\frac{1}{2}\right)^n + \left(\frac{2}{3}\right)^n} = \frac{\frac{1}{2}\left(\frac{3}{4}\right)^n + \frac{2}{3}}{1 + \left(\frac{3}{4}\right)^n} \underset{n\to\infty}{\to} \frac{2}{3} < 1$

6. No, since $\sqrt[n]{a_n} = \sqrt[n]{\frac{1}{n^2}} \to 1$ as $n \to \infty$.

Section 7.3

1. Clearly $\frac{1}{n\sqrt{n}} > \frac{1}{(n+1)\sqrt{n+1}}$ and $\lim_{n\to\infty} \frac{1}{n\sqrt{n}} = 0$, i.e., convergence follows by Theorem 7.3.1.

3. (a) The series is not absolutely convergent ($\int_{2}^{\infty}\frac{dx}{x\ln(x)}$ does not exist).

   (b) The series is not absolutely convergent ($\lim_{x\to 0}\frac{\sin(x)}{x}=1$ and $\sum\left(\frac{1}{2}\right)^n$ converges).

5. Before each negative term insert k (which varies) positive terms whose sum exceeds 1.

Section 7.4

2. Cauchy product is $1+\frac{3}{4}+\frac{35}{72}+\frac{11}{36}+\ldots$

Section 7.5

1. The radius of convergence is $R=1$. The first two derivatives are $f'(x)=\frac{1}{1-x}$ and $f''(x)=\frac{1}{(1-x)^2}$.

2. The Taylor series is $x-\frac{x^2}{2}+\frac{x^3}{3}-\frac{x^4}{4}+\ldots$ and the radius of convergence is $R=1$.

6. The Taylor series is $1+x+\frac{x^2}{2!}+\frac{x^3}{3!}+\ldots$ and it converges uniformly for $|x|\le 100$ since for an arbitrary $\varepsilon>0$ we can find n such that

$$\frac{e^{100}\cdot 100^{n+1}}{(n+1)!}<\varepsilon$$

A similar argument holds for $|x|\le 200$ (see Example 7.5.7 and Eq. (10.2.8)).

Chapter 8

Section 8.1

1. (i) Clearly $I=\int_{a}^{a}f(x)dx=\int_{0}^{a}f(x+T)dx=\int_{0}^{a}f(x+T)d(x+T)$. By substituting $u=x+T$ we get $I=\int_{T}^{a+T}f(u)du=\int_{T}^{a+T}f(x)dx$.

2. $f(x) = 2(\sin(x) - \frac{1}{2}\sin(2x) + \frac{1}{3}\sin(3x) - \frac{1}{4}\sin(4x) + \ldots)$

6. (d) $a_n = 0$, $n \geq 0$; $b_n = \frac{1}{\pi}\int_0^\pi \cos^2(x)\sin(nx)dx$, $n \geq 1$

Section 8.2

4. The function $f(x) = \sqrt{x}$, $0 < x < 2$ does not satisfy the requirements since $f'(x) = \frac{1}{2\sqrt{x}} \to \infty$ as $x \to 0$. The function $f(x) = \sqrt{x}$, $0.001 < x < 2$ satisfies both requirements of Theorem 8.2.1.

Section 8.3

1. (a) even (c) odd (d) neither
2. $f'(0) = \lim_{h\to 0}\frac{f(h) - f(-h)}{h} = \lim_{h\to 0}\frac{f(h) - f(h)}{h} = 0$

   $g(-x) = -g(x) \Rightarrow g(x) + g(-x) = 0 \Rightarrow 2g(0) = 0 \Rightarrow g(0) = 0$
3. Consider an even function $f(x)$. Then

$$\frac{f(x+h) - f(x)}{h} = -\frac{f(-x-h) - f(-x)}{-h}$$

   and if $h \to 0$ the left-hand side converges to $f'(x)$ while the right-hand side converges to $-f'(x)$.

Chapter 9

Section 9.1

1. We discuss only real solutions. **Step 1.** Calculate $A = b^2 - 4ac$. **Step 2.** If $A < 0$ there are no real solutions – stop; If $A = 0$ get the single solution $x_1 = \frac{-b}{2a}$ and stop; Otherwise get the two different solutions $x_1 = \frac{-b + \sqrt{A}}{2a}$, $x_2 = \frac{-b - \sqrt{A}}{2a}$ and stop.
5. 5 iterations.

Section 9.2

1. Consider the function

$$f(x) = \begin{cases} 2x\,, 0 \le x < \dfrac{1}{2} \\ 1 - x\,, \dfrac{1}{2} \le x \le 1 \end{cases}$$

   Clearly it does not satisfy either condition, yet $f\left(\frac{1}{2}\right) = \frac{1}{2}$.

5. There is an indication that $|f(s)| > 1$.

Section 9.3

1. 5 iterations, $x_5 = 1.718771927$. 3. The Newton method provides the particular solution $s = 0$ where $f'(s) = 0$. This is not the case at the other solution $s = 3$.

Section 9.4

2. The maximum error is bounded by $6 \cdot 10^{-2}$.

Section 9.5

3. The quadratic least-squares approximation is $p_2(x) = x$ since the data is an exact straight line.
4. (b) The data is almost linear which indicates a small coefficient for x^2.
5. The least-squares approximation is $y = 4.671x - 2.143$.

Section 9.6

1. 3.116 2. The exact value is π and Eq. (9.6.4) provides $f''(\xi) = -1.006$. It is easily seen that that the second derivative $f''(x) = \sin(x) + x\cos(x)$ obtains this value somewhere between $\frac{\pi}{2}$ and π. 5. The approximated value is 8.3894 vs. the exact value 8.3891. 6. The integrand's fourth derivative is bounded by $6e^2$. Hence $\frac{6e^2 \cdot 2 \cdot h^4}{180} \le 10^{-8}$ which implies $h \le 0.0119$.

Chapter 10

Section 10.1

1. If $\sqrt{e}=\frac{m}{n}$ where m,n integers than $e=\frac{m^2}{n^2}$ is rational which contradicts Theorem 10.1.1. Similarly, if $\sqrt{\frac{e+1}{e-1}}=r$ for some rational r than $e=\frac{r+1}{r-1}$, i.e., rational.

Section 10.2

1. $\binom{2n}{n}\approx\frac{2^{2n}}{\sqrt{\pi n}}$, $n\to\infty$ 5. The approximate value is $\sqrt{\frac{81}{39}}\approx 1.446$. The exact value is 1.441…

Section 10.3.1

1. (b) Definition: the whole plane, discontinuity: both axes.
2. (a) $\frac{\partial f}{\partial x}=-\frac{1}{x^2}+\frac{x}{\sqrt{x^2+y^2}}$, $x\neq 0$, $\frac{\partial f}{\partial y}=\frac{y}{\sqrt{x^2+y^2}}$, $(x.y)\neq(0,0)$

Section 10.3.2

1. $x_1=x_2=\frac{\sqrt{2}}{2}$, $x_1=x_2=-\frac{\sqrt{2}}{2}$ 3. (b) $x=\frac{ab^2}{a^2+b^2}$, $y=\frac{a^2b}{a^2+b^2}$

Index

Abel's test, 202
Abel's theorem, 209
absolute convergence, 193–203, 206, 208
absolute value, 30
absolutely convergent, 195–196, 198–199, 200, 203, 205, 207
Aitken's method, 241–242, 244
algebraic functions, 48–54
algebraic number, 51, 54
algorithm, 233–236, 239, 242, 247, 256, 259
alternating series, 194–196
Archimedes, 3, 5, 235
Archimedes' axiom, 25
Aristotle, 2
arithmetic mean, 33–34

Barrow, 6
Bernoulli's inequality, 22, 33, 60, 67
Bessel's inequality, 223
binomial theorem, 27–28, 62, 69, 75
Bolzano – Weierstrass theorem, 80–81, 83, 96–97, 100
bounded sequence, 58, 81

Cantor, 2
Cartesian product, 37, 39
Cauchy, 2, 6
Cauchy condensation test, 189, 190
Cauchy product, 204–207, 209–210
Cauchy Sequences, 74, 82–86
Cauchy's ratio test, 193
Cauchy's root test, 191–193, 212
Cauchy-Schwarz inequality, 32, 35
Cavalieri, 5
closed interval, 38, 72, 80
comparison test, 186–187, 190–191, 196–197, 207
composite function, 95–96, 117–118, 120, 127–128
conditionally convergent, 198
continuous functions, 87–106
contractive sequence, 83–85
convergence rate, 79
convergence test, 234–235
convergence theorem, 227–228, 230–231
convergent sequence, 56, 63, 69, 77, 82–83
cubic splines, 250–253

D'Alembert's Ratio Test, 191
Darboux integrals, 149–155
Darboux sums, 148, 150–151, 154
Dedekind cuts, 23–24
Descartes, 5
differentiable functions, 107–146
dirichlet's test, 201, 268
divergent sequence, 56, 62, 69

elementary function, 175–177
euclid, 3, 5
Eudoxus, 2, 3
Euler, 76
Euler's summation formula, 264–266, 268
even functions, 53–54, 229–230, 232
exponential function, 100, 103, 127–129

Fermat, 6
fourier, 217
fourier coefficients, 220–221, 223, 229–230
fourier series, 217–232

fourier sine series, 230–232
fundamental period, 218
Fundamental Theorem of Calculus, 165–171

Gaussian integration, 259–261
geometric means, 33–34
Gerschgorin's theorem, 51
Goethe, 9

harmonic series, 184, 196, 202
Hermite, 263
Hieron, 3
Hipocrates, 2
Huygens, 7–8

implicit relation, 120–124, 130
improper integrals, 179–182
infimum, 42–44, 72–73, 96–97
infinite products, 203–210
infinite series, 183–216
inflection point, 135, 137- 140
integrable functions, 155–158, 174
integral test, 188, 190, 193
integration by parts, 176, 178
integration by substitution, 177
interpolation, 247, 251
interpolator, 247- 251, 253, 257
inverse function, 42, 119, 127–128, 130
irrational numbers, 30
iterative process, 235

jump discontinuity, 91

Kepler, 5
Kronecker, 14

lagrange interpolation, 250
lagrange interpolator, 249, 257
lagrange multipliers, 263, 270–281
lagrange polynomial, 247–250, 252, 257
least – squares approximations, 252–256
left derivative, 110, 114, 131
Leibniz, 2, 3, 6–9, 107, 117–118
Leibniz's Alternating Series Test, 194

length of a curve, 170
L'Hospital's Rules, 141–146
limit point, 70–73, 80–81, 83
linear least – squares method, 253–254
linear least-squares approximation, 255
lipschitz condition, 93, 106, 237, 239, 240, 242
local maximum, 120–122, 135–138
local minimum, 120, 135–139
logarithmic functions, 125, 127–128, 130

mathematical induction, 15, 28
mean-value theorems, 122–125, 134, 141, 168, 171–172, 174, 238
method of exhaustion, 3
Minkowsky inequality, 35
monotone decreasing, 75–76, 86
monotone functions, 42, 45
monotone increasing, 75–78, 81, 85–86
monotone sequences, 74–77
negative integers, 19

Newton, 2, 3, 6–9, 107, 109, 132
Newton – Raphson Method (NRM), 242–247
nonnegative integers, 28
numerical methods, 233–261

odd functions, 53–54, 229–232
open interval, 38, 72
order of convergence, 240

Parmenides, 1
Parseval's identity, 224
partial derivative, 270- 273, 275
peano axioms, 15
periodic function, 53, 217, 220, 225, 228–230
piecewise continuous, 225–227
plato, 1, 2
plutarch, 3
positive integers, 11, 19, 25
positive numbers, 13
power function, 119
power series, 203, 204, 210–214

quadratic least – squares, 254–256
quadratic least-squares approximation, 255–256
quadratic least-squares method, 255

radius of convergence, 212–213, 216
rational numbers, 17, 19–26, 29–30
real-valued function, 39
Riemann integrable, 147, 151, 153, 155–159, 161, 163, 165, 171
Riemann integrable functions, 158–159, 160, 162–163, 165, 172
Riemann Integral, 147–155, 158–167, 256
Riemann sum, 151, 153
right derivative, 110, 131
Roberval, 5
Rolle's theorem, 121–122, 134
Root Test, 192

sawtooth function, 227
Schiller, 9
secant method, 247
SIM, 236, 239–241, 244
Simpson's rule, 257–259, 261
Socrates, 1
stationary points, 135, 139–141
stirling's formula, 266–269
strictly monotone, 99–100, 103
supremum, 42–44, 71–73, 96

Taylor series, 210, 213–216
Taylor's theorem, 131, 133–135, 138, 141, 146, 214
The Irrationality of e, 263–264
theorem, 238
transcendental number, 51–52
trapezoidal rule, 256–257, 261
triangle inequality, 30–31
triangular wave, 220–221, 230
Trichotomy law, 16
trigonometric series, 217–224, 218

uniform continuity, 104–106

Wallis, 6
Wallis' formula, 267, 269

Zeno, 1, 2